# Atherosclerosis and Vascular Imaging

## Special Issue Editor
Michael Henein

MDPI • Basel • Beijing • Wuhan • Barcelona • Belgrade

**MDPI**

*Special Issue Editor*
Michael Henein
Umea University
Sweden

*Editorial Office*
MDPI AG
St. Alban-Anlage 66
Basel, Switzerland

This edition is a reprint of the Special Issue published online in the open access journal *International Journal of Molecular Sciences* (ISSN 1422-0067) from 2014–2015 (available at:

http://www.mdpi.com/journal/ijms/special_issues/vascular_imaging).

For citation purposes, cite each article independently as indicated on the article page online and as indicated below:

Author 1; Author 2. Article title. *Journal Name* **Year**, *Article number*, page range.

ISBN 978-3-03842-434-5 (Pbk)
ISBN 978-3-03842-435-2 (PDF)

# Table of Contents

# About the Special Issue Editor

**Michael Henein** holds the Chair of Cardiology at Umea University, Sweden, and is a visiting Professor at the St. George University, London, and Brunel University, Middlesex. After 15 years of clinical and academic work at the Royal Brompton Hospital and Imperial College London, he started an international career establishing academic activities in Sweden, Italy, UK, Kosovo, and Egypt. He currently focuses on international cardiology research programs in the field of coronary artery atherosclerosis and calcification, incorporating imaging techniques, biochemistry, metabolomics and genetics. In addition, Professor Henein has a special interest in other areas of cardiology including heart failure and valve disease. He has published over 300 papers in peer reviewed journals, written four books and 15 chapters. He has supervised 15 PhD theses and has lectured in many international conferences. Professor Henein founded two new journals in the field of cardiology, the International Journal of Cardiology–Heart and Vasculature and the International Cardiovascular Forum, and has served as Associate Editor for the International Journal of Cardiology for 8 years. He has also served for two years as the Guest Editor for IJMS.

# Preface to "Atherosclerosis and Vascular Imaging"

Cardiovascular disease is the main cause of death in the West, and vascular disease is the most common cardiovascular clinical problem. The disease results in serious morbidity and mortality, and carries economic cost implications. While conventional risk factors are well established, and their biomarkers regularly monitored, patients may continue to suffer subclinical active disease, even in the absence of risk factors, until they present with sudden cardiac death or stroke. Early disease detection using direct imaging has shown to be more accurate in identifying vulnerable patients and unstable plaques than conventional risk factors. This IJMS issue deals with the current opinion concerning the state-of-the-art imaging technologies available for clinical applications and their unique value over the sole use of conventional risk factor analysis, in identifying vulnerable patients, recommending aggressive treatments, prognosticating, and in assessing related nutritional and environmental issues.

Michael Henein
*Special Issue Editor*

International Journal of
*Molecular Sciences*

MDPI

Case Report

# Imaging of a Cilioretinal Artery Embolisation

Marion R. Munk [1,2], Rukhsana G. Mirza [1] and Lee M. Jampol [1,*]

[1] Department of Ophthalmology, Northwestern University, Feinberg School of Medicine, Chicago, IL 60611, USA; marion.munk@northwestern.edu (M.R.M.); r-mirza@northwestern.edu (R.G.M.)
[2] Department of Ophthalmology, Medical University of Vienna, Vienna 1090, Austria
* Correspondence: l-jampol@northwestern.edu; Tel.: +1-312-908-8152; Fax: +1-312-503-8152

Received: 31 July 2014; in revised form: 29 August 2014; Accepted: 1 September 2014;
Published: 4 September 2014

**Abstract:** Retinal artery occlusion can be the first indicator of a significant cardiovascular disorder and the need for treatment. We present the case of a 69-year-old man with a cilioretinal artery occlusion and retinal ischemia. Retinal imaging, in particular fundus autofluorescence, highlighted an intraluminal hyperautofluorescent lesion which led to the diagnosis of retinal emboli. Subsequently a severe, previously undiagnosed carotid occlusive disease was discovered. The patient underwent prompt endarterectomy.

**Keywords:** retinal artery occlusion; spectral domain optical coherence tomography (SD-OCT); fundus autofluorescence; retinal ischemia; endarterectomy; cardiovascular disorder; atherosclerosis

## 1. Introduction

Retinal artery occlusion and retinal ischemia can be the first signs of a significant and symptomatic cardiovascular disorder and indicators of an urgent need for treatment. We show a case of a patient who presented with visual complaints and was found to have two focal areas of retinal whitening. Due to a complicated systemic history, his presentation led initially to a workup by the urgent care clinic for infectious retinitis and septic emboli. However, after exclusion of an infectious origin, the intraluminal hyperautofluorescent plaques in the macular arterioles were detected on retinal imaging, which led to the correct diagnosis of retinal emboli and associated retinal ischemia. These specific features led to the finding of severe, previously undiagnosed carotid occlusive disease. The patient underwent prompt endarterectomy, thus preventing further events.

## 2. Case Report

A 69-year-old male awoke with a "foggy spot the size of a pencil eraser" in the center of his vision of the right eye. His ocular history was unremarkable. His medical history was notable for hypertension and hyperlipidemia, which were treated with Amlodipine 5 mg and Simvastatin 40 mg once a day. He admitted ethanol abuse (about 1 bottle of bourbon per day). In addition, he reported a recent episode of cough and night sweats. He also complained of green sputum and chills.

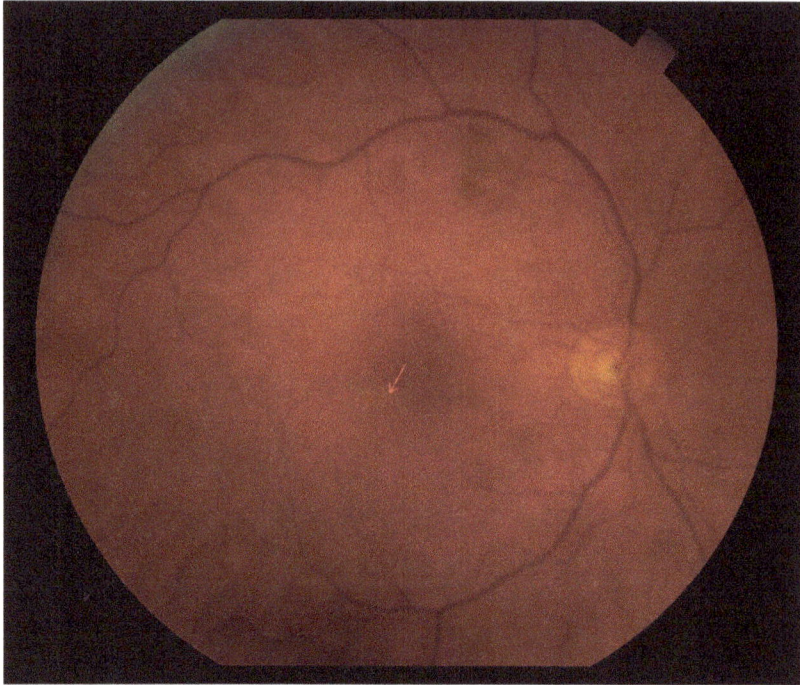

**Figure 1.** Color fundus image OD reveals two retinal parafoveal whitish lesions. The tiny, yellowish emboli in the small afferent arterioles are obscured by the retinal whitening (arrow).

At presentation to the urgent care clinic, his best corrected visual acuity was 20/20 OU and the anterior segment was unremarkable. Ophthalmoscopy revealed two parafoveal white retinal lesions (Figure 1). No vitreous cells/haze was noted. Due to his systemic complaints, QuantiFERON Gold, rapid plasma reagin (RPR), HIV Enzyme-linked Immunosorbent Assay and toxoplasmosis serology testing were obtained and found to be normal. An echocardiogram was obtained to assess for possible embolic retinitis and was unremarkable. The retina service was consulted, and retinal imaging was performed. At the level of the two retinal lesions, spectral domain optical coherence tomography (SD-OCT) displayed corresponding hyperreflectivity in the inner retinal layers (Figure 2A,B). Red-free, short wave 488 nm fundus autofluorescence (FAF) as well as infrared imaging highlighted these two lesions (Figures 2 and 3). Fluorescein angiography (FA) revealed normal filling and transit time. Beside the two hypo-autofluorescent lesions FAF showed hyperautofluorescent emboli in the terminal vascular bed of the cilioretinal artery (Figure 3). This latter finding prompted ordering carotid Doppler sonography and subsequently computer tomography angiography, which revealed >90% stenosis within the right carotid bulb primarily related to an atherosclerotic plaque. It was noted that there was passage of only a thin trickle of contrast material with a 1 mm residual lumen diameter. There was a collapse of the distal right internal carotid artery (ICA), related to hemodynamic compromise. Based on these findings, the vascular surgeon performed a successful right carotid endarterectomy with patch graft angioplasty.

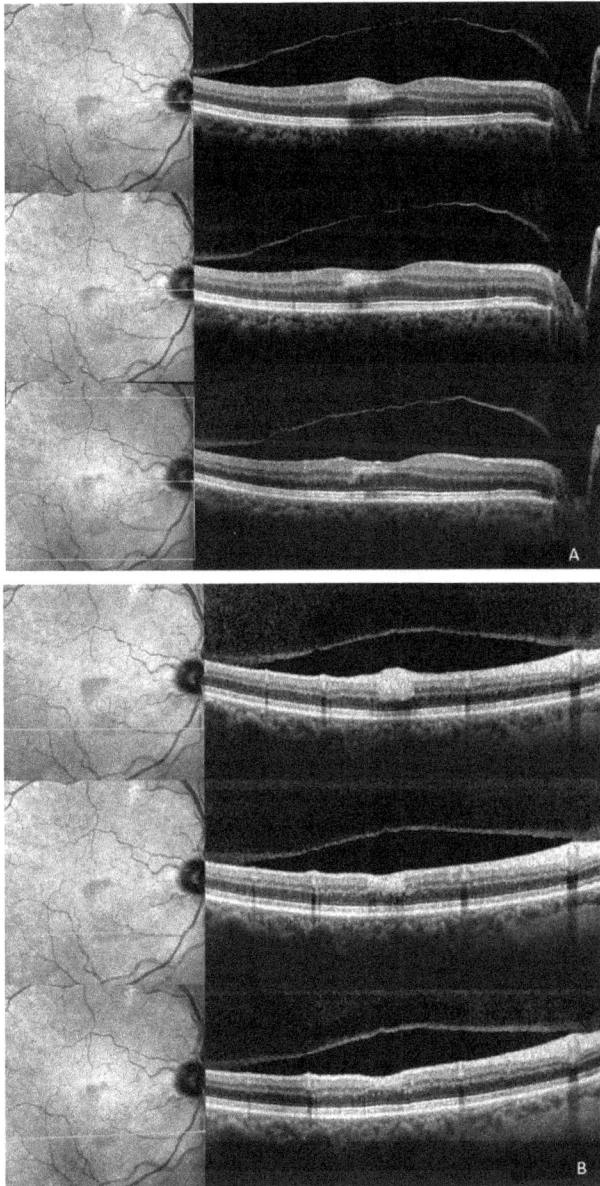

**Figure 2. Top**: The ischemic superior (**A**) and inferior (**B**) lesions show corresponding hyperreflectivity in the outer plexiform (OPL), inner nuclear (INL), inner plexiform (IPL), ganglion cell (GCL), and retinal nerve fiber layer (RNFL) on spectral domain optical coherence tomography (SD-OCT). On infrared image the ischemic areas are hyporeflective due to the scattering and/or absorbance of light. The four hyperreflective bands corresponding to the retinal pigment epithelium and the photoreceptors seem intact; **Middle**: Two weeks after initial presentation the hyperreflectivity is still visible and thinning of the respective layers is already appreciable. Hyporeflectivity already starts to fade on infrared imaging; **Bottom**: One month after initial presentation hyperreflectivity on infrared is still present. Affected layers revealed further thinning and the inner nuclear layer is not identifiable any more.

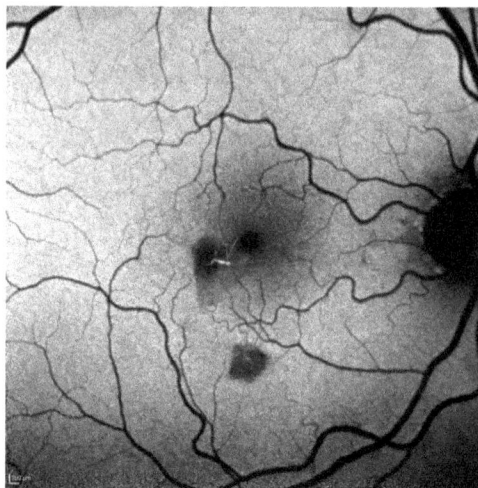

**Figure 3.** Short wave fundus autofluorescence shows two hypoautofluorescent areas (scattered and/or absorbed light due to swollen inner retina). The hyperautofluorescent emboli in the afferent arterioles are clearly visible (arrows). Scale bar: 200 μm.

One month after presentation, the whitish lesions were nearly invisible. On FAF and IR, the hypoautofluorescent and hyporeflective lesions were still visible but were fading (Figures 2 and 4) and SD-OCT revealed continuous thinning of the retinal nerve fiber layer (RNFL), ganglion cell layer (GCL), inner plexiform layer (IPL) and inner nuclear layer (INL) with a relative thickening of the outer nuclear layer (ONL) (Figure 2). Two months thereafter a subtle hyporeflectivity was still visible on IR, and FAF highlighted the unchanged hyperautofluorescent emboli, while SD-OCT revealed further thinning of respective layers.

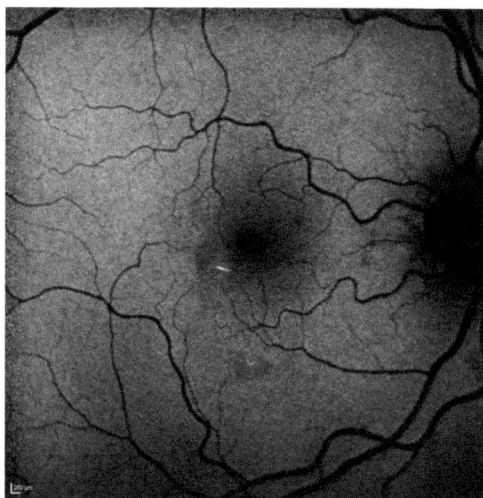

**Figure 4.** After one month the hypoautofluorescence has faded (as the inner retina atrophies) on short wave blue 488 nm autofluorescence. The hyperautofluorescent emboli are still clearly visible. Scale bar: 200 μm.

## 3. Discussion

Retinal artery occlusions and retinal ischemia can be the first signs of cardiovascular disorders. The case presented here demonstrates that retinal imaging can be crucial for the detection and diagnosis of a cardiovascular disease leading to immediate important therapy.

The retinal blood vessels are responsible for the blood supply of the inner retinal layers, whereas the outer retina is nourished by the choroidal vascular system, with the outer plexiform layer acting as a watershed zone [1,2]. Retinal ischemia due to central, branch, or cilioretinal artery occlusion may present with whitening in the affected non-perfused area, which can be seen in ophthalmoscopy and color photography. The whitening represents swelling and, acutely, is seen as hyperreflectivity throughout all inner retinal layers on SD-OCT. Based on which retinal layers present this localized hyperreflectivity on SD-OCT, the location and level of the thromboembolic event can be presumed [3]. Subsequently, as these areas "die" we see progressive thinning of respective layers over time [4]. The ischemic area that is white and swollen initially scatters and/or absorbs light and therefore is hypoautofluorescent on FAF and is hyporeflective on infrared imaging. As this retina atrophies, these changes fade over time concordant to the subsequent retinal thinning [2,5–7]. Although hyporeflectivity on IR is a nonspecific sign for absorbance and/or scattering of light, focal, hyporeflective lesions on IR are, in conjunction with localized hyperreflectivity on SD-OCT and whitening on color photography, indicative of ischemic events. This hyporeflectivity on IR is found irrespective of the severity of the occlusion, *i.e.*, proximal, larger vessel involvement may lead to retinal artery occlusions, affecting all inner retinal layers. Small microvasculature thromboembolic events of the capillary network in turn cause localized ischemic lesions, limited to particular retinal layers [2–4,8]. However, the most important finding that led to the diagnosis and subsequently to the work-up and therapy of this patient was the hyperautofluorescent emboli found on FAF. The appearance of such emboli and the usefulness of FAF to detect and illustrate retinal artery occlusions was only noted recently [7]. This previous report assumed that hyperautofluorescence may be present in only certain types of emboli such as calcified emboli [7]. However, in our case the arteriosclerotic plaques in the right carotid bulb visualized by CT angiography and carotid Doppler sonography were described as cholesterol and not calcium containing thromboembolic plaques. In clinical studies of coronary plaques using color fluorescent angioscopy to evaluate fluorescence of the major components of atherosclerotic plaques, it became evident that excitation using short wavelength filters results in an emission spectrum of blue, light blue and white autofluorescence for collagen I, IV and calcium containing plaques, respectively. Cholesterol and cholesteryl esters in turn exhibited a yellow and orange fluorescence [9]. This indicates that the composition of a thromboembolic plaque may be differentiable based on their emission spectrum [9]. The usefulness and applicability of fluorescence to identify retinal emboli need further studies. These findings may then also help guide the differential diagnosis of underlying etiologies [10,11].

In summary, retinal imaging can reveal characteristic findings of retinal ischemia and embolic plaques. This may help to identify patients with cardiovascular diseases in need of life-saving intervention.

**Acknowledgments:** This work was supported in part by an unrestricted grant from Research to Prevent Blindness Inc., New York to Northwestern University. The authors want to thank Evica Simjanoski for taking and processing the images.

**Author Contributions:** Study concept and design: Marion R. Munk, Lee M Jampol and Rukhsana G. Mirza; Acquisition of data: Marion R. Munk; Analysis and interpretation of the data: Marion R. Munk, Lee M Jampol and Rukhsana G. Mirza; Drafting the manuscript: Marion R. Munk; Critical revision of the manuscript: Lee M Jampol and Rukhsana G. Mirza; Obtaining funding: Lee M Jampol.

**Conflicts of Interest:** The authors declare no conflict of interest.

## References

1. Toussaint, D.; Kuwabara, T.; Cogan, D.G. Retinal vascular patterns. II. Human retinal vessels studied in three dimensions. *Arch. Ophthalmol.* **1961**, *65*, 575–581. [CrossRef]

2. Sarraf, D.; Rahimy, E.; Fawzi, A.A.; Sohn, E.; Barbazetto, I.; Zacks, D.N.; Mittra, R.A.; Klancnik, J.M.; Mrejen, S.; Goldberg, N.R.; *et al.* Paracentral acute middle maculopathy: A new variant of acute macular neuroretinopathy associated with retinal capillary ischemia. *JAMA Ophthalmol.* **2013**, *131*, 1275–1287. [CrossRef]

3. Aleman, T.S.; Tapino, P.J.; Brucker, A.J. Evidence of recurrent microvascular occlusions associated with acute branch retinal artery occlusion demonstrated with spectral-domain optical coherence tomography. *Retina* **2012**, *32*, 1687–1688.

4. Ritter, M.; Sacu, S.; Deak, G.G.; Kircher, K.; Sayegh, R.G.; Pruente, C.; Schmidt-Erfurth, U.M. *In vivo* identification of alteration of inner neurosensory layers in branch retinal artery occlusion. *Br. J. Ophthalmol.* **2012**, *96*, 201–207. [CrossRef]

5. Mathew, R.; Papavasileiou, E.; Sivaprasad, S. Autofluorescence and high-definition optical coherence tomography of retinal artery occlusions. *Clin. Ophthalmol.* **2010**, *4*, 1159–1163.

6. Coady, P.A.; Cunningham, E.T., Jr.; Vora, R.A.; McDonald, H.R.; Johnson, R.N.; Jumper, J.M.; Fu, A.D.; Haug, S.J.; Williams, S.L.; Lujan, B.J. Spectral domain optical coherence tomography findings in eyes with acute ischaemic retinal whitening. *Br. J. Ophthalmol.* **2014**. [CrossRef]

7. Siddiqui, A.A.; Paulus, Y.M.; Scott, A.W. Use of fundus autofluoresecence to evaluate retinal artery occlusion. *Retina* **2014**, in press.

8. Fawzi, A.A.; Pappuru, R.R.; Sarraf, D.; Le, P.P.; McCannel, C.A.; Sobrin, L.; Goldstein, D.A.; Honowitz, S.; Walsh, A.C.; Sadda, S.R.; *et al.* Acute macular neuroretinopathy: Long-term insights revealed by multimodal imaging. *Retina* **2012**, *32*, 1500–1513.

9. Uchida, Y.; Uchida, Y.; Kawai, S.; Kanamaru, R.; Sugiyama, Y.; Tomaru, T.; Maezawa, Y.; Kameda, N. Detection of vulnerable coronary plaques by color fluorescent angioscopy. *JACC Cardiovasc. Imaging* **2010**, *3*, 398–408.

10. Ramakrishna, G.; Malouf, J.F.; Younge, B.R.; Connolly, H.M.; Miller, F.A. Calcific retinal embolism as an indicator of severe unrecognised cardiovascular disease. *Heart* **2005**, *91*, 1154–1157. [CrossRef]

11. Mughal, M.M.; Khan, M.K.; DeMarco, J.K.; Majid, A.; Shamoun, F.; Abela, G.S. Symptomatic and asymptomatic carotid artery plaque. *Expert Rev. Cardiovasc. Ther.* **2011**, *9*, 1315–1330. [CrossRef]

International Journal of
*Molecular Sciences*

MDPI

*Review*

# Coronary CT Angiography in Managing Atherosclerosis

Joachim Eckert *, Marco Schmidt, Annett Magedanz, Thomas Voigtländer and Axel Schmermund

Cardioangiologisches Centrum Bethanien, Im Prüfling 23, D-60389 Frankfurt, Germany;
m.schmidt@ccb.de (M.S.); a.magedanz@ccb.de (A.M.); t.voigtlaender@ccb.de (T.V.);
a.schmermund@ccb.de (A.S.)
* Correspondence: j.eckert@ccb.de; Tel.: +49-69-9450-280; Fax: +49-69-4616-13

Academic Editor: Michael Henein
Received: 4 January 2015; Accepted: 4 February 2015; Published: 9 February 2015

**Abstract:** Invasive coronary angiography (ICA) was the only method to image coronary arteries for a long time and is still the gold-standard. Technology of noninvasive imaging by coronary computed-tomography angiography (CCTA) has experienced remarkable progress during the last two decades. It is possible to visualize atherosclerotic lesions in the vessel wall in contrast to "lumenography" performed by ICA. Coronary artery disease can be ruled out by CCTA with excellent accuracy. The degree of stenoses is, however, often overestimated which impairs specificity. Atherosclerotic lesions can be characterized as calcified, non-calcified and partially calcified. Calcified plaques are usually quantified using the Agatston-Score. Higher scores are correlated with worse cardiovascular outcome and increased risk of cardiac events. For non-calcified or partially calcified plaques different angiographic findings like positive remodelling, a large necrotic core or spotty calcification more frequently lead to myocardial infarctions. CCTA is an important tool with increasing clinical value for ruling out coronary artery disease or relevant stenoses as well as for advanced risk stratification.

**Keywords:** atherosclerosis; coronary plaques; coronary computed-tomography angiography (CCTA); coronary calcium; cardiac events

## 1. Background

Recent developments of CT scanners have improved accuracy especially regarding the visualization of the coronary arteries. A better spatial and temporal resolution makes it possible to scan the heart and the coronary arteries free of motion and to detect vascular plaques and stenoses. Still, heart rates below 60–65/min are preferable to achieve high quality images with a low radiation exposure using prospective ECG (electrocardiographic)-gating. Common nomenclature distinguishes between different types of plaque: calcified, noncalcified and predominant calcified or predominant noncalcified [1]. Calcified plaques are visualized and quantified by CT scans without injection of contrast agent (calcium scanning). For detecting different types of plaque as well as determining possible coronary stenoses, intravenous contrast agent must be injected prior to the scan (CT-angiography, CTA).

## 2. Coronary Plaque Morphology and Pathophysiology

On the basis of the CT images, coronary plaques are classified as calcified and noncalcified or as "mixed" plaques containing both aspects. Pathophysiologically, subendothelial lipoprotein retention triggers inflammatory responses via macrophages and T-cells with chronic maladaptive progression of atherosclerotic lesions [2]. Looking at plaques on a cellular basis, early atherosclerotic changes can be classified into 3 types [3] which reflect microscopic changes like accumulation of macrophages (type I) and which are already seen in infant arteries. Later, fatty streaks, foam cells and deposits of lipid inside smooth-muscle cells can be found (type II). These lesions tend to start to develop in puberty. Type III lesions mark the border where these microscopic changes become visible to the eye. Macroscopic changes begin, and the so-called "atheroma" is formed. Advanced lesions can again be classified into 3 types (types IV–VI) [4]. Type IV lesions encompass the lipid core which is called atheroma. As soon as fibrous tissue grows the lesion is classified type V ("fibroatheroma"). If a thrombus or hemorrhage develops on the atheroma or fibroatheroma the lesion is regarded "complicated" (type VI) and, hence, patients can become symptomatic. Lesions IV and V can be asymptomatic due to maintenance of the vessel diameter. Glagov *et al.* first described adaptive changes of arterial size in the course of plaque formation [5]. The entire vessel grows with increasing plaque volume so that the lumen diameter is maintained. Furthermore, a frequent pathology seen in myocardial infarctions due to plaque rupture is the thin cap fibroatheroma (TCFA), which is characterized by a necrotic core covered by a fibrous cap measuring <65 μm [6]. Even though the classifications cannot be directly compared, the TCFA corresponds to a subgroup of the Stary type V lesion. Speckled calcification can be visualized in the majority of ruptured plaques. TCFA seems to be the precursor lesion of plaque rupture. It is frequently associated with expansive remodeling. These changes cannot be detected in invasive angiography because the vessel wall is invisible and only the lumen, which may appear normal, is displayed. Coronary CT-angiography (CCTA) may fill this diagnostic gap, since changes of the vessel wall can directly be visualized.

### 3. Coronary Calcification

Coronary artery calcification (CAC) is a frequent pathology seen in CT scans (Figure 1). The amount of calcium is quantified using the Agatston-Score [7]. It is correlated with the extent of atherosclerotic plaque burden [8]. In most patients presenting with acute coronary syndromes or sustaining sudden cardiac death, calcifications in the coronary artery wall can be detected [9,10]. A high amount of calcium, however, does not necessarily correlate with angiographic luminal stenoses, nor is there a fixed relationship with vulnerability of plaques [11]. *Vice versa*, a lack of coronary calcium makes stenotic lesions unlikely, but it is not possible to definitely rule out coronary stenoses [12,13].

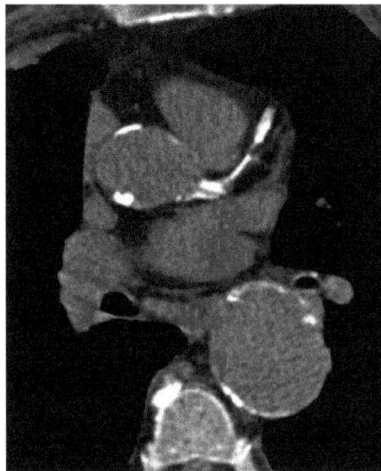

**Figure 1.** Native calcium scan with severe calcification of left main, left anterior descending (LAD) and the aorta.

There are still debates on the mechanisms of coronary artery calcification. Studies could show that calcification is not a mere passive response to injury but an active process similar to bone formation [14,15]. This process already starts in the second decade of life [16]. Mostly, calcifications are part of atherosclerotic changes, share the same risk factors, and can predominantly be found in advanced lesions [17].

The amount of calcium is influenced by gender, ethnicity and age [18]. Different data exist concerning the possible individual modification of coronary calcium. Lifestyle changes and aggressive medical therapy (especially with "statins") might slow the progress of calcification [19,20]. Interestingly, recent data show that the progression of calcification is mainly driven by genetic conditions and to a minor extent by classical risk factors such as hypertension or LDL cholesterol [21,22]. It is, however, important that although progression of calcification seems to be inevitable this does not hold true for the clinical outcome and adverse cardiac events of patients on lifestyle changes or medication for risk-factor modification.

## 4. Clinical Implication and Prognosis of Coronary Artery Calcium

Studies have demonstrated that cardiovascular events are low and the overall prognosis is good in the absence of coronary calcifications [23]. Coronary calcium scoring in combination with assessment of the Framingham Score in asymptomatic people can improve risk stratification especially in individuals with risks between 10% and 19% in 10 years according to the Framingham Score [24]. High calcium scores are associated with future cardiovascular events and worse survival outcome. Cardiovascular risk increases proportionally to the amount of calcium and is highest with Agatston-Scores above 400. An annual progression of more than 15% enhances the risk of myocardial infarctions [17,19,25]. Patients after myocardial infarctions have higher CAC progressions than subjects who remained event-free [26]. Positive predictive values of CAC progression as a marker of risk are, however, low [17]. Repeated CAC scans can therefore not be recommended as a control of adequate medical therapies or lifestyle changes. Single calcium scores are recommended in asymptomatic persons with intermediate risk (Framingham risk score 10%–20%) as support for clinical decisions whether to start aggressive medical therapy. In high or low risk populations, CAC scoring does not necessarily add relevant information.

## 5. Coronary CT Angiography

For calcium scoring, a native CT scan is sufficient. To gain information on coronary stenoses and plaque morphology, contrast media (50–100 mL) must be injected and the scan timed in the phase of maximal contrast enhancement. In contrast to invasive coronary angiography (ICA), CCTA offers the advantage of visualizing the vessel wall. Thus, it is possible to detect atherosclerotic lesions despite a preserved vessel lumen as well as lesions causing a coronary stenosis (Figure 2), even in revascularized patients (Figures 3 and 4).

For clinical purposes, CCTA performs best in individuals who are at low to intermediate risk of coronary artery disease (CAD) [27]. For high-risk individuals, the diagnostic performance of CCTA is lower; patients frequently need ICA afterwards due to suspected high-grade stenoses in CCTA or severe calcifications.

Using the latest CT scanners (at least $2 \times 128$ slices), CCTA can be performed with a radiation exposure of <1 mSv. High pitch spiral mode with iterative reconstruction is able to visualize the whole heart in a single diastole with excellent image quality [28–30]. To obtain images with low radiation exposure and little motion artifacts, patients' heart rate should be <60–65/min. Beta blockers are often administered prior to the scan.

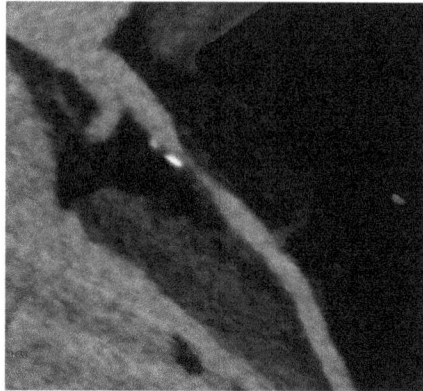

**Figure 2.** Predominantly noncalcified plaque with high-grade stenosis of LAD.

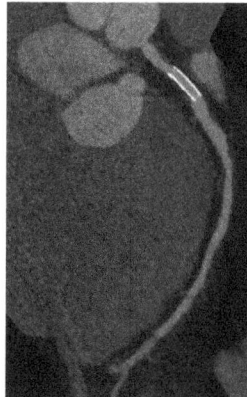

**Figure 3.** LAD after revascularization with a patent drug eluting stent showing a very good result 18 months after implantation.

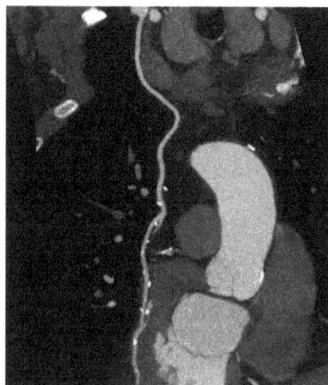

**Figure 4.** Patent right mammary artery bypass graft (free transplant with end-to-side anastomosis on left mammary artery) with anastomosis on obtuse marginal branch.

## 6. Imaging of Coronary Plaques and Stenoses

When performing CCTA in patients with intermediate risk for CAD, a substantial portion of the patients show coronary plaques (Figure 5). Hausleiter *et al.* assessed 161 patients of whom almost 30% had noncalcified plaques; most had both noncalcified and calcified plaques. In this group, 6% had plaques without any calcification [31]. Several studies compared the diagnostic accuracy of detecting coronary artery stenoses compared to invasive angiography [32], some additionally with intravascular ultrasonography (IVUS) [33–35]. Sensitivity for detection of plaques range above 90%, negative predictive values approach 100% in patients with low to intermediate probabilities of CAD. CCTA is a reliable method, especially for ruling out relevant plaques and stenoses in coronary arteries (Figure 6). One major limitation is a reduced ability to reliably quantify the degree of stenoses [36] which is the reason for lower positive predictive values and specificity due to the fact that stenoses tend to be overestimated in CCTA especially in calcified lesions. Specificity ranges between 64% and 87%, depending on patient characteristics such as obesity or calcification [32,35,37]. A recent meta-analysis comprised 42 studies in which CCTA was compared to IVUS for detection of any plaques. Sensitivity and specificity were 93% and 92%, respectively [34]. Furthermore, imaging artifacts can lead to misinterpretation. Most of the existing studies were, however, performed using 64-slice CCTA. Technology has remarkably improved in the last decade so that dual-source scanners with $2 \times 128$ slices and more are the technical standard at present. In a meta-analysis by Voros *et al.*, it could be demonstrated that sensitivity improves from 84% to 94% when images are obtained with 64-slice scanners compared to 16-slice scanners [38]. Still, different attenuation values inside the same plaques (fibrous, lipid-rich, necrotic and calcified) make the classification and reproducibility of lesions challenging.

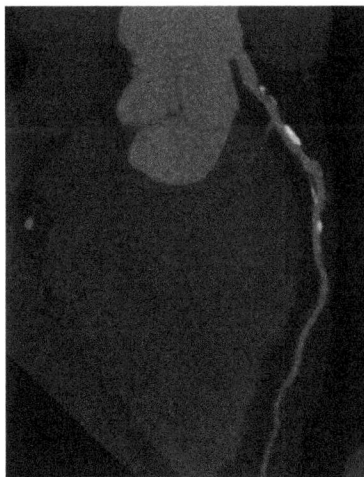

**Figure 5.** Calcified and noncalcified plaques in LAD.

Cheng *et al.* demonstrated that visual detection of plaque presence is reproducible [39]. Intraobserver, interobserver and interscan variability were excellent, but large differences in agreement existed regarding total plaque volume. The reason is probably the problem of quantifying small coronary plaques by CCTA due to technical limitations in spatial resolution. Moderate reproducibility of plaque burden and degree of coronary stenoses was also reported by Leber *et al.* using 64-slice CT scanners [36,40]. Interobserver variability depends on image quality. Pflederer *et al.* showed that in the left anterior descending coronary artery (LAD), where image quality was best, interobserver variability was significantly lower than in the left circumflex (LCX) or right coronary artery (RCA) (17% in LAD *versus* 32% RCA) [41].

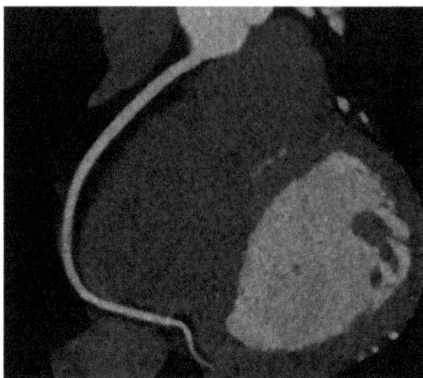

**Figure 6.** Normal right coronary artery.

A commonly used standardized score for quantification of coronary calcification is the Agatston-Score [7]. A standardized, reproducible tool for quantification of noncalcified plaques does not exist. One important reason for that is the limited spatial resolution of CCTA that makes small plaques difficult to detect. Furthermore, noncalcified plaques can show a wide range of attenuation values due to differences in morphology. An automated quantification using a special software to assess minimal lumen area, plaque burden, percentage luminal stenosis and degree of remodeling was used by Boogers *et al.* in 51 patients [42]. Plaque quantification was feasible and reproducible, and significant correlations could be demonstrated for all parameters. Minimal lumen area was, however, underestimated, and lumen area stenosis overestimated compared to IVUS, especially in calcified lesions.

Furthermore, CCTA is a helpful planning tool for revascularization of chronic total coronary occlusions (CTO). Rolf *et al.* could show that CCTA prior to percutaneous coronary intervention (PCI) could significantly improve success rates of the intervention in 30 patients [43]. Three-dimensional images derived from CCTA guided the advance of the guide wire during the intervention and could thus lead to a success rate of 90% *versus* 63% in the control group without CT. Another study with 100 patients could demonstrate significantly fewer complications, such as coronary perforations, but no improvement in success [44]. Severe coronary calcification seems to be an independent predictor of failure of revascularization of a CTO lesion [45].

### 7. Qualitative Plaque Characterization

IVUS is considered the gold standard for *in vivo* plaque quantification and characterization [46]. Normal intimal thickness of healthy young subjects measures 0.15 mm [47], below the spatial resolution of CCTA, which is 0.24–0.3 mm for the latest generation of CT scanners. The vessel wall becomes thickened in atherosclerotic lesions which makes it a diagnostic target for non invasive CT evaluation.

Attenuation values inside noncalcified plaques vary according to histological findings. Fibrotic tissue is associated with higher CT density whereas a necrotic core is negatively correlated to density, although with a wide range of overlap [33,48]. Contrast flow rates and concentration inside the vessel as well as microcalcifications often seen around the necrotic core affect density measurements [49,50]. Furthermore, slice thickness and convolution kernels hamper reproducibility of density measurements. Dual-energy CT might enhance the differentiation between the necrotic core and fibrous tissue, however, with a loss of temporal resolution [51].

Atherosclerotic lesions can lead to acute coronary events and death. The morphological characterization of plaques being prone to complications is of eminent interest, and some studies have been reported on aspects of plaque morphology in the context of acute coronary syndromes. Characteristics of ruptured plaques include expansive remodeling, a large necrotic core, thin cap fibroatheroma (TCFA) and macrophage infiltration [52]. Motoyama *et al.* demonstrated that in patients with acute coronary syndromes ACS, CT can identify plaques showing expansive remodeling, low atheroma attenuation values, and spotty calcifications [53]. In another study, 1059 patients underwent CCTA; 45 of these patients showed expansive remodeling and low attenuation plaques [54]. Twenty-two percent of the patients harboring both pathologies developed myocardial infarction in the follow-up period. On the other hand, only four of 820 patients with neither sign had a cardiac event. Hoffmann *et al.* suggested that culprit lesions tend to have greater noncalcified areas, whereas largely calcified plaques indicate more stability [55].

A special pathology in atherosclerotic lesions is the so-called "napkin-ring-sign" which has a high specificity and positive predictive value for advanced lesions [56]; it can be visualized by using CCTA. The napkin-ring-sign is characterized by a plaque core with low attenuation surrounded by a rim-like area of higher attenuation, potentially representing TCFA. Recently, Otsuka *et al.* could demonstrate that the presence of the napkin-ring-sign is strongly associated with acute coronary syndromes [57].

Another interesting approach of detecting vulnerable plaques might be $^{18}$F-sodium fluoride and $^{18}$F-FDG uptake diagnosed by PET-CT. Dweck *et al.* demonstrated that $^{18}$F-NaF uptake was significantly higher in individuals with coronary atherosclerosis (defined by calcium score > 0) in contrast to subjects without (calcium score 0) [58]. Uptake of $^{18}$F-NaF seems to be related to inflammation and active calcification with similarities to bone metabolism. Recently, evaluation of patients with myocardial infarction could show that >90% of the patients had increased uptakes in the culprit lesions [59]. In other plaques with increased uptake, high-risk factors such as expansive remodelling, microcalcifications, and a larger necrotic core could be seen on IVUS. It has yet to be demonstrated that increased uptake will translate into future cardiac events.

### 8. Hemodynamic Relevance of Angiographic Stenoses

Frequently, intermediate stenoses (30%–70%) of coronary arteries are detected on CCTA and it is not evident if these lesions cause ischemia. For intermediate stenoses diagnosed on invasive angiography it is recommended to perform Fractional Flow Reserve (FFR) measurements to assess the functional relevance. De Bruyne *et al.* demonstrated that patients having lesions with FFR of less than 0.8 benefit from revascularization whereas stenoses with FFR of more than 0.8 should be treated conservatively with medical therapy alone [60]. An FFR < 0.8 is thus considered to cause ischemia. Sensitivity for diagnosing high grade stenoses on CCTA is excellent; specificity is, however, poor due to false positive results because of overestimation of stenoses [61]. There was a weak correlation between significant coronary lesions on CCTA and ICA combined with FFR < 0.75; diagnostic accuracy was only 49%. Relying solely on the visual aspect might lead to unnecessary revascularizations and be potentially harmful. Hence, the concept of measuring FFR(CT) noninvasively by CCTA was perceived during the last years. Min *et al.* could show in a multicenter trial that diagnostic accuracy for FFR(CT) was superior to CCTA alone although specificity was still poor [62]. FFR(CT) measurements seem to be reproducible [63], and can be calculated from the normal CCTA dataset without additional image acquisition by using special equations of fluid dynamics [64].

### 9. CCTA in the Emergency Department

CCTA can be used to rule out relevant CAD in the emergency department for patients presenting with symptoms such as angina without signs of myocardial infarction (ST(segment)-elevation on ECG, positive cardiac enzymes). Frequently, patients presenting with chest pain are admitted to hospital, or stay in the emergency department for many hours. One trial showed a significant reduction of time to diagnosis from 15 h in the control group to 3.4 h in the CCTA group [65]. Both approaches were safe, but CCTA appeared cost effective, and patients who had a CT scan required less subsequent diagnostic workup for recurrent chest pain symptoms.

The ROMICAT-Trial described an excellent sensitivity (100%) of diagnosing CAD in patients presenting with chest pain at low to intermediate pretest probability [66]. Fifty percent of all patients had no CAD at all. Patients were followed-up for two years regarding major adverse cardiac events (MACE) [67]. Patients with no CAD in CCTA had no risk for MACE in the following two years whereas risk was 4.6% in nonobstructive CAD and 30.3% in obstructive CAD. A limitation was that almost 10% of patients were lost to follow-up.

CCTA seems to be a useful diagnostic tool with good safety in the early triage of patients in the emergency department.

## 10. Prognostic Data of CCTA

As for CAC many studies evaluated the prognostic implication of CCTA in symptomatic patients (Figure 7). Al-Mallah *et al.* followed up 8627 patients with suspected CAD concerning outcomes of death and myocardial infarction [68]. CCTA results added discriminatory power to the Agatston-Score regarding outcomes. This additional value was highest in patients with moderate calcium scores (Agatston 1–100). There is strong evidence supported by many trials that individuals without any signs of CAD in CCTA have an excellent prognosis [69–73]. Rates of MACE approached 0% in the years of follow-up in the studies. CCTA has not only incremental value over Calcium-Scoring but also over routine risk factors of cardiovascular disease [71,74–76].

On the other hand, individuals with signs of CAD on the images can be stratified regarding the risk of cardiovascular events according to different findings. Ahmadi *et al.* examined 3499 symptomatic patients of which 1102 had nonobstructive CAD; these patients were followed-up for 10 years [74]. Among the patients with plaques, event-free survival was best for patients with calcified plaques (98.6%) and decreased in mixed plaques (96.7%) and further decreased in non-calcified plaques (90.4%). Mortality rose proportionally to the amount of diseased vessels (1-, 2- or 3-vessel disease). Hou *et al.* did a follow-up on 5007 patients for myocardial infarction, death or coronary revascularization (MACE) [76]. MACE occurred in 0.8% with no plaque, 3.7% with nonobstructive disease, 27.6% with 1-vessel, 35.5% with 2-vessel and 57.7% with 3-vessel-disease.

No standardized score—such as the Agatston-Score for calcified plaques—exists for quantifying non-calcified plaques. Such a score which comprises the numbers of coronary segments with different morphology and amount of plaque is only used in studies [68,75]. In these, cardiac events are related to higher scores.

To conclude, rates of cardiovascular risk and MACE go hand in hand with the amount of plaque in the coronary arteries. Plaque morphology may play an important role, but it cannot be diagnosed on native calcium scans. A non-negligible number of individuals have non-calcified plaques in the absence of calcium, ranging from 4% to 25% according to selection of study population [68,78]. These patients might benefit most from CCTA for risk stratification over CAC although no data exist to prove that, and therefore CCTA is not recommended for that purpose.

| | Absolute Events CCTA+ | Absolute Events CCTA- | | -LR (95% CI) |
|---|---|---|---|---|
| **Any CAD vs No CAD** | n = 6035 | n = 3557 | | |
| MACE (18 studies) | 497 | 23 | | 0.008 (0.0004, 0.17) |
| Death (17 studies) | 161 | 21 | | 0.12 (0.02, 0.76) |
| | | | | |
| **Non-obs. vs No CAD** | n = 3185 | n = 3557 | | |
| MACE (16 studies) | 90 | 23 | | 0.20 (0.04, 0.96) |
| Death (14 studies) | 58 | 21 | | 0.34 (0.06, 1.90) |
| | | | | |
| **Obstr. vs Non-obstr. CAD** | n = 2772 | n = 3185 | | |
| MACE (16 studies)* | 381 | 90 | | 0.21 (0.12, 0.36) |
| Death (14 studies) | 101 | 58 | | 0.64 (0.24, 1.68) |
| Death or MI (10 studies) | 150 | 65 | | 0.46 (0.23, 0.92) |
| MI (10 studies) | 49 | 7 | | 0.39 (0.10, 1.53) |
| Revasc (9 studies) | 199 | 14 | | 0.13 (0.07, 0.25) |

Negative Likelihood Ratio: 0 .5 1 1.5 2

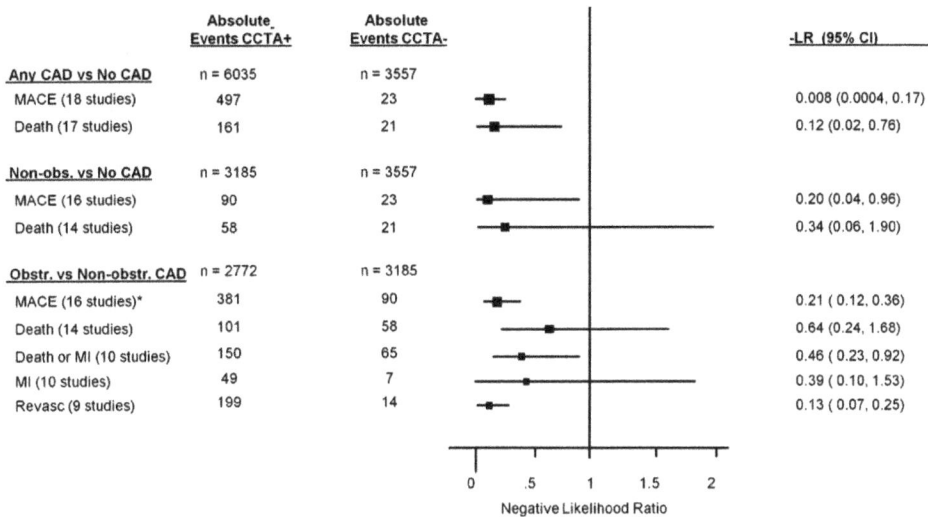

**Figure 7.** Pooled lifetime risk (LR) for major adverse cardiac events (MACE), death, death or myocardial infarctions (MI), MI and revascularization stratified by Coronary CT Angiography (CCTA) findings. Reproduced from [77] with permission from Hulten, *J. Am. Coll. Cardiol.*; published by Elsevier, 2011.

## 11. Summary and Future Directions

CCTA is a very reliable diagnostic tool for proving and ruling out obstructive CAD. Still, and in spite of remarkable improvements in image quality due to progress in technology, there are factors such as severe obesity or calcifications that impede diagnostic accuracy. Studies have demonstrated that the type and amount of plaque are related to cardiac events independently of the remaining lumen diameter. Important prognostic implications have been proven especially for calcium scoring. Different types of plaque can be visualized by CCTA, and high-risk lesions being prone to acute coronary syndromes have been described. This ability of CCTA may be able to provide important prognostic information, particularly compared with ICA. However, it remains to be demonstrated whether specific treatment of morphological high-risk plaques on CCTA will translate into fewer future cardiac events and improvements in prognosis.

**Conflicts of Interest:** The authors declare no conflict of interest.

## References

1. Leipsic, J.; Abbara, S.; Achenbach, S.; Cury, R.; Earls, J.P.; Mancini, G.J.; Nieman, K.; Pontone, G.; Raff, G.L. SCCT guidelines for the interpretation and reporting of coronary CT angiography: A report of the Society of Cardiovascular Computed Tomography Guidelines Committee. *J. Cardiovasc. Comput. Tomogr.* **2014**, *8*, 342–358. [CrossRef] [PubMed]
2. Tabas, I.; Williams, K.J.; Borén, J. Subendothelial lipoprotein retention as the initiating process in atherosclerosis: Update and therapeutic implications. *Circulation* **2007**, *116*, 1832–1844. [CrossRef] [PubMed]
3. Stary, H.C.; Chandler, A.B.; Glagov, S.; Guyton, J.R.; Insull, W.; Rosenfeld, M.E.; Schaffer, S.A.; Schwartz, C.J.; Wagner, W.D.; Wissler, R.W. A definition of initial, fatty streak, and intermediate lesions of atherosclerosis: A report from the Committee on Vascular Lesions of the Council on Arteriosclerosis, American Heart Association. *Arterioscler. Thromb. Vasc. Biol.* **1994**, *14*, 840–856. [CrossRef]

4. Stary, H.C.; Chandler, A.B.; Dinsmore, R.E.; Fuster, V.; Glagov, S.; Insull, W.; Rosenfeld, M.E.; Schwartz, C.J.; Wagner, W.D.; Wissler, R.W. A definition of advanced types of atherosclerotic lesions and a histological classification of atherosclerosis: A report from the Committee on Vascular Lesions of the Council on Arteriosclerosis, American Heart Association. *Circulation* **1995**, *92*, 1355–1374. [CrossRef] [PubMed]
5. Glagov, S.; Weisenberg, E.; Zarins, C.K.; Stankunavicius, R.; Kolettis, G.J. Compensatory enlargement of human atherosclerotic coronary arteries. *N. Engl. J. Med.* **1987**, *316*, 1371–1375. [CrossRef] [PubMed]
6. Virmani, R.; Burke, A.P.; Farb, A.; Kolodgie, F.D. Pathology of the vulnerable plaque. *J. Am. Coll. Cardiol.* **2006**, *47*, C13–C18. [CrossRef] [PubMed]
7. Agatston, A.S.; Janowitz, W.R.; Hildner, F.J.; Zusmer, N.R.; Viamonte, M., Jr.; Detrano, R. Quantification of coronary artery calcium using ultrafast computed tomography. *J. Am. Coll. Cardiol.* **1990**, *15*, 827–832. [CrossRef]
8. O'Rourke, R.A.; Brundage, B.H.; Froelicher, V.F.; Greenland, P.; Grundy, S.M.; Hachamovitch, R.; Pohost, G.M.; Shaw, L.J.; Weintraub, W.S.; Winters, W.L.; *et al.* American college of cardiology/American heart association expert consensus document on electron-beam computed tomography for the diagnosis and prognosis of coronary artery disease committee members. *J. Am. Coll. Cardiol.* **2000**, *36*, 326–340. [CrossRef] [PubMed]
9. Pohle, K.; Ropers, D.; Mäffert, R.; Geitner, P.; Moshage, W.; Regenfus, M.; Kusus, M.; Daniel, W.G.; Achenbach, S. Coronary calcifications in young patients with first, unheralded myocardial infarction: A risk factor matched analysis by electron beam tomography. *Heart* **2003**, *89*, 625–628. [CrossRef] [PubMed]
10. Schmermund, A.; Schwartz, R.S.; Adamzik, M.; Sangiorgi, G.; Pfeifer, E.A.; Rumberger, J.A.; Burke, A.P.; Farb, A.; Virmani, R. Coronary atherosclerosis in unheralded sudden coronary death under age 50: Histo-pathologic comparison with "healthy" subjects dying out of hospital. *Atherosclerosis* **2001**, *155*, 499–508. [CrossRef] [PubMed]
11. Davies, M.J. The composition of coronary-artery plaques. *N. Engl. J. Med.* **1997**, *336*, 1312–1314. [CrossRef] [PubMed]
12. Marwan, M.; Ropers, D.; Pflederer, T.; Daniel, W.G.; Achenbach, S. Clinical characteristics of patients with obstructive coronary lesions in the absence of coronary calcification: An evaluation by coronary CT angiography. *Heart* **2009**, *95*, 1056–1060. [CrossRef]
13. Gottlieb, I.; Miller, J.M.; Arbab-Zadeh, A.; Dewey, M.; Clouse, M.E.; Sara, L.; Niinuma, H.; Bush, D.E.; Paul, N.; Vavere, A.L.; *et al.* The absence of coronary calcification does not exclude obstructive coronary artery disease or the need for revascularization in patients referred for conventional coronary angiography. *J. Am. Coll. Cardiol.* **2010**, *55*, 627–634. [CrossRef] [PubMed]
14. Boström, K.; Watson, K.E.; Horn, S.; Wortham, C.; Herman, I.M.; Demer, L.L. Bone morphogenetic protein expression in human atherosclerotic lesions. *J. Clin. Investig.* **1993**, *91*, 1800–1809. [CrossRef] [PubMed]
15. Jeziorska, M.; McCollum, C.; Wooley, D.E. Observations on bone formation and remodelling in advanced atherosclerotic lesions of human carotid arteries. *Virchows Arch.* **1998**, *433*, 559–565. [CrossRef] [PubMed]
16. Stary, H.C. The sequence of cell and matrix changes in atherosclerotic lesions of coronary arteries in the first forty years of life. *Eur. Heart J.* **1990**, *11* (Suppl. E), 3–19. [CrossRef]
17. Greenland, P.; Bonow, R.O.; Brundage, B.H.; Budoff, M.J.; Eisenberg, M.J.; Grundy, S.M.; Lauer, M.S.; Post, W.S.; Raggi, P.; Redberg, R.F.; *et al.* ACCF/AHA 2007 clinical expert consensus document on coronary artery calcium scoring by computed tomography in global cardiovascular risk assessment and in evaluation of patients with chest pain. *J. Am. Coll. Cardiol.* **2007**, *49*, 378–402. [CrossRef] [PubMed]
18. McClelland, R.L.; Chung, H.; Detrano, R.; Post, W.; Kronmal, R.A. Distribution of coronary artery calcium by race, gender, and age: Results from the Multi-Ethnic Study of Atherosclerosis (MESA). *Circulation* **2006**, *113*, 30–37. [CrossRef] [PubMed]
19. Raggi, P.; Cooil, B.; Ratti, C.; Callister, T.Q.; Budoff, M. Progression of coronary artery calcium and occurrence of myocardial infarction in patients with and without diabetes mellitus. *Hypertension* **2005**, *46*, 238–243. [CrossRef] [PubMed]
20. Goh, V.K.; Lau, C.P.; Mohlenkamp, S.; Rumberger, J.A.; Achenbach, S.; Budoff, M.J. Outcome of coronary plaque burden: A 10-year follow-up of aggressive medical management. *Cardiovasc. Ultrasound* **2010**, *8*, 5. [CrossRef] [PubMed]
21. Cassidy-Bushrow, A.E.; Bielak, L.F.; Sheedy, P.F.; Turner, S.T.; Kullo, I.J.; Lin, X.; Peyser, P.A. Coronary artery calcification progression is heritable. *Circulation* **2007**, *116*, 25–31. [CrossRef] [PubMed]

22. Erbel, R.; Lehmann, N.; Churzidse, S.; Rauwolf, M.; Mahabadi, A.A.; Möhlenkamp, S.; Moebus, S.; Bauer, M.; Kälsch, H.; Budde, T.; *et al.* Heinz Nixdorf Recall study investigators progression of coronary artery calcification seems to be inevitable, but predictable—Results of the Heinz Nixdorf Recall (HNR) study. *Eur. Heart J.* **2014**, *35*, 2960–2971. [CrossRef] [PubMed]

23. Sarwar, A.; Shaw, L.J.; Shapiro, M.D.; Blankstein, R.; Hoffmann, U.; Hoffman, U.; Cury, R.C.; Abbara, S.; Brady, T.J.; Budoff, M.J.; *et al.* Diagnostic and prognostic value of absence of coronary artery calcification. *JACC Cardiovasc. Imaging* **2009**, *2*, 675–688. [CrossRef] [PubMed]

24. Greenland, P.; LaBree, L.; Azen, S.P.; Doherty, T.M.; Detrano, R.C. Coronary artery calcium score combined with Framingham score for risk prediction in asymptomatic individuals. *JAMA* **2004**, *291*, 210–215. [CrossRef] [PubMed]

25. Budoff, M.J.; Hokanson, J.E.; Nasir, K.; Shaw, L.J.; Kinney, G.L.; Chow, D.; Demoss, D.; Nuguri, V.; Nabavi, V.; Ratakonda, R.; *et al.* Progression of coronary artery calcium predicts all-cause mortality. *JACC Cardiovasc. Imaging* **2010**, *3*, 1229–1236. [CrossRef] [PubMed]

26. Raggi, P.; Callister, T.Q.; Shaw, L.J. Progression of coronary artery calcium and risk of first myocardial infarction in patients receiving cholesterol-lowering therapy. *Arterioscler. Thromb. Vasc. Biol.* **2004**, *24*, 1272–1277. [CrossRef] [PubMed]

27. Meijboom, W.B.; van Mieghem, C.A.G.; Mollet, N.R.; Pugliese, F.; Weustink, A.C.; van Pelt, N.; Cademartiri, F.; Nieman, K.; Boersma, E.; de Jaegere, P.; *et al.* 64-Slice computed tomography coronary angiography in patients with high, intermediate, or low pretest probability of significant coronary artery disease. *J. Am. Coll. Cardiol.* **2007**, *50*, 1469–1475. [CrossRef] [PubMed]

28. Achenbach, S.; Goroll, T.; Seltmann, M.; Pflederer, T.; Anders, K.; Ropers, D.; Daniel, W.G.; Uder, M.; Lell, M.; Marwan, M. Detection of coronary artery stenoses by low-dose, prospectively ECG-triggered, high-pitch spiral coronary CT angiography. *JACC Cardiovasc. Imaging* **2011**, *4*, 328–337. [CrossRef] [PubMed]

29. Kröpil, P.; Rojas, C.A.; Ghoshhajra, B.; Lanzman, R.S.; Miese, F.R.; Scherer, A.; Kalra, M.; Abbara, S. Prospectively ECG-triggered high-pitch spiral acquisition for cardiac CT angiography in routine clinical practice: Initial results. *J. Thorac. Imaging* **2012**, *27*, 194–201. [CrossRef] [PubMed]

30. Yin, W.H.; Lu, B.; Hou, Z.H.; Li, N.; Han, L.; Wu, Y.J.; Niu, H.X.; Silverman, J.R.; Nicola De Cecco, C.; Schoepf, U.J. Detection of coronary artery stenosis with sub-milliSievert radiation dose by prospectively ECG-triggered high-pitch spiral CT angiography and iterative reconstruction. *Eur. Radiol.* **2013**, *23*, 2927–2933. [CrossRef] [PubMed]

31. Hausleiter, J.; Meyer, T.; Hadamitzky, M.; Kastrati, A.; Martinoff, S.; Schömig, A. Prevalence of noncalcified coronary plaques by 64-slice computed tomography in patients with an intermediate risk for significant coronary artery disease. *J. Am. Coll. Cardiol.* **2006**, *48*, 312–318. [CrossRef] [PubMed]

32. Budoff, M.J.; Dowe, D.; Jollis, J.G.; Gitter, M.; Sutherland, J.; Halamert, E.; Scherer, M.; Bellinger, R.; Martin, A.; Benton, R.; *et al.* Diagnostic performance of 64-multidetector row coronary computed tomographic angiography for evaluation of coronary artery stenosis in individuals without known coronary artery disease: results from the prospective multicenter ACCURACY (Assessment by Coronary Computed Tomographic Angiography of Individuals Undergoing Invasive Coronary Angiography) trial. *J. Am. Coll. Cardiol.* **2008**, *52*, 1724–1732. [CrossRef] [PubMed]

33. Choi, B.J.; Kang, D.K.; Tahk, S.J.; Choi, S.Y.; Yoon, M.H.; Lim, H.S.; Kang, S.J.; Yang, H.M.; Park, J.S.; Zheng, M.; *et al.* Comparison of 64-slice multidetector computed tomography with spectral analysis of intravascular ultrasound backscatter signals for characterizations of noncalcified coronary arterial plaques. *Am. J. Cardiol.* **2008**, *102*, 988–993. [CrossRef] [PubMed]

34. Fischer, C.; Hulten, E.; Belur, P.; Smith, R.; Voros, S.; Villines, T.C. Coronary CT angiography *versus* intravascular ultrasound for estimation of coronary stenosis and atherosclerotic plaque burden: A meta-analysis. *J. Cardiovasc. Comput. Tomogr.* **2013**, *7*, 256–266. [CrossRef] [PubMed]

35. Meijboom, W.B.; Meijs, M.F.L.; Schuijf, J.D.; Cramer, M.J.; Mollet, N.R.; van Mieghem, C.A.G.; Nieman, K.; van Werkhoven, J.M.; Pundziute, G.; Weustink, A.C.; *et al.* Diagnostic accuracy of 64-slice computed tomography coronary angiography: A prospective, multicenter, multivendor study. *J. Am. Coll. Cardiol.* **2008**, *52*, 2135–2144. [CrossRef] [PubMed]

36. Leber, A.W.; Knez, A.; von Ziegler, F.; Becker, A.; Nikolaou, K.; Paul, S.; Wintersperger, B.; Reiser, M.; Becker, C.R.; Steinbeck, G.; *et al.* Quantification of obstructive and nonobstructive coronary lesions by 64-slice computed tomography: A comparative study with quantitative coronary angiography and intravascular ultrasound. *J. Am. Coll. Cardiol.* **2005**, *46*, 147–154. [CrossRef] [PubMed]

37. Alkadhi, H.; Scheffel, H.; Desbiolles, L.; Gaemperli, O.; Stolzmann, P.; Plass, A.; Goerres, G.W.; Luescher, T.F.; Genoni, M.; Marincek, B.; *et al.* Dual-source computed tomography coronary angiography: Influence of obesity, calcium load, and heart rate on diagnostic accuracy. *Eur. Heart J.* **2008**, *29*, 766–776. [CrossRef] [PubMed]

38. Voros, S.; Rinehart, S.; Qian, Z.; Joshi, P.; Vazquez, G.; Fischer, C.; Belur, P.; Hulten, E.; Villines, T.C. Coronary atherosclerosis imaging by coronary CT angiography: Current status, correlation with intravascular interrogation and meta-analysis. *JACC Cardiovasc. Imaging* **2011**, *4*, 537–548. [CrossRef] [PubMed]

39. Cheng, V.Y.; Nakazato, R.; Dey, D.; Gurudevan, S.; Tabak, J.; Budoff, M.J.; Karlsberg, R.P.; Min, J.; Berman, D.S. Reproducibility of coronary artery plaque volume and composition quantification by 64-detector row coronary computed tomographic angiography: An intraobserver, interobserver, and interscan variability study. *J. Cardiovasc. Comput. Tomogr.* **2009**, *3*, 312–320. [CrossRef] [PubMed]

40. Leber, A.W.; Becker, A.; Knez, A.; von Ziegler, F.; Sirol, M.; Nikolaou, K.; Ohnesorge, B.; Fayad, Z.A.; Becker, C.R.; Reiser, M.; *et al.* Accuracy of 64-slice computed tomography to classify and quantify plaque volumes in the proximal coronary system: A comparative study using intravascular ultrasound. *J. Am. Coll. Cardiol.* **2006**, *47*, 672–677. [CrossRef] [PubMed]

41. Pflederer, T.; Schmid, M.; Ropers, D.; Ropers, U.; Komatsu, S.; Daniel, W. G.; Achenbach, S. Interobserver variability of 64-slice computed tomography for the quantification of non-calcified coronary atherosclerotic plaque. *Rofo* **2007**, *179*, 953–957. [CrossRef] [PubMed]

42. Boogers, M.J.; Broersen, A.; van Velzen, J.E.; de Graaf, F.R.; El-Naggar, H.M.; Kitslaar, P.H.; Dijkstra, J.; Delgado, V.; Boersma, E.; de Roos, A.; *et al.* Automated quantification of coronary plaque with computed tomography: Comparison with intravascular ultrasound using a dedicated registration algorithm for fusion-based quantification. *Eur. Heart J.* **2012**, *33*, 1007–1016. [CrossRef] [PubMed]

43. Rolf, A.; Werner, G.S.; Schuhbäck, A.; Rixe, J.; Möllmann, H.; Nef, H.M.; Gundermann, C.; Liebetrau, C.; Krombach, G.A.; Hamm, C.W.; *et al.* Preprocedural coronary CT angiography significantly improves success rates of PCI for chronic total occlusion. *Int. J. Cardiovasc. Imaging* **2013**, *29*, 1819–1827. [CrossRef] [PubMed]

44. Ueno, K.; Kawamura, A.; Onizuka, T.; Kawakami, T.; Nagatomo, Y.; Hayashida, K.; Yuasa, S.; Maekawa, Y.; Anzai, T.; Jinzaki, M.; *et al.* Effect of preoperative evaluation by multidetector computed tomography in percutaneous coronary interventions of chronic total occlusions. *Int. J. Cardiol.* **2012**, *156*, 76–79. [CrossRef] [PubMed]

45. Soon, K.H.; Cox, N.; Wong, A.; Chaitowitz, I.; Macgregor, L.; Santos, P.T.; Selvanayagam, J.B.; Farouque, H.M.O.; Rametta, S.; Bell, K.W.; *et al.* CT coronary angiography predicts the outcome of percutaneous coronary intervention of chronic total occlusion. *J. Interv. Cardiol.* **2007**, *20*, 359–366. [CrossRef] [PubMed]

46. Mintz, G.S.; Nissen, S.E.; Anderson, W.D.; Bailey, S.R.; Erbel, R.; Fitzgerald, P.J.; Pinto, F.J.; Rosenfield, K.; Siegel, R.J.; Tuzcu, E.M.; *et al.* American college of cardiology clinical expert consensus document on standards for acquisition, measurement and reporting of Intravascular Ultrasound Studies (IVUS): A report of the American college of cardiology task force on clinical expert consensus documents. *J. Am. Coll. Cardiol.* **2001**, *37*, 1478–1492. [CrossRef] [PubMed]

47. Nissen, S.E.; Yock, P. Intravascular ultrasound: Novel pathophysiological insights and current clinical applications. *Circulation* **2001**, *103*, 604–616. [CrossRef] [PubMed]

48. Marwan, M.; Taher, M.A.; El Meniawy, K.; Awadallah, H.; Pflederer, T.; Schuhbäck, A.; Ropers, D.; Daniel, W.G.; Achenbach, S. *In vivo* CT detection of lipid-rich coronary artery atherosclerotic plaques using quantitative histogram analysis: A head to head comparison with IVUS. *Atherosclerosis* **2011**, *215*, 110–115. [CrossRef] [PubMed]

49. Schroeder, S.; Flohr, T.; Kopp, A.F.; Meisner, C.; Kuettner, A.; Herdeg, C.; Baumbach, A.; Ohnesorge, B. Accuracy of density measurements within plaques located in artificial coronary arteries by X-ray multislice CT: Results of a phantom study. *J. Comput. Assist. Tomogr.* **2001**, *25*, 900–906. [CrossRef] [PubMed]

50. Cademartiri, F.; Mollet, N.R.; Runza, G.; Bruining, N.; Hamers, R.; Somers, P.; Knaapen, M.; Verheye, S.; Midiri, M.; Krestin, G.P.; *et al.* Influence of intracoronary attenuation on coronary plaque measurements using multislice computed tomography: Observations in an ex vivo model of coronary computed tomography angiography. *Eur. Radiol.* **2005**, *15*, 1426–1431. [CrossRef] [PubMed]

51. Obaid, D.R.; Calvert, P.A.; Gopalan, D.; Parker, R.A.; West, N.E.J.; Goddard, M.; Rudd, J.H.F.; Bennett, M.R. Dual-energy computed tomography imaging to determine atherosclerotic plaque composition: A prospective study with tissue validation. *J. Cardiovasc. Comput. Tomogr.* **2014**, *8*, 230–237. [CrossRef] [PubMed]

52. Narula, J.; Finn, A.V.; Demaria, A.N. Picking plaques that pop . . . . *J. Am. Coll. Cardiol.* **2005**, *45*, 1970–1973. [CrossRef]

53. Motoyama, S.; Kondo, T.; Sarai, M.; Sugiura, A.; Harigaya, H.; Sato, T.; Inoue, K.; Okumura, M.; Ishii, J.; Anno, H.; *et al.* Multislice computed tomographic characteristics of coronary lesions in acute coronary syndromes. *J. Am. Coll. Cardiol.* **2007**, *50*, 319–326. [CrossRef] [PubMed]

54. Motoyama, S.; Sarai, M.; Harigaya, H.; Anno, H.; Inoue, K.; Hara, T.; Naruse, H.; Ishii, J.; Hishida, H.; Wong, N.D.; *et al.* Computed tomographic angiography characteristics of atherosclerotic plaques subsequently resulting in acute coronary syndrome. *J. Am. Coll. Cardiol.* **2009**, *54*, 49–57. [CrossRef]

55. Hoffmann, U.; Moselewski, F.; Nieman, K.; Jang, I.K.; Ferencik, M.; Rahman, A.M.; Cury, R.C.; Abbara, S.; Joneidi-Jafari, H.; Achenbach, S.; *et al.* Noninvasive assessment of plaque morphology and composition in culprit and stable lesions in acute coronary syndrome and stable lesions in stable angina by multidetector computed tomography. *J. Am. Coll. Cardiol.* **2006**, *47*, 1655–1662. [CrossRef] [PubMed]

56. Maurovich-Horvat, P.; Schlett, C.L.; Alkadhi, H.; Nakano, M.; Otsuka, F.; Stolzmann, P.; Scheffel, H.; Ferencik, M.; Kriegel, M.F.; Seifarth, H.; *et al.* The napkin-ring sign indicates advanced atherosclerotic lesions in coronary CT angiography. *JACC Cardiovasc. Imaging* **2012**, *5*, 1243–1252. [CrossRef] [PubMed]

57. Otsuka, K.; Fukuda, S.; Tanaka, A.; Nakanishi, K.; Taguchi, H.; Yoshikawa, J.; Shimada, K.; Yoshiyama, M. Napkin-ring sign on coronary CT angiography for the prediction of acute coronary syndrome. *JACC Cardiovasc. Imaging* **2013**, *6*, 448–457. [CrossRef] [PubMed]

58. Dweck, M.R.; Chow, M.W.L.; Joshi, N.V.; Williams, M.C.; Jones, C.; Fletcher, A.M.; Richardson, H.; White, A.; McKillop, G.; van Beek, E.J.R.; *et al.* Coronary arterial [18]F-sodium fluoride uptake: A novel marker of plaque biology. *J. Am. Coll. Cardiol.* **2012**, *59*, 1539–1548. [CrossRef] [PubMed]

59. Joshi, N.V.; Vesey, A.T.; Williams, M.C.; Shah, A.S.V.; Calvert, P.A.; Craighead, F.H.M.; Yeoh, S.E.; Wallace, W.; Salter, D.; Fletcher, A.M.; *et al.* [18]F-fluoride positron emission tomography for identification of ruptured and high-risk coronary atherosclerotic plaques: A prospective clinical trial. *Lancet* **2014**, *383*, 705–713. [CrossRef] [PubMed]

60. De Bruyne, B.; Fearon, W.F.; Pijls, N.H.J.; Barbato, E.; Tonino, P.; Piroth, Z.; Jagic, N.; Mobius-Winckler, S.; Rioufol, G.; Witt, N.; *et al.* Fractional flow reserve-guided PCI for stable coronary artery disease. *N. Engl. J. Med.* **2014**, *371*, 1208–1217. [CrossRef] [PubMed]

61. Meijboom, W.B.; van Mieghem, C.A.G.; van Pelt, N.; Weustink, A.; Pugliese, F.; Mollet, N.R.; Boersma, E.; Regar, E.; van Geuns, R.J.; de Jaegere, P.J.; *et al.* Comprehensive assessment of coronary artery stenoses: Computed tomography coronary angiography *versus* conventional coronary angiography and correlation with fractional flow reserve in patients with stable angina. *J. Am. Coll. Cardiol.* **2008**, *52*, 636–643. [CrossRef] [PubMed]

62. Min, J.K.; Leipsic, J.; Pencina, M.J.; Berman, D.S.; Koo, B.K.; van Mieghem, C.; Erglis, A.; Lin, F.Y.; Dunning, A.M.; Apruzzese, P.; *et al.* Diagnostic accuracy of fractional flow reserve from anatomic CT angiography. *JAMA* **2012**, *308*, 1237–1245. [CrossRef] [PubMed]

63. Gaur, S.; Bezerra, H.G.; Lassen, J.F.; Christiansen, E.H.; Tanaka, K.; Jensen, J.M.; Oldroyd, K.G.; Leipsic, J.; Achenbach, S.; Kaltoft, A.K.; *et al.* Fractional flow reserve derived from coronary CT angiography: Variation of repeated analyses. *J. Cardiovasc. Comput. Tomogr.* **2014**, *8*, 307–314. [CrossRef] [PubMed]

64. Taylor, C.A.; Fonte, T.A.; Min, J.K. Computational fluid dynamics applied to cardiac computed tomography for noninvasive quantification of fractional flow reserve: Scientific basis. *J. Am. Coll. Cardiol.* **2013**, *61*, 2233–2241. [CrossRef] [PubMed]

65. Goldstein, J.A.; Gallagher, M.J.; O'Neill, W.W.; Ross, M.A.; O'Neil, B.J.; Raff, G.L. A randomized controlled trial of multi-slice coronary computed tomography for evaluation of acute chest pain. *J. Am. Coll. Cardiol.* **2007**, *49*, 863–871. [CrossRef] [PubMed]

66. Hoffmann, U.; Bamberg, F.; Chae, C.U.; Nichols, J.H.; Rogers, I.S.; Seneviratne, S.K.; Truong, Q.A.; Cury, R.C.; Abbara, S.; Shapiro, M.D.; *et al.* Coronary computed tomography angiography for early triage of patients with acute chest pain: The ROMICAT (Rule Out Myocardial Infarction using Computer Assisted Tomography) trial. *J. Am. Coll. Cardiol.* **2009**, *53*, 1642–1650. [CrossRef] [PubMed]

67. Schlett, C.L.; Banerji, D.; Siegel, E.; Bamberg, F.; Lehman, S.J.; Ferencik, M.; Brady, T.J.; Nagurney, J.T.; Hoffmann, U.; Truong, Q.A. Prognostic value of CT angiography for major adverse cardiac events in patients with acute chest pain from the emergency department: 2-Year outcomes of the ROMICAT trial. *JACC Cardiovasc. Imaging* **2011**, *4*, 481–491. [CrossRef] [PubMed]

68. Al-Mallah, M.H.; Qureshi, W.; Lin, F.Y.; Achenbach, S.; Berman, D.S.; Budoff, M.J.; Callister, T.Q.; Chang, H.J.; Cademartiri, F.; Chinnaiyan, K.; *et al.* Does coronary CT angiography improve risk stratification over coronary calcium scoring in symptomatic patients with suspected coronary artery disease? Results from the prospective multicenter international CONFIRM registry. *Eur. Heart J. Cardiovasc. Imaging* **2014**, *15*, 267–274. [CrossRef] [PubMed]

69. Andreini, D.; Pontone, G.; Mushtaq, S.; Bartorelli, A.L.; Bertella, E.; Antonioli, L.; Formenti, A.; Cortinovis, S.; Veglia, F.; Annoni, A.; *et al.* A long-term prognostic value of coronary CT angiography in suspected coronary artery disease. *JACC Cardiovasc. Imaging* **2012**, *5*, 690–701. [CrossRef] [PubMed]

70. Hadamitzky, M.; Täubert, S.; Deseive, S.; Byrne, R.A.; Martinoff, S.; Schömig, A.; Hausleiter, J. Prognostic value of coronary computed tomography angiography during 5 years of follow-up in patients with suspected coronary artery disease. *Eur. Heart J.* **2013**, *34*, 3277–3285. [CrossRef] [PubMed]

71. Hadamitzky, M.; Freissmuth, B.; Meyer, T.; Hein, F.; Kastrati, A.; Martinoff, S.; Schömig, A.; Hausleiter, J. Prognostic value of coronary computed tomographic angiography for prediction of cardiac events in patients with suspected coronary artery disease. *JACC Cardiovasc. Imaging* **2009**, *2*, 404–411. [CrossRef] [PubMed]

72. Min, J.K.; Shaw, L.J.; Devereux, R.B.; Okin, P.M.; Weinsaft, J.W.; Russo, D.J.; Lippolis, N.J.; Berman, D.S.; Callister, T.Q. Prognostic value of multidetector coronary computed tomographic angiography for prediction of all-cause mortality. *J. Am. Coll. Cardiol.* **2007**, *50*, 1161–1170. [CrossRef] [PubMed]

73. Pundziute, G.; Schuijf, J.D.; Jukema, J.W.; Boersma, E.; de Roos, A.; van der Wall, E.E.; Bax, J.J. Prognostic value of multislice computed tomography coronary angiography in patients with known or suspected coronary artery disease. *J. Am. Coll. Cardiol.* **2007**, *49*, 62–70. [CrossRef] [PubMed]

74. Ahmadi, N.; Nabavi, V.; Hajsadeghi, F.; Flores, F.; French, W.J.; Mao, S.S.; Shavelle, D.; Ebrahimi, R.; Budoff, M. Mortality incidence of patients with non-obstructive coronary artery disease diagnosed by computed tomography angiography. *Am. J. Cardiol.* **2011**, *107*, 10–16. [CrossRef] [PubMed]

75. Chow, B.J.W.; Wells, G.A.; Chen, L.; Yam, Y.; Galiwango, P.; Abraham, A.; Sheth, T.; Dennie, C.; Beanlands, R.S.; Ruddy, T.D. Prognostic value of 64-slice cardiac computed tomography severity of coronary artery disease, coronary atherosclerosis, and left ventricular ejection fraction. *J. Am. Coll. Cardiol.* **2010**, *55*, 1017–1028. [CrossRef] [PubMed]

76. Hou, Z.; Lu, B.; Gao, Y.; Jiang, S.; Wang, Y.; Li, W.; Budoff, M.J. Prognostic value of coronary CT angiography and calcium score for major adverse cardiac events in outpatients. *JACC Cardiovasc. Imaging* **2012**, *5*, 990–999. [CrossRef] [PubMed]

77. Hulten, E.A.; Carbonaro, S.; Petrillo, S.P.; Mitchell, J.D.; Villines, T.C. Prognostic value of cardiac computed tomography angiography: A systematic review and meta-analysis. *J. Am. Coll. Cardiol.* **2011**, *57*, 1237–1247. [CrossRef] [PubMed]

78. Choi, E.K.; Choi, S.I.; Rivera, J.J.; Nasir, K.; Chang, S.A.; Chun, E.J.; Kim, H.K.; Choi, D.J.; Blumenthal, R.S.; Chang, H.J. Coronary computed tomography angiography as a screening tool for the detection of occult coronary artery disease in asymptomatic individuals. *J. Am. Coll. Cardiol.* **2008**, *52*, 357–365. [CrossRef] [PubMed]

International Journal of
*Molecular Sciences*

MDPI

Article

# A Variant in the Osteoprotegerin Gene Is Associated with Coronary Atherosclerosis in Patients with Rheumatoid Arthritis: Results from a Candidate Gene Study

Cecilia P. Chung [1], Joseph F. Solus [1], Annette Oeser [1], Chun Li [2], Paolo Raggi [3,*], Jeffrey R. Smith [1] and C. Michael Stein [1]

[1] Departments of Medicine and Biostatistics, Vanderbilt University, Nashville, TN 37232, USA; c.chung@vanderbilt.edu (C.P.C.); joseph.f.solus@vanderbilt.edu (J.F.S.); annette.oeser@vanderbilt.edu (A.O.); jeffrey.smith@vanderbilt.edu (J.R.S.); mike.stein@vanderbilt.edu (C.M.S.)
[2] Biostatistics, Vanderbilt University, Nashville, TN 37232, USA; Chun.li3@case.edu
[3] Mazankowski Alberta Heart Institute, Department of Medicine, University of Alberta, Edmonton, AB T6G 2B7, Canada
* Correspondence: raggi@ualberta.ca; Tel.: +1-780-407-4575; Fax: +1-780-407-7834

Academic Editor: Michael Henein
Received: 14 December 2014; Accepted: 6 February 2015; Published: 11 February 2015

**Abstract:** Objective: Patients with rheumatoid arthritis (RA) have accelerated atherosclerosis, but there is limited information about the genetic contribution to atherosclerosis in this population. Therefore, we examined the association between selected genetic polymorphisms and coronary atherosclerosis in patients with RA. Methods: Genotypes for single-nucleotide polymorphisms (SNPs) in 152 candidate genes linked with autoimmune or cardiovascular risk were measured in 140 patients with RA. The association between the presence of coronary artery calcium (CAC) and SNP allele frequency was assessed by logistic regression with adjustment for age, sex, and race. To adjust for multiple comparisons, a false discovery rate (FDR) threshold was set at 20%. Results: Patients with RA were $54 \pm 11$ years old and predominantly Caucasian (89%) and female (69%). CAC was present in 70 patients (50%). A variant in rs2073618 that encodes an Asn3Lys missense substitution in the osteoprotegerin gene (*OPG, TNFRSF11B*) was significantly associated with the presence of CAC (OR = 4.09, $p < 0.00026$) and withstands FDR correction. Conclusion: Our results suggest that a polymorphism of the *TNFRSF11B* gene, which encodes osteoprotegerin, is associated with the presence of coronary atherosclerosis in patients with RA. Replication of this finding in independent validation cohorts will be of interest.

**Keywords:** rheumatoid arthritis; genetic variation; SNP; atherosclerosis; rs2073618

## 1. Introduction

Patients with rheumatoid arthritis (RA) have accelerated atherosclerosis, with a two-fold increased risk in cardiovascular events [1]. However, traditional cardiovascular risk factors and inflammatory markers do not fully explain this increased risk [2–5]. Therefore, efforts to identify additional risk factors and improve our ability to predict cardiovascular risk in this patient population are important.

Rheumatoid arthritis and atherosclerosis are chronic inflammatory conditions and both have significant genetic susceptibility. Because heritability accounts for up to 60% of the risk in RA and 30%–60% of the risk in cardiovascular disease [6] there has been increasing interest in attempting to identify genetic markers of atherosclerosis in RA.

*Int. J. Mol. Sci.* **2015**, *16*, 3885–3894

Recent studies have described the role of a few single nucleotide polymorphisms (SNPs) in single genes, such as *MTHFR*, *TNFA*, and *ZC3HC1*, *NFKB* or other SNPs such as rs964184 on atherosclerosis in RA [7–11]. In this study, we undertook a more extensive candidate gene approach and selected polymorphisms located in 152 genes involved in inflammation, autoimmunity and cardiovascular disease (CVD). We examined the association of these selected genetic polymorphisms with coronary atherosclerosis in patients with RA.

## 2. Results

Patients with RA were $54 \pm 11$ years old on average and predominantly Caucasian (89%) and female (69%). Patients had a mean disease activity score based on the 28 joint count assessment (DAS28) of $3.6 \pm 1.6$ units and coronary calcium was detected in 70 patients (50%). Patients with coronary calcium were older, had longer disease duration, higher systolic blood pressure. There were more female patients in the group with coronary calcium than in the group without coronary calcium (Table 1).

**Table 1.** Clinical characteristics of patients with rheumatoid arthritis with and without coronary calcium.

| Patient Characteristics | With Coronary Calcium ($n$ = 70) | Without Coronary Calcium ($n$ = 70) | $p$-Value |
|---|---|---|---|
| Age (years) | $60 \pm 9$ | $48 \pm 10$ | <0.001 |
| Female sex (%) | 59% | 80% | 0.01 |
| Caucasian | 90% | 87% | 0.40 |
| Disease duration (years) | $11 \pm 11$ | $6 \pm 8$ | 0.03 |
| Systolic blood pressure | $138 \pm 21$ | $129 \pm 17$ | 0.02 |
| Diastolic blood pressure | $76 \pm 11$ | $75 \pm 10$ | 0.24 |
| Body mass index | $29 \pm 6$ | $30 \pm 7$ | 0.64 |
| Total cholesterol | $184 \pm 38$ | $184 \pm 41$ | 0.67 |
| LDL cholesterol | $113 \pm 27$ | $109 \pm 32$ | 0.42 |

After adjustment for age, gender and race, SNPs in the genes tumor necrosis factor receptor superfamily member 11 b (*TNFRSF11B*), matrix metalloproteinase-3 (*MMP3*), interleukin-12 (*IL12*), matrix metalloproteinase-9 (*MMP9*), nucleotide-binding oligomerization domain-2 (*NOD2*), C-reactive protein (*CRP*), myeloperoxidase (*MPO*), resistin (*RETN*), interferon regulatory factor 5 (*IRF5*) and Fcχ receptor 2A (*FCGR2A*) were significantly associated with the presence of CAC (Table 2). *MMP9* contained more than one significant SNP; examination of pairwise LD (Linkage disequilibrium), defined as the correlations among neighboring alleles [12], between these is consistent with a single genetic effect per gene (pairwise $R^2$ range 0.622–0.965 for all showing association). Strong linkage disequilibrium (LD) was observed between rs3918249 and rs17576 (chromosome 20, $R^2$ = 0.965). All other pairs had an $R^2 < 0.8$.

Of all the variants initially associated with the presence of coronary atherosclerosis, rs2073618, which encodes an Asn3Lys missense change in the osteoprotegerin gene (*TNFRSF11B*), was most highly associated with the presence of coronary artery calcium (CAC) (OR = 4.09, $p$ = 0.00026). This association remained after false discovery rate (FDR) correction.

In the 139 patients in whom rs2073618 genotypes were available, 37 (26.6%) had the CC, 62 (44.6%) the CG, and 40 (28.8%) the GG genotypes. Figure 1 depicts the prevalence of coronary calcium by genotype. Among RA patients with CC genotype, 75.7% had coronary calcium; as compared with 43.6% in those with CG genotype and with 37.5% in those with GG genotype ($p$ = 0.001). However, in a *post-hoc* analysis examining the association of rs2073618 genotypes and serum osteoprotegerin concentrations, differences were non-significant [13]. (Median (interquartile range, IQR) concentrations were 1548 (1042–2509), 1497 (1100–1698), and 1657 (1132–2105) pg/mL, respectively, $p$ = 0.38).

**Table 2.** Genetic association with coronary atherosclerosis in patients with rheumatoid arthritis (RA).

| SNP | Gene | Major, Minor Allele | Minor Allele Frequency | Odds Ratio * (95% C.I.) | *p*-Value |
|---|---|---|---|---|---|
| rs2073618 ** | *TNFRSF11B* | G,C | 0.36 | 4.09 (1.93–8.70) | 0.00026 |
| rs522616 | *MMP3* | T,C | 0.29 | 4.43 (1.77–11.10) | 0.001 |
| rs2853694 | *IL12B* | T,C | 0.29 | 3.09 (1.53–6.24) | 0.002 |
| rs3918249 | *MMP9* | T,C | 0.49 | 0.36 (0.18–0.69) | 0.002 |
| rs17576 | *MMP9* | A,G | 0.45 | 0.34 (0.17–0.68) | 0.002 |
| rs3918253 | *MMP9* | C,T | 0.28 | 0.38 (0.20–0.72) | 0.003 |
| rs2274756 | *MMP9* | G,A | 0.16 | 0.24 (0.09–0.64) | 0.004 |
| rs751271 | *NOD2* | T,G | 0.49 | 2.57 (1.33–4.96) | 0.005 |
| rs650108 | *MMP3* | G,A | 0.41 | 2.82 (1.29–6.17) | 0.009 |
| rs1800947 | *CRP* | C,G | 0.04 | 5.93 (1.43–24.71) | 0.014 |
| rs9562414 | *TNFSF11* | A,G | 0.06 | 0.25 (0.08–0.79) | 0.019 |
| rs2107545 | *MPO* | T,C | 0.26 | 0.42 (0.20–0.89) | 0.023 |
| rs3745367 | *RETN* | G,A | 0.39 | 0.45 (0.22–0.91) | 0.026 |
| rs10954213 | *IRF5* | G,A | 0.47 | 0.50 (0.27–0.94) | 0.030 |
| rs633137 | *TNFSF11* | T,A | 0.08 | 0.34 (0.13–0.91) | 0.031 |
| rs2243828 | *MPO* | A,G | 0.23 | 0.43 (0.20–0.96) | 0.040 |
| rs1801274 | *FCGR2A* | A,G | 0.43 | 0.52 (0.28–1.00) | 0.049 |

* Odds ratios for the comparison between minor and major allele; SNP = single-nucleotide polymorphism; C.I. = confidence interval; ** rs2073618 is the only SNP that remains significant after false discovery rate (FDR) correction.

## A Variant in *TNFRSF11B (rs2073618)* is Associated with the Presence of Coronary Atherosclerosis in RA

**Figure 1.** Prevalence of coronary calcium by genotype. The *p*-value was obtained using the Fisher's exact test.

## 3. Discussion

To the best of our knowledge, this is the first study examining the relationship between relevant published genetic variants in a range of genes of biological relevance to coronary atherosclerosis and inflammation in patients with RA. The major finding of this study is that a polymorphism of the *TNFRSF11B* gene, which encodes osteoprotegerin, is associated with the presence of coronary atherosclerosis in patients with RA.

Osteoprotegerin is a member of the TNF receptor family and acts as a soluble decoy receptor for receptor activator of nuclear factor ΚB ligand (RANKL) [14]. Initially, it was recognized as a mediator preventing osteoclast differentiation, activation and survival, and thus, a regulator of bone resorption [14]. More recently, in addition to its role in bone metabolism, increased osteoprotegerin concentrations have been associated with the presence and severity of atherosclerosis in the general population [15] and with inflammatory diseases, such as RA [16].

We previously showed that in patients with RA, the concentration of osteoprotegerin was higher than in control subjects, and increased osteoprotegerin concentrations were associated with both coronary artery calcification and sedimentation rate [13]. Furthermore, osteoprotegerin is associated with carotid atherosclerosis and endothelial activation in patients with RA [17]. Therefore, osteoprotegerin and the gene encoding it are potential mechanistic links between inflammation and atherosclerosis in RA [13].

In a recent study, Genre *et al.* examined the role of three functional polymorphisms (rs3134063, rs2073618 and rs3134069) located in the gene encoding osteoprotegerin. The study showed a protective effect of the osteoprotegerin (OPG) CGA haplotype on the risk of cardiovascular disease in patients who were anti-cyclic citrullinated peptide (anti-CCP) negative. Further examination of the role of individual SNPs suggested that the GG genotype in rs2073618 was associated with a significant reduction of cerebrovascular, but not cardiovascular events [18]. Concordant with our main findings in this study, two recent studies showed that the same polymorphism in the gene encoding osteoprotegerin, rs2073618, was associated with atherosclerosis in other patient populations. The same CC genotype was associated with a three-fold increased risk of stroke in patients with diabetes [19] and the frequency was two-fold higher in patients who underwent carotid endarterectomy compared to control subjects. This last association was even stronger when the results were adjusted for age, sex, hypertension, hypercholesterolemia, diabetes, coronary artery disease, peripheral artery disease and smoking [20].

However, despite consistent results showing that the CC genotype in rs2073618 confers an increased risk of atherosclerosis in different patient populations, the mechanisms underlying this association remain unclear. We and others have found that increased osteoprotegerin concentrations independently predict coronary atherosclerosis, endothelial activation, carotid atherosclerosis and cardiovascular disease [14,21,22]. Therefore, one possible explanation for our findings was that osteoprotegerin concentrations were affected by rs2073618 variants. However, that was not the case, as we did not find an association between osteoprotegerin concentrations and rs2073618 genotype. Other potential explanations could be hypothesized. For example, considering that this polymorphism causes a substitution from asparagine to lysine, it is possible that the variant alters function of the osteoprotegerin protein rather than its concentration. Both genome evolutionary rate profiling and functionally conserved sequence element alignments suggest that this substitution might be deleterious. Thus, this substitution may have an effect on either protein or transcriptional activity.

In addition to rs2073618, our data suggest that SNPs located in genes including *MMP3*, *IL12*, *MMP9*, *NOD2*, *CRP*, *MPO*, *RETN*, *IRF5*, and *FCGR2A* could be of interest. Although the results did not remain significant after adjustment for multiple comparisons, the magnitude of the effect appears large.

Other genetic variants, such as the rheumatoid arthritis shared epitope, have been proposed as markers of endothelial dysfunction and coronary atherosclerosis in RA [23–25]. Our study did not find an association between the SNPs tagging *HLA-DRB-1* and *CAC*. In concordance with our other findings, there are reports of no association between genetic variants in genes regulating IL-1, interferon-gamma, toll like receptor-4, and IL-6 and atherosclerosis in RA [26–29]. Most of these studies were limited to examining single genetic variants.

Recent studies in other populations also examined the association between multiple genetic variants and CAC. A meta-analysis of genome-wide association studies for CAC in community cohort studies identified SNPs near *CDKN2A* and *CDKN2B*, *PHACTR1*, *MRAS*, *COL4A1/COL4A2*, and *SORT1* as significantly associated with CAC [30]. Another study, using a candidate gene approach to evaluate patients with chronic kidney disease, identified rs13260 (*COL4A1*) and rs7964239 (*BCAT1*) as SNPs associated with the presence of coronary calcium scores >0 [31], but because the selection of our candidate SNPs was done before these studies were published, we did not examine them.

Our results need to be interpreted in the light of some limitations. First, the sample size is small for genetic association studies and our results should be seen as hypothesis generating. Although rs2073618 reached the predefined 20% FDR threshold adjustment for multiple comparisons, it is possible that some other relevant SNPs did not reach statistical significance due to limitations in power. Second, this study examined the role of individual SNPs as markers for atherosclerosis; but the development of risk scores with multiple validated SNPs may have even better potential as tool to improve cardiovascular risk estimation. Finally, given the potential for false positive associations [32] and noted differences with the study by Genre *et al.* [18] replication of our findings in other cohorts will be necessary.

## 4. Methods

One hundred and forty patients with RA, who have participated in studies to evaluate the prevalence and risk factors of coronary atherosclerosis in RA, were included. Details of the protocol have been described previously [3]. In summary, patients fulfilled the 1987 American College of Rheumatology classification criteria for RA [33] and were 18 years or older. Clinical information was obtained by medical record review, structured interview, and physical examination. Blood was collected and erythrocyte sedimentation rate (ESR) measured, and the disease activity score based on the 28 joint count assessment (DAS28) [34] was calculated.

As described previously, [3] the presence of coronary artery calcium (CAC), a non-invasive measurement of coronary atherosclerosis, was detected by electron beam computed tomography (EBCT) scan with an Imatron C-150 scanner (GE/Imatron, South San Francisco, CA, USA). All scans were read by a single investigator (PR), who was unaware of the subjects' clinical status.

After review of the literature, 653 SNPs in 152 genes involved in inflammation, CVD and autoimmune diseases (see Table S1), were selected. Tagging SNP markers were selected based upon HapMap CEU subject data ($R^2 \geq 0.8$, maf > 0.05) for genes encompassing 3–20 kb upstream to 1–10 kb downstream of each respective coding region or were represented by a single SNP selected from published disease association studies. Deoxyribonucleic acid (DNA) was extracted from blood samples using Gentra Puregene DNA Purification reagents and standard protocols on an Autopure LS instrument (Qiagen, Valencia, CA, USA). Quantification was done with a Nanodrop 2000 instrument (Thermo Scientific, Waltham, MA, USA). Samples were diluted to 50 ng/μL and genotyping was done by GoldenGate assay on a Beadstation 500GX instrument (Illumina Corp, San Diego, CA, USA); 7.5% of the SNPs failed in genotyping or were monomorphic in our RA population and were not included in analysis. No SNPs were excluded for deviation from Hardy–Weinberg Equilibrium. All samples were called in >90% of the remaining SNPs and were included in subsequent analysis.

The association between rs2073618 genotypes and osteoprotegerin concentrations was examined in a *post-hoc* analysis using the Kruskal–Wallis test. As previously reported [13], serum osteoprotegerin was measured using enzyme-linked immunosorbent assay method using commercial kits (R & D Systems, Minneapolis, MN, USA).

Baseline characteristics in patients with RA are presented as mean ($\pm$SD), median (interquartile range), or frequency (%). Baseline characteristics between patients with and without coronary calcium were compared with the use of Wilcoxon rank sum test or with Fisher's exact test, as appropriate. The association between the presence of CAC and individual SNPs was assessed with logistic regression with an additive effect for each allele. Results were adjusted for age, sex and race, covariates that were chosen a priori. A two-sided *p*-value of <0.05 was considered significant. To adjust for multiple comparisons, a pre-specified false discovery rate (FDR) threshold was set at 20%.

We conducted the analyses using STATA 12.0 (Stata Corp., College Station, TX, USA). SNAP (SNP annotation and proxy search; www.broadinstitute.org) was used to estimate pair-wise linkage disequilibrium (LD) between individual SNPs within the same gene.

All study participants provided written informed consent and the study was approved by the Institutional Review Board of Vanderbilt University on 10 August 2014 (IRB# 000567).

**Supplementary Materials:** Supplementary materials can be found at http://www.mdpi.com/1422-0067/16/02/3885/s1.

*Int. J. Mol. Sci.* **2015**, *16*, 3885–3894

**Acknowledgments:** This study was supported by grants (P60AR056116, K23AR064768, and GM5M01-RR00095) from the National Institutes of Health, Grant UL1 RR024975-01, now at the National Center for Advancing Translational Sciences, Grant 2 UL1 TR000445-06, and the Vanderbilt Physician Scientist Development Award.

**Author Contributions:** Cecilia P. Chung, Joseph F. Solus, Paolo Raggi, Chun Li, Jeffrey R. Smith, and C. Michael Stein conceived and designed the study; Joseph F. Solus and Annette Oeser performed the experiments; Cecilia P. Chung, Joseph F. Solus, Chun Li, and C. Michael Stein analyzed the data; Cecilia P. Chung, Joseph F. Solus, Annette Oeser, Chun Li, Paolo Raggi, Jeffrey R. Smith, and C. Michael Stein contributed reagents/materials/analysis tools; Cecilia P. Chung and C. Michael Stein wrote the initial draft. All authors critically reviewed the manuscript and approved its final version.

**Conflicts of Interest:** The authors declare no conflict of interest.

## References

1. Solomon, D.H.; Karlson, E.W.; Rimm, E.B.; Cannuscio, C.C.; Mandl, L.A.; Manson, J.E.; Stampfer, M.J.; Curhan, G.C. Cardiovascular morbidity and mortality in women diagnosed with rheumatoid arthritis. *Circulation* **2003**, *107*, 1303–1307. [CrossRef] [PubMed]

2. Chung, C.P.; Oeser, A.; Avalos, I.; Gebretsadik, T.; Shintani, A.; Raggi, P.; Sokka, T.; Pincus, T.; Stein, C.M. Utility of the Framingham risk score to predict the presence of coronary atherosclerosis in patients with rheumatoid arthritis. *Arthritis Res. Ther.* **2006**, *8*, R186. [CrossRef] [PubMed]

3. Chung, C.P.; Oeser, A.; Raggi, P.; Gebretsadik, T.; Shintani, A.K.; Sokka, T. Increased coronary-artery atherosclerosis in rheumatoid arthritis: Relationship to disease duration and cardiovascular risk factors. *Arthritis Rheumatol.* **2005**, *52*, 3045–3053. [CrossRef]

4. Del Rincon, I.D.; Willliams, K.; Stern, M.P.; Freeman, G.L.; Escalante, A. High incidence of cardiovascular events in a rheumatoid arthritis cohort not explained by traditional cardiac risk factors. *Arthritis Rheumatol.* **2001**, *44*, 2737–2745. [CrossRef]

5. Gonzalez-Gay, M.A.; Gonzalez-Juanatey, C.; Pineiro, A.; Garcia-Porrua, C.; Testa, A.; Llorca, J. High-grade C-reactive protein elevation correlates with accelerated atherogenesis in patients with rheumatoid arthritis. *J. Rheumatol.* **2005**, *32*, 1219–1223. [PubMed]

6. Rodriguez-Rodriguez, L.; Lopez-Mejias, R.; Garcia-Bermudez, M.; Gonzalez-Juanatey, C.; Gonzalez-Gay, M.A.; Martin, J. Genetic markers of cardiovascular disease in rheumatoid arthritis. *Mediat. Inflamm.* **2012**, *2012*, 574817.

7. Palomino-Morales, R.; Gonzalez-Juanatey, C.; Vazquez-Rodriguez, T.R.; Rodriguez, L.; Miranda-Filoy, J.A.; Llorca, J.; Martin, J.; Gonzalez-Gay, M.A. A1298C polymorphism in the *MTHFR* gene predisposes to cardiovascular risk in rheumatoid arthritis. *Arthritis Res. Ther.* **2010**, *12*, R71. [CrossRef] [PubMed]

8. Rodriguez-Rodriguez, L.; Gonzalez-Juanatey, C.; Palomino-Morales, R.; Vázquez-Rodríguez, T.R.; Miranda-Filloy, J.A.; Fernández-Gutiérrez, B.; Llorca, J.; Martin, J.; González-Gay, M.A. *TNFA-308* (rs1800629) polymorphism is associated with a higher risk of cardiovascular disease in patients with rheumatoid arthritis. *Atherosclerosis* **2011**, *216*, 125–130. [CrossRef] [PubMed]

9. Lopez-Mejias, R.; Genre, F.; Garcia-Bermudez, M.; Corrales, A.; González-Juanatey, C.; Llorca, J.; Miranda-Filloy, J.A.; Rueda-Gotor, J.; Blanco, R.; Castañeda, S.; *et al.* The *ZC3HC1* rs11556924 polymorphism is associated with increased carotid intima-media thickness in patients with rheumatoid arthritis. *Arthritis Res. Ther.* **2013**, *15*, R152.

10. Lopez-Mejias, R.; Genre, F.; Garcia-Bermudez, M.; Castañeda, S.; González-Juanatey, C.; Llorca, J.; Corrales, A.; Miranda-Filloy, J.A.; Rueda-Gotor, J.; Gómez-Vaquero, C.; Rodríguez-Rodríguez, L.; *et al.* The 11q23.3 genomic region—rs964184—is associated with cardiovascular disease in patients with rheumatoid arthritis. *Tissue Antigens* **2013**, *82*, 344–347.

11. Lopez-Mejias, R.; Garcia-Bermudez, M.; Gonzalez-Juanatey, C.; Castañeda, S.; Miranda-Filloy, J.A.; Gómez-Vaquero, C.; Fernández-Gutiérrez, B.; Balsa, A.; Pascual-Salcedo, D.; Blanco, R.; *et al.* *NFKB1-94ATTG* ins/del polymorphism (rs28362491) is associated with cardiovascular disease in patients with rheumatoid arthritis. *Atherosclerosis* **2012**, *224*, 426–429.

12. Asanuma, Y.; Chung, C.P.; Oeser, A.; Solus, J.F.; Avalos, I.; Gebretsadik, T.; Shintani, A.; Raggi, P.; Sokka, T.; Pincus, T.; Stein, C.M. Serum osteoprotegerin is increased and independently associated with coronary-artery atherosclerosis in patients with rheumatoid arthritis. *Atherosclerosis* **2007**, *195*, e135–e141. [CrossRef] [PubMed]

13. Simonet, W.S.; Lacey, D.L.; Dunstan, C.R.; Kelley, M.; Chang, M.S.; Luthy, R.; Nguyen, H.Q.; Wooden, S.; Bennett, L.; Boone, T.; *et al.* Osteoprotegerin: A novel secreted protein involved in the regulation of bone density. *Cell* **1997**, *89*, 309–319.

14. Reich, D.E.; Cargill, M.; Bolk, S.; Ireland, J.; Sabeti, P.C.; Richter, D.J.; Lavery, T.; Kouyoumjian, R.; Farhadian, S.F.; Ward, R.; *et al.* Linkage disequilibrium in the human genome. *Nature* **2001**, *411*, 199–204.

15. Jono, S.; Ikari, Y.; Shioi, A.; Mori, K.; Miki, T.; Hara, K.; Nishizawa, Y. Serum osteoprotegerin levels are associated with the presence and severity of coronary artery disease. *Circulation* **2002**, *106*, 1192–1194. [CrossRef] [PubMed]

16. Ziolkowska, M.; Kurowska, M.; Radzikowska, A.; Luszczykiewicz, G.; Wiland, P.; Dziewczopolski, W.; Filipowicz-Sosnowska, A.; Pazdur, J.; Szechinski, J.; Kowalczewski, J.; *et al.* High levels of osteoprotegerin and soluble receptor activator of nuclear factor kappa B ligand in serum of rheumatoid arthritis patients and their normalization after anti-tumor necrosis factor alpha treatment. *Arthritis Rheumatol.* **2002**, *46*, 1744–1753.

17. Genre, F.; Lopez-Mejias, R.; Miranda-Filloy, J.A.; Ubilla, B.; Carnero-Lopez, B.; Palmou-Fontana, N.; Gomez-Acebo, I.; Blanco, R.; Rueda-Gotor, J.; Pina, T.; *et al.* Osteoprotegerin correlates with disease activity and endothelial activation in non-diabetic ankylosing spondylitis patients undergoing TNF-α antagonist therapy. *Clin. Exp. Rheumatol.* **2014**, *32*, 640–646.

18. Genre, F.; Lopez-Mejias, R.; Garcia-Bermudez, M.; Castaneda, S.; Gonzalez-Juanatey, C.; Llorca, J.; Corrales, A.; Ubilla, B.; Miranda-Filloy, J.A.; Pina, T.; *et al.* Osteoprotegerin CGA haplotype protection against cerebrovascular complications in anti-CCP negative patients with rheumatoid arthritis. *PLoS One* **2014**, *9*, e106823.

19. Biscetti, F.; Straface, G.; Giovannini, S.; Santoliquido, A.; Angelini, F.; Santoro, L.; Porreca, C.F.; Pecorini, G.; Ghirlanda, G.; Flex, A.; *et al.* Association between *TNFRSF11B* gene polymorphisms and history of ischemic stroke in Italian diabetic patients. *Hum. Genet.* **2013**, *132*, 49–55.

20. Straface, G.; Biscetti, F.; Pitocco, D.; Bertoletti, G.; Misuraca, M.; Vincenzoni, C.; Snider, F.; Arena, V.; Stigliano, E.; Angelini, F.; *et al.* Assessment of the genetic effects of polymorphisms in the osteoprotegerin gene, *TNFRSF11B*, on serum osteoprotegerin levels and carotid plaque vulnerability. *Stroke* **2011**, *42*, 3022–3028.

21. Lopez-Mejias, R.; Ubilla, B.; Genre, F.; Corrales, A.; Hernandez, J.L.; Ferraz-Amaro, I.; Tsang, L.; Llorca, J.; Blanco, R.; Gonzalez-Juanatey, C.; *et al.* Osteoprotegerin concentrations relate independently to established cardiovascular disease in rheumatoid arthritis. *J. Rheumatol.* **2015**, *42*, 39–45.

22. Dessein, P.H.; Lopez-Mejias, R.; Gonzalez-Juanatey, C.; Genre, F.; Miranda-Filloy, J.A.; Llorca, J.; Gonzalez-Gay, M.A. Independent relationship of osteoprotegerin concentrations with endothelial activation and carotid atherosclerosis in patients with severe rheumatoid arthritis. *J. Rheumatol.* **2014**, *41*, 429–436. [CrossRef] [PubMed]

23. Gonzalez-Juanatey, C.; Testa, A.; Garcia-Castelo, A.; Garcia-Porrua, C.; Llorca, J.; Vidan, J.; Hajeer, A.H.; Ollier, W.E.; Mattey, D.L.; Gonzalez-Gay, M.A. HLA-DRB1 status affects endothelial function in treated patients with rheumatoid arthritis. *Am. J. Med.* **2003**, *114*, 647–652. [CrossRef] [PubMed]

24. Gonzalez-Gay, M.A.; Gonzalez-Juanatey, C.; Llorca, J.; Ollier, W.E.; Martin, J. Contribution of HLA-DRB1 shared epitope alleles and chronic inflammation to the increased incidence of cardiovascular disease in rheumatoid arthritis: Comment on the article by Farragher *et al.*. *Arthritis Rheumatol.* **2008**, *58*, 2584–2585. [CrossRef]

25. Farragher, T.M.; Goodson, N.J.; Naseem, H.; Silman, A.J.; Thomson, W.; Symmons, D.; Barton, A. Association of the *HLA-DRB1* gene with premature death, particularly from cardiovascular disease, in patients with rheumatoid arthritis and inflammatory polyarthritis. *Arthritis Rheumatol.* **2008**, *58*, 359–369. [CrossRef]

26. Lopez-Mejias, R.; Garcia-Bermudez, M.; Gonzalez-Juanatey, C.; Castaneda, S.; Perez-Esteban, S.; Miranda-Filloy, J.A.; Gomez-Vaquero, C.; Fernandez-Gutierrez, B.; Balsa, A.; Pascual-Salcedo, D.; *et al.* Lack of association between IL6 single nucleotide polymorphisms and cardiovascular disease in Spanish patients with rheumatoid arthritis. *Atherosclerosis* **2011**, *219*, 655–658.

27. Garcia-Bermudez, M.; Lopez-Mejias, R.; Gonzalez-Juanatey, C.; Castaneda, S.; Miranda-Filloy, J.A.; Blanco, R.; Fernandez-Gutierrez, B.; Balsa, A.; Gonzalez-Alvaro, I.; Gomez-Vaquero, C.; *et al.* Lack of association between *TLR4* rs4986790 polymorphism and risk of cardiovascular disease in patients with rheumatoid arthritis. *DNA Cell Biol.* **2012**, *31*, 1214–1220.

28. Garcia-Bermudez, M.; Lopez-Mejias, R.; Gonzalez-Juanatey, C.; Corrales, A.; Castaneda, S.; Ortiz, A.M.; Miranda-Filloy, J.A.; Gomez-Vaquero, C.; Fernandez-Gutierrez, B.; Balsa, A.; *et al. CARD8* rs2043211 (p.C10X) polymorphism is not associated with disease susceptibility or cardiovascular events in Spanish rheumatoid arthritis patients. *DNA Cell Biol.* **2013**, *32*, 28–33.

29. Garcia-Bermudez, M.; Lopez-Mejias, R.; Gonzalez-Juanatey, C.; Corrales, A.; Robledo, G.; Castaneda, S.; Miranda-Filloy, J.A.; Blanco, R.; Fernandez-Gutierrez, B.; Balsa, A.; *et al.* Analysis of the interferon gamma (rs2430561, +874T/A) functional gene variant in relation to the presence of cardiovascular events in rheumatoid arthritis. *PLoS One* **2012**, *7*, e47166.

30. O'Donnell, C.J.; Kavousi, M.; Smith, A.V.; Kardia, S.L.; Feitosa, M.F.; Hwang, S.J.; Sun, Y.V.; Province, M.A.; Aspelund, T.; Dehghan, A.; *et al.* Genome-wide association study for coronary artery calcification with follow-up in myocardial infarction. *Circulation* **2011**, *124*, 2855–2864.

31. Ferguson, J.F.; Matthews, G.J.; Townsend, R.R.; Raj, D.S.; Kanetsky, P.A.; Budoff, M.; Fischer, M.J.; Rosas, S.E.; Kanthety, R.; Rahman, M.; *et al.* Candidate gene association study of coronary artery calcification in chronic kidney disease: Findings from the CRIC study (Chronic Renal Insufficiency Cohort). *J. Am. Coll. Cardiol.* **2013**, *62*, 789–798.

32. Lopez-Mejias, R.; Genre, F.; Corrales, A.; Gonzalez-Juanatey, C.; Ubilla, B.; Llorca, J.; Miranda-Filloy, J.A.; Pina, T.; Blanco, R.; Castaneda, S.; *et al.* Investigation of a *PON1* gene polymorphism (rs662 polymorphism) as predictor of subclinical atherosclerosis in patients with rheumatoid arthritis. *Ann. Rheum. Dis.* **2014**, *73*, 1749–1750.

33. Arnett, F.C.; Edworthy, S.M.; Bloch, D.A.; Mc Shane, D.J.; Fries, J.F.; Cooper, N.S.; Healey, L.A.; Kaplan, S.R.; Liang, M.H.; Luthra, H.S. The American Rheumatism Association 1987 revised criteria for the classification of rheumatoid arthritis. *Arthritis Rheumatol.* **1988**, *31*, 315–324. [CrossRef]

34. Prevoo, M.L.; van't Hof, M.A.; Kuper, H.H.; van Leeuwen, M.A.; van de Putte, L.B.; van Riel, P.L. Modified disease activity scores that include twenty-eight-joint counts. Development and validation in a prospective longitudinal study of patients with rheumatoid arthritis. *Arthritis Rheumatol.* **1995**, *38*, 44–48.

International Journal of
*Molecular Sciences*

MDPI

*Review*

# Coronary Artery Calcium Screening: Does it Perform Better than Other Cardiovascular Risk Stratification Tools?

Irfan Zeb [1],* and Matthew Budoff [2]

[1] Department of Medicine, Bronx-Lebanon Hospital Center, 1650 Grand Concourse, Bronx, NY 10457, USA
[2] Department of Cardiology, Los Angeles Biomedical Research Institute at Harbor-UCLA Medical Center, Torrance, CA 90502, USA; mbudoff@labiomed.org
* Correspondence: izeb82@gmail.com; Tel.: +1-310-972-8978

Academic Editor: Michael Henein
Received: 26 January 2015; Accepted: 5 March 2015; Published: 23 March 2015

**Abstract:** Coronary artery calcium (CAC) has been advocated as one of the strongest cardiovascular risk prediction markers. It performs better across a wide range of Framingham risk categories (6%–10% and 10%–20% 10-year risk categories) and also helps in reclassifying the risk of these subjects into either higher or lower risk categories based on CAC scores. It also performs better among population subgroups where Framingham risk score does not perform well, especially young subjects, women, family history of premature coronary artery disease and ethnic differences in coronary risk. The absence of CAC is also associated with excellent prognosis, with 10-year event rate of 1%. Studies have also compared with other commonly used markers of cardiovascular disease risk such as Carotid intima-media thickness and highly sensitive C-reactive protein. CAC also performs better compared with carotid intima-media thickness and highly sensitive C-reactive protein in prediction of coronary heart disease and cardiovascular disease events. CAC scans are associated with relatively low radiation exposure (0.9–1.1 mSv) and provide information that can be used not only for risk stratification but also can be used to track the progression of atherosclerosis and the effects of statins.

**Keywords:** coronary artery calcium; risk stratification; cardiovascular risk

## 1. Introduction

Atherosclerosis coronary artery disease is among the leading cause of morbidity and mortality in the Western world. The shear burden of cardiovascular disease on healthcare costs is enormous, with an estimate of 475 billion US dollars spent in the year 2009 alone [1]. By 2030, real total direct medical costs of cardiovascular disease (CVD) are projected to increase to $\approx$ \$ 918 billion [2]. In order to drive the cost down, emphasis is on preventive measures and earlier detection of cardiovascular disease. There is a number of risk factors algorithms, biomarkers and imaging studies to screen for cardiovascular diseases. Framingham risk scores (FRS), Reynolds risk score, highly sensitive C-reactive protein (hs-CRP), carotid intima media thickness (CIMT) and coronary artery calcium (CAC) are among the various measures that can be used for screening of cardiovascular disease among asymptomatic population. CAC score has emerged as one of the strongest risk prediction tools. It represents calcific atherosclerosis in the coronary arteries and correlates well with the overall burden of atherosclerosis in the coronary arteries. The current review article will compare CAC score with the remaining risk stratification tools.

## 2. Framingham Risk Score and Coronary Artery Calcium

Framingham risk score is the most commonly used cardiovascular risk stratification tool in the general population due to its ease of use. It takes into account major cardiovascular risk factors including age, sex, dyslipidemia, smoking and hypertension [3]. This risk scoring system predicts an estimate of population risk for CVD events over 10-years and categorizes individual risk for developing CVD to low (10-year risk of <10%), intermediate (10-year risk of 10%–20%), and high (10-year risk of >20%) risk. FRS is easy to calculate and provides a good overall estimate of patient risk for future cardiovascular problems. However, there are several limitations to the use of this classification for risk assessment in the general population.

Shaw *et al.* [4] followed up a cohort of 10,377 asymptomatic individuals for mean duration of five years and compared cardiac risk factors such as family history of coronary artery disease, hypercholesterolemia, hypertension, smoking and diabetes mellitus with CAC score (Figure 1). CAC was found to be an independent predictor of mortality ($p < 0.001$) and it provided improved discrimination value compared to the cardiac risk factors alone (concordance index 0.78 *vs.* 0.72, $p < 0.001$) in a multivariable model for prediction of death. CAC was also superior to estimated Framingham risk score in outcome classification ability (Area under the ROC curve = 0.73 *vs.* 0.67, $p < 0.001$).

Greenland *et al.* [5] evaluated predictive value of CAC across increasing levels of FRS categories (0%–9%, 10%–15%, 16%–20% and ≥21%). For low risk FRS categories (0%–9%), CAC >300 did not provide an increased risk. CAC >100 was associated with increased risk among FRS categories 10%–15% and higher. The risk associated with FRS category 10%–15% and CAC score >300 (hazard ratio 17.6, $p < 0.001$) was comparable with the risk associated with FRS category >21% and CAC >300 (hazard ratio 19.1, $p < 0.001$). 2010 American College of Cardiology (ACC) and American Heart Association (AHA) guidelines [6] have incorporated CAC for cardiovascular risk assessment in asymptomatic adults at intermediate risk (10%–20% 10-year risk) (Class II a indication). The current clinical practice guidelines [6] do not recommend the use of CAC for low risk population risk assessment. Okwuosa *et al.* [7] evaluated the prevalence of CAC in low risk FRS categories and the number needed to screen to detect one individual with CAC score >300. There were 1.7% and 4.4% of the population found to have CAC score ≥300 within FRS categories of 0%–2.5% and 2.6%–5.0%, respectively. The respective number needed to screen were 59.7 and 22.7. The number needed to screen to detect persons with CAC score ≥300 decreased with increasing FRS categories; with only 4.2 and 3.3 individuals needed to be screened with FRS of 15.1%–20% and >20%. Okwuosa *et al.* [8] evaluated role of novel markers for CAC score progression among patients within low risk FRS category (<10%). They looked at the following variables: LDL particle number (LDLpn), urine albumin, CRP using a high-sensitivity assay, D-dimer, factor VIIIc, total homocysteine (tHcy), fibrinogen, cystatin C, soluble intercellular adhesion molecule-1 (sICAM-1) and carotid intima media thickness (CIMT). The study showed significant association of most of these risk factors with CAC progression in univariate and age adjusted models. This statistical significance was lost after adjustment for traditional cardiovascular risk factors. The study also evaluated predictive value of various combinations of novel markers with CAC progression compared with a base model composed of traditional risk factors. These combination models showed little or no improvement over the base model in discrimination and informativeness. Taylor *et al.* [9] performed a study evaluating the predictive value of CAC in 2000 healthy young men and women aged 40–50 years of age. A total of nine coronary events occurred, with four in men with 10-year FRS risk <6% and five men with 10-year FRS risk of 6%–10%. None of the events occurred in men with 10-year FRS risk above 10%. Seven out of nine men had CAC among those who suffered coronary heart disease (CHD) events. This study showed that the presence of any CAC was associated with an 11.8-fold increased risk of acute CHD ($p = 0.002$) after controlling for the FRS. 2010 ACC/AHA guidelines [6] suggest use of CAC for cardiovascular risk assessment of individuals with 10-year FRS risk of 6%–10% (Class II b indication). The current clinical practice guidelines do not recommend the use of CAC for low risk individuals (10-year FRS risk <6%) for cardiovascular risk assessment (Class III indication). The study performed by Greenland *et al.* [5] did not show any benefit of cardiovascular risk assessment with CAC score in the 10-year FRS category of less than 10%. The mean age of population at the time of computed tomography (CT) scan in this study was 65.7 years. The population groups in both Taylor *et al.* [9] (mean age 43 years) and Okwuosa *et al.* [7] were younger (mean age: 60.9 years, 53% women). It is already known that the FRS performs poorly in younger patients and women [10–12]. In a multi-ethnic study of atherosclerosis, 90% of the women were classified as low risk based on FRS risk classification [13]. There were 32% of low risk women who were found to have CAC score of more than 0 (four percent had CAC score ≥300). In these women, CAC score of >300 were highly predictive of future CHD and CVD events (6.7% and 8.6% absolute CHD and CVD risk, respectively) over a 3.75 year period. Ninety-five percent of the US women younger than 70 are found to be at low risk for coronary heart disease according to the data from Third National Health and Nutrition Examination Survey (NHANES III) [10] and, thereby, do not qualify for more aggressive medical management for standard risk factors according to the National Cholesterol Education Program Expert Panel on Detection, Evaluation, and Treatment of High Blood Cholesterol in Adults (Adult Treatment Panel III) (NCEP/ATP III) [14]. These persons otherwise may be classified as low-risk, and may not benefit from preventive treatments that can otherwise be offered to them based on CAC scores.

Elias-Smale *et al.* [15] evaluated the effect of CAC in risk assessment of elderly population (2028 subjects, age 69.6 ± 6.2 years and showed that CAC was able to reclassify 52% of men and women in intermediate risk category.

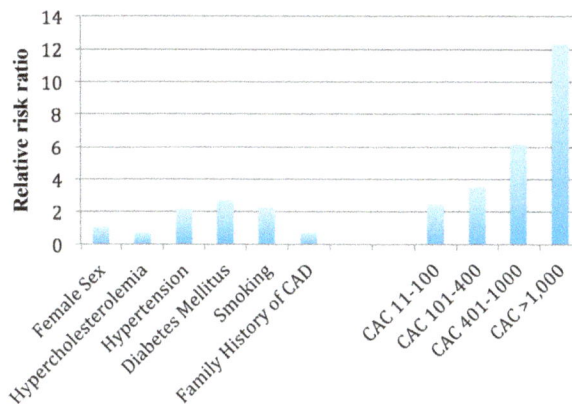

**Figure 1.** Predictive value of coronary artery calcium [4].

Family history of premature coronary artery disease represents a unique situation. The persons may suffer from cardiovascular events at an earlier age. The usual risk stratification tools such as FRS may perform poorly in these persons. Reynolds risk score includes parental family history of premature CAD and high-sensitivity C-reactive protein (hs-CRP) to traditional risk factors. CAC use has been assessed in this situation as well [16–18]. Nasir *et al.* [16] showed that the age-, gender- and race-adjusted prevalence of CAC >0 was significantly higher with presence of any family history of premature CHD than those with no family history of premature CHD among individuals classified as low risk (35% *vs.* 23%, *p* < 0.0001) and among those at intermediate risk (70% *vs.* 60%, *p* = 0.01). Similar results were seen with CAC ≥75th percentile among individuals at low risk (24% *vs.* 14%, *p* = 0.0003) and intermediate-risk (34% *vs.* 20%, *p* < 0.001) respectively. A *post hoc* analysis from St. Francis Heart study [17] revealed benefit of statin treatment in individuals with a family history of premature CAD and CAC above 80th percentile.

CAC has been shown to perform better across a wide range of FRS categories (6%–10% 10-year risk and 10%–20% 10-year risk). FRS is a population-based risk of 10-year event rates; however, individual persons may not be at the same level of risk in each category as determined by FRS. The individual risk can be further stratified with the use of various markers like hs-CRP and CAC. Polonski *et al.* [19] showed that addition of CAC to traditional risk factors improves risk prediction significantly compared to traditional risk factors alone. Traditional risk factors were able to classify 69% of the cohort into the highest or lowest risk categories, whereas addition of CAC to this model increased it to 77%. CAC along with traditional risk factors was able to reclassify 23% of those who had cardiac events into high risk and 13% without cardiac events into low risk.

Coronary atherosclerosis burden differs between different ethnic groups [20]. Framingham risk score does not account for the ethnicities. Detrano *et al.* [20] performed coronary calcium scans on population-based sample of 6722 men and women, of whom 38.6% were white, 27.6% were black, 21.9% were Hispanic, and 11.9% were Chinese. They found that adjusted risk for any coronary events for CAC score categories 1–100, 101–300 and >300 increased by a factor of 3.61-fold, 7.73-fold and 9.67-fold when compared with CAC score of zero ($p < 0.001$, respectively). The adjusted risk for any coronary events for ethnic groups Whites, Chinese, Black and Hispanics was 1.22-fold ($p < 0.001$), 1.36-fold ($p < 0.005$), 1.39-fold ($p < 0.001$) and 1.18-fold ($p < 0.001$), respectively. Across four ethnic groups, a doubling of CAC increased the risk of major coronary event by 15%–35% and the risk of any coronary event by 18%–39%. The areas under the receiver-operating-characteristic curve showed superior predictive value of CAC compared to the standard risk factors for both major and any coronary events.

The predictive value of increasing CAC score is well established. The protective value of CAC score zero is also very promising. A number of studies have consistently reported better outcomes associated with the absence of CAC (Table 1). These persons may require less intensive treatments and further diagnostic testing due to very low event rates in this population group. Blaha *et al.* [21] evaluated annualized all-cause mortality rates in 44,052 consecutive asymptomatic patients referred for CAC testing. They found that 45% had no CAC on screening and there were 104 deaths in those with CAC score zero. Patients with CAC score of 0 have an excellent prognosis, with a 10-year event rate of about 1%.

**Table 1.** Predictive value of zero coronary artery calcium score.

| Author | Year | Total No. of Participants | Participants with CAC Score 0 | Event Rate | Follow up Period |
|--------|------|---------------------------|-------------------------------|------------|-------------------|
| Blaha *et al.* [21] | 2009 | 44,052 | 19,898 | 0.87/1000 person-years | Mean follow-up of 5.6 ± 2.6 years |
| Arad *et al.* [22] | 2005 | 4613 | 1504 | 1/1000 person-years | 4.3 years |
| Taylor *et al.* [9] | 2005 | 2000 | 1263 | 0.6/1000 person-years | Mean follow-up of 3.0 ±1.4 years |
| Budoff *et al.* [23] | 2007 | 25,253 | 11,046 | 0.6/1000 person-years | Mean follow-up of 6.8 ± 3 years |
| Detrano *et al.* [20] | 2009 | 6722 | 3409 | 0.6/1000 person-years | 3.7 years |
| Shaw *et al.* [4] | 2003 | 10,377 | 5067 | 1.5 events/1000 person-years | Mean follow-up of 5.0 years |
| LaMonte *et al.* [24] | 2005 | 10,746 | 2692 | 1.6 events/1000 person-years | 3.5 years |

The value of CAC in cardiac disease prediction has been well established. Herman *et al.* [25] evaluated association of CAC with stroke and showed that log10 (CAC + 1) is an independent predictor of stroke (hazards ratio, 1.52 (95% confidence interval, 1.19–1.92); $p = 0.001$) in addition to age, systolic blood pressure and smoking in subjects at low or intermediate vascular risk. This association was independent of atrial fibrillation detected at baseline and follow up exams in prediction of stroke (fully adjusted hazard ratio, 1.31 (1.00–1.71); $p = 0.049$) that may be the underlying cause for cryptogenic strokes. O'Neal *et al.* [26] showed that higher CAC scores were associated with increased risk for atrial fibrillation (CAC = 1–100: Hazard Ratio 1.4, 95% CI 1.01–2.0; CAC = 101–300: HR 1.6, 95% CI 1.1–2.4; CAC > 300: HR 2.1, 95% CI 1.4–2.9) in a model adjusted for sociodemographics, cardiovascular risk factors, and potential confounders.

The recently published ACC/AHA guidelines on the assessment of cardiovascular risk [27] used a different risk tool to estimate 10-year and long term risk of atherosclerotic cardiovascular disease (ASCVD) that incorporates age, sex, race, HDL-cholesterol, total cholesterol, diabetes, systolic blood pressure or treatment for hypertension and smoking status. This new risk stratification tool may replace FRS for risk prediction. However there are currently no studies available that compare CAC with 10-year ASCVD risk tool.

### 3. Reynolds Risk Score and Coronary Artery Calcium

Reynolds risk score (RRS) includes traditional risk factors used in FRS and adds parental family history of premature CAD and high-sensitivity C-reactive protein (hs-CRP) [28]. It provides improved prediction of CVD events in both men and women and has been proposed as an alternative to FRS in risk stratification [28,29]. DeFillepis *et al.* [30] compared the performance of FRS and RRS in predicting CAC incidence and progression among 5140 individuals in MESA population. Both FRS and RRS were predictive of incident CAC (relative risk: 1.40 and 1.41 per 5% increase in risk, respectively) and CAC progression (relative risk 6.92 and 6.82 per 5% increase in risk, respectively). There was discordance in risk category classification (<10% and >10% per 10-year HD risk) in 13.7% of patients. RRS was found superior to FRS in providing additional predictive information for incidence and progression of CAC when discordance between the scoring systems existed. Desai *et al.* [31] evaluated whether CAC leads to reclassification of RRS risk assessment in patients without prior CHD. There were 72% of patients with CAC score >400 and 88% of patients with high CAC percentile had low or intermediate risk by RRS risk classification. CAC can potentially improve risk reclassification for RRS risk assessment.

### 4. Carotid Intima Media Thickness and Coronary Artery Calcium

Carotid intima-media thickness (CIMT) represents carotid atherosclerosis and is being used as a surrogate marker for atherosclerosis. Its utility relies on its ability to predict future adverse cardiovascular events [32]. Lorenz *et al.* [33] performed a meta-analysis of eight studies where they showed that the age- and sex-adjusted overall estimates of the relative risk of 1.26 (95% CI 1.21–1.30) for myocardial infarction and1.32 (95% CI 1.27–1.38) for stroke for each one-standard deviation difference of common carotid IMT. CIMT does predict future CHD events but is a slightly better predictor for cerebrovascular events. Folsom *et al.* [34] compared CAC with CIMT in a cohort of 6698 subjects aged 45–84 years of age in four ethnic groups enrolled in MESA study and showed that the hazard of CVD increased 2.1 (95% CI 1.8–2.5) fold for log-transformed CAC *vs.* 1.3-fold (95% CI 1.1–1.4) for maximum CIMT for each standard deviation increment of log-transformed CAC and maximum CIMT. For coronary heart disease, the hazard ratios per 1-SD increment increased 2.5-fold (95% CI 2.1–3.1) for CAC score and 1.2-fold (95% CI 1.0–1.4) for IMT. The study also showed CAC to be a superior predictor of incident CVD compared with CIMT (area under the curve 0.81 *vs.* 0.78), respectively. Terry *et al.* [35] compared the performance of CAC and CIMT for non-invasive detection of prevalent CAD ($\geq$50% stenosis in one or more coronary arteries identified on coronary angiography). CAC performed better than CIMT in terms of prevalent CAD for each quartile increase in both measures (odds ratio: 8-fold *vs.* 1.7-fold, respectively).

CAC seems to be better than CIMT in patients with atherosclerosis CAD. However, among patients with coronary micro-vascular dysfunction, CIMT seem to perform better than CAC score. Danad *et al.* [36] evaluated 120 patients without documented evidence of CAD to compare CAC with CIMT for prediction of minimal coronary vascular resistance, a measure used to detect coronary micro-vascular disease [37]. Using bivariate and multivariable regression analysis using backward elimination revealed that only CIMT was a predictor of minimal coronary vascular resistance. Naqvi *et al.* [38] showed that CIMT based mean vascular age ($61.6 \pm 11.4$ years) was significantly higher than coronary calcium age ($58.3 \pm 11.1$ years) ($p = 0.001$); both of these were significantly higher than mean chronological age ($p < 0.0001$ and $p < 0.04$, respectively). CIMT was able to upgrade or downgrade FRS by >5% compared with CAC (42% of CIMT cases *vs.* 17% of CAC cases).

Studies comparing CAC with CIMT in younger population groups, where the prevalence of CAC is very low, CIMT may perform better in those population groups. Davis *et al.* [39] performed a study in young population (33–42 years of age) evaluating the relationship of CAC with CIMT. The multivariate model showed low density cholesterol-C ($p < 0.001$), pack-year smoking ($p < 0.005$) and CAC ($p < 0.05$) were significant in men; and low density cholesterol-C ($p < 0.001$), systolic blood pressure ($p < 0.01$) and CAC ($p < 0.05$) were significant in women and were predictive of CIMT in this population.

Lester *et al.* [40] performed a study on relatively younger population, aged 36–59 years of age, comparing CAC and CIMT to detect subclinical atherosclerosis. A large proportion of population had a CAC score of zero (75%). There was evidence of carotid atherosclerosis in 47% of the population with CAC score of zero, showing the role of CIMT in low risk, relatively younger population group with CAC score of zero.

## 5. C-Reactive Protein and Coronary Artery Calcium

AHA/ACC recommends measurement of C-reactive protein (CRP) in men 50 years of age or older or women 60 years of age or older with low-density lipoprotein cholesterol less than 130 mg/dL can be useful in selection of patients for statin therapy [41]. These recommendations are based on JUPITER trial [42] assessing the benefit of statin treatment in apparently healthy persons with hyperlipidemia (LDL-C < 130 mg/dL) but with elevated high-sensitivity CRP levels (hs-CRP $\geq$ 2 mg/dL). Measurement of hs-CRP has been shown as an independent prediction of future vascular events and improves global classification of risk, regardless of the LDL cholesterol level [43–49]. Arad *et al.* [22] performed a study evaluating 4903 apparently healthy middle-aged persons comparing CAC to standard coronary disease risk factors and CRP that were followed for a period of 4.3 years. CAC was shown to perform better than CRP and to highly correlate with future CAD events in this study. Park *et al.* [50] performed a study evaluating 1461 subjects without coronary heart disease comparing CAC and CRP in predicting cardiovascular events in non-diabetic individuals. The study showed that both CAC was predictive of non-fatal myocardial infarction or coronary death and any cardiovascular events ($p < 0.005$) whereas CRP was only predictive of cardiovascular events ($p = 0.03$). Risk group analysis that are defined by increasing tertiles for CAC (<3.7, 3.7–142.1, >142.1) and the 75th percentile for CRP (>4.05 mg/L) were associated with increasing risk with increasing CAC and CRP. Relative risks for the medium-calcium/low-CRP risk group to high-calcium/high-CRP risk group ranged from 1.8 to 6.1 for MI/coronary death ($p = 0.003$) and 2.8–7.5 for any cardiovascular event ($p < 0.001$). Lakoski *et al.* [51] examined the MESA participants to determine how many individuals risk can be reclassified with the use of CAC and CRP. The study showed that 30% of the intermediate-risk subjects by FRS had CRP concentration greater than 3 mg/L and 33% had CAC scores more than 100 AU. In gender specific analysis, 49% of intermediate-risk women and 27% of intermediate-risk men had CRP >3 mg/L, compared with 33% of intermediate-risk men and women had CAC >100 AU. When gender specific cut points for CRP and CAC were use, the same percentage of intermediate risk women and men had CRP above 75th percentile (28% and 27%, respectively) whereas more intermediate-risk women 40%, CAC > 50) than men (25%, CAC > 180) had a high CAC score.

Blaha *et al.* [52] performed a study evaluating whether CAC score may further risk stratify a JUPITER-eligible population (LDL-C < 130 mg/dL and hs-CRP $\geq$ 2.0 mg/dL) in participants enrolled in MESA study. For CHD events, the five-year number needed to treat for CAC scores of 0, 1–100 and >100 was 549, 94 and 24, respectively. In JUPITOR trial, the five-year number needed to prevent the occurrence of one primary end point was 25 [42]. The presence of CAC was associated with 4.3-fold increased CHD (95% CI 2.0–9.3) and 2.6-fold increased CVD (95% CI 1.5–4.5), while hs-CRP was not associated with either CHD or CVD after multivariable adjustment [52].

Möhlenkamp *et al.* [53] performed a study involving 3966 subjects without known CAD or acute inflammation, followed for 5.1 $\pm$ 0.3 years, to determine whether combined presence of CAC and hs-CRP improves discrimination and stratification of coronary events and all-cause mortality. For coronary events, net reclassification improvement (NRI) was 23.8% ($p$ = 0.0007) for CAC and 10.5% ($p$ = 0.026) for hs-CRP. Addition of CAC to Framingham risk variables and hs-CRP improved discrimination of coronary risk but not *vice versa*. Among persons with CAC score of zero, hs-CRP >3 mg/L was associated with a significantly higher coronary event risk compared with hs-CRP <3.0 mg/L ($p$ = 0.006). Reilly *et al.* [54] performed a study involving 914 participants from the Study of Inherited Risk of Coronary Atherosclerosis (SIRCA) who were free of clinical CAD and had a family history of premature CAD. The study showed that median CAC scores increased across ordinal CRP categories in women but not in men (Krushal-Wallis $\chi^2$ = 22.5, $p$ < 0.001 in women, $\chi^2$ = 2.5, $p$ < 0.29 in men, respectively). CRP levels were found to be significant predictors of CAC scores in women after adjusting for traditional risk factors, which was lost after adjustment for body mass index. There was no such association found between CRP and CAC in men.

## 6. Coronary Artery Calcium and Other Imaging Parameters for Risk Prediction

Various imaging markers that can be easily obtained from coronary artery calcium scan have been proposed for prediction of CHD and CVD events. Pericardial fat is present around the heart—encompassing the coronary arteries—has been shown to be a predictor of coronary events [55]. Budoff *et al.* [56] showed that thoracic aortic wall calcification to be a significant predictor of incident coronary events in women, independent of CAC. Zeb *et al.* [57] showed liver fat measured by CT scans to be an independent predictor of incident coronary heart disease events. Yeboah *et al.* [58] performed a study to assess whether addition of computed tomography risk markers thoracic aorta calcium (TAC), aortic valve calcification (AVC), mitral annular calcification (MAC), pericardial adipose tissue volume (PAT), and liver attenuation (LA) to FRS + CAC provide improved discrimination for incident CHD and incident CVD events. CAC, TAC, AVC and MAC were all significantly associated with incident CVD/CHD/mortality among intermediate risk subjects; CAC had the strongest association. The addition of CAC to the FRS provides superior discrimination especially in intermediate-risk individuals. Among participants with intermediate FRS risk, the addition of TAC, AVC, MAC, PAT, or LA to FRS + CAC resulted in a significant reduction in area of the curve for incident CHD (0.712 *vs.* 0.646, 0.655, 0.652, 0.648 and 0.569; all $p$ < 0.01, respectively). The addition of CAC to FRS resulted in a superior discrimination for incident CHD in the intermediate-risk groups compared to when TAC, AVC, MAC, PAT and LA were added to FRS + CAC (0.024, 0.026, 0.019, 0.012 and 0.012, respectively). These risk markers are unlikely to be useful for improving cardiovascular risk prediction.

## 7. Conclusions

CAC score has been shown as the strongest predictor of incident coronary events and is able to reclassify low-to-intermediate risk groups and certain subgroups, especially women and young adults most of which may be classified as low risk by FRS risk stratification. Most of the adult population will have CAC scores of zero (Table 1). The prevalence of CAC scores increases with increasing age. There is a number of different cutoffs that are used to denote increased CHD risk [20,59], however, the most commonly used cutoffs of increased CHD risk are CAC score 1–100, CAC 101–300 and CAC > 300 [20]. Studies have shown that subjects who undergo CAC scans tend to have better compliance rates with medications compared with other groups, and promote a healthy behavior in these patients (a picture equals a thousand words) [60,61]. CAC scans can be used to evaluate the efficacy of treatment regimens over time. The accelerated progression of CAC is associated with adverse cardiovascular events [62–65]. There are certain concerns associated with the use of CAC scores. Among them, the foremost concern associated with CAC use is radiation exposure. The recent advances in imaging techniques has lowered the radiation exposure associated with CAC scans and is comparable to mammograms in terms in radiation exposure (0.9 to 1.1 mSv) [66,67]. The scan can be easily performed in an outpatient setting without any prior preparation. Medicare is currently reimbursing the cost of CAC scans for selected population groups. Another issue that can arise with the use of CAC scans are incidental findings which can drive the cost upward due to further downstream testing as a result of these incidental findings. The prevalence of incidental findings can range from 4% to 8% [68–70]. There is a number of cost effectiveness analysis studies that are looking at the cost associated with the use of CAC scans. However, the information obtained with CAC can be very useful for earlier detection, institution of statins and prevention of adverse cardiovascular events.

**Author Contributions:** Irfan Zeb and Matthew Budoff contributed to the preparation of this review article.

**Conflicts of Interest:** The authors declare no conflict of interest.

## References

1. Lloyd-Jones, D.; Adams, R.; Carnethon, M.; de Simone, G.; Ferguson, T.B.; Flegal, K.; Ford, E.; Furie, K.; Go, A.; Greenlund, K.; *et al.* Heart disease and stroke statistics—2009 Update: A report from the American Heart Association Statistics Committee and Stroke Statistics Subcommittee. *Circulation* **2009**, *119*, 480–486. [CrossRef] [PubMed]
2. Go, A.S.; Mozaffarian, D.; Roger, V.L.; Benjamin, E.J.; Berry, J.D.; Blaha, M.J.; Dai, S.; Ford, E.S.; Fox, C.S.; Franco, S.; *et al.* Heart disease and stroke statistics—2014 Update: A report from the American Heart Association. *Circulation* **2014**, *129*, e28–e292. [CrossRef] [PubMed]
3. Grundy, S.M.; Pasternak, R.; Greenland, P.; Smith, S., Jr.; Fuster, V. Assessment of cardiovascular risk by use of multiple-risk-factor assessment equations: A statement for healthcare professionals from the American Heart Association and the American College of Cardiology. *Circulation* **1999**, *100*, 1481–1492. [CrossRef] [PubMed]
4. Shaw, L.J.; Raggi, P.; Schisterman, E.; Berman, D.S.; Callister, T.Q. Prognostic value of cardiac risk factors and coronary artery calcium screening for all-cause mortality. *Radiology* **2003**, *228*, 826–833. [CrossRef] [PubMed]
5. Greenland, P.; LaBree, L.; Azen, S.P.; Doherty, T.M.; Detrano, R.C. Coronary artery calcium score combined with Framingham score for risk prediction in asymptomatic individuals. *JAMA* **2004**, *291*, 210–215. [CrossRef] [PubMed]
6. Greenland, P.; Alpert, J.S.; Beller, G.A.; Benjamin, E.J.; Budoff, M.J.; Fayad, Z.A.; Foster, E.; Hlatky, M.A.; Hodgson, J.M.; Kushner, F.G.; *et al.* 2010 ACCF/AHA guideline for assessment of cardiovascular risk in asymptomatic adults: A report of the American College of Cardiology Foundation/American Heart Association task force on practice guidelines. *J. Am. Coll. Cardiol.* **2010**, *56*, e50–e103. [CrossRef] [PubMed]
7. Okwuosa, T.M.; Greenland, P.; Ning, H.; Liu, K.; Bild, D.E.; Burke, G.L.; Eng, J.; Lloyd-Jones, D.M. Distribution of coronary artery calcium scores by Framingham 10-year risk strata in the MESA (Multi-Ethnic Study of Atherosclerosis) potential implications for coronary risk assessment. *J. Am. Coll. Cardiol.* **2011**, *57*, 1838–1845. [CrossRef] [PubMed]
8. Okwuosa, T.M.; Greenland, P.; Burke, G.L.; Eng, J.; Cushman, M.; Michos, E.D.; Ning, H.; Lloyd-Jones, D.M. Prediction of coronary artery calcium progression in individuals with low Framingham Risk Score: The Multi-Ethnic Study of Atherosclerosis. *JACC Cardiovasc. Imaging* **2012**, *5*, 144–153. [CrossRef] [PubMed]
9. Taylor, A.J.; Bindeman, J.; Feuerstein, I.; Cao, F.; Brazaitis, M.; O'Malley, P.G. Coronary calcium independently predicts incident premature coronary heart disease over measured cardiovascular risk factors: Mean three-year outcomes in the Prospective Army Coronary Calcium (PACC) project. *J. Am. Coll. Cardiol.* **2005**, *46*, 807–814. [CrossRef] [PubMed]

10. Ford, E.S.; Giles, W.H.; Mokdad, A.H. The distribution of 10-year risk for coronary heart disease among US adults: Findings from the National Health and Nutrition Examination Survey III. *J. Am. Coll. Cardiol.* **2004**, *43*, 1791–1796. [CrossRef] [PubMed]

11. Berry, J.D.; Lloyd-Jones, D.M.; Garside, D.B.; Greenland, P. Framingham risk score and prediction of coronary heart disease death in young men. *Am. Heart J.* **2007**, *154*, 80–86. [CrossRef] [PubMed]

12. Akosah, K.O.; Schaper, A.; Cogbill, C.; Schoenfeld, P. Preventing myocardial infarction in the young adult in the first place: How do the National Cholesterol Education Panel III guidelines perform? *J. Am. Coll. Cardiol.* **2003**, *41*, 1475–1479. [CrossRef] [PubMed]

13. Lakoski, S.G.; Greenland, P.; Wong, N.D.; Schreiner, P.J.; Herrington, D.M.; Kronmal, R.A.; Liu, K.; Blumenthal, R.S. Coronary artery calcium scores and risk for cardiovascular events in women classified as "low risk" based on Framingham risk score: The Multi-Ethnic Study of Atherosclerosis (MESA). *Arch. Intern. Med.* **2007**, *167*, 2437–2442. [CrossRef] [PubMed]

14. Expert Panel on Detection, Evaluation, and Treatment of High Blood Cholesterol in Adult. Executive summary of the third report of the National Cholesterol Education Program (NCEP) expert panel on detection, evaluation, and treatment of high blood cholesterol in adults (Adult Treatment Panel III). *JAMA* **2001**, *285*, 2486–2497.

15. Elias-Smale, S.E.; Proenca, R.V.; Koller, M.T.; Kavousi, M.; van Rooij, F.J.; Hunink, M.G.; Steyerberg, E.W.; Hofman, A.; Oudkerk, M.; Witteman, J.C. Coronary calcium score improves classification of coronary heart disease risk in the elderly: The Rotterdam study. *J. Am. Coll. Cardiol.* **2010**, *56*, 1407–1414. [CrossRef] [PubMed]

16. Nasir, K.; Budoff, M.J.; Wong, N.D.; Scheuner, M.; Herrington, D.; Arnett, D.K.; Szklo, M.; Greenland, P.; Blumenthal, R.S. Family history of premature coronary heart disease and coronary artery calcification: Multi-Ethnic Study of Atherosclerosis (MESA). *Circulation* **2007**, *116*, 619–626. [CrossRef] [PubMed]

17. Mulders, T.A.; Sivapalaratnam, S.; Stroes, E.S.; Kastelein, J.J.; Guerci, A.D.; Pinto-Sietsma, S.J. Asymptomatic individuals with a positive family history for premature coronary artery disease and elevated coronary calcium scores benefit from statin treatment: A *post hoc* analysis from the St. Francis Heart Study. *JACC Cardiovasc. Imaging* **2012**, *5*, 252–260. [CrossRef] [PubMed]

18. Pandey, A.K.; Blaha, M.J.; Sharma, K.; Rivera, J.; Budoff, M.J.; Blankstein, R.; Al-Mallah, M.; Wong, N.D.; Shaw, L.; Carr, J.; et al. Family history of coronary heart disease and the incidence and progression of coronary artery calcification: Multi-Ethnic Study of Atherosclerosis (MESA). *Atherosclerosis* **2014**, *232*, 369–376. [CrossRef] [PubMed]

19. Polonsky, T.S.; McClelland, R.L.; Jorgensen, N.W.; Bild, D.E.; Burke, G.L.; Guerci, A.D.; Greenland, P. Coronary artery calcium score and risk classification for coronary heart disease prediction. *JAMA* **2010**, *303*, 1610–1016. [CrossRef] [PubMed]

20. Detrano, R.; Guerci, A.D.; Carr, J.J.; Bild, D.E.; Burke, G.; Folsom, A.R.; Liu, K.; Shea, S.; Szklo, M.; Bluemke, D.A.; et al. Coronary calcium as a predictor of coronary events in four racial or ethnic groups. *New Engl. J. Med.* **2008**, *358*, 1336–1345. [CrossRef] [PubMed]

21. Blaha, M.; Budoff, M.J.; Shaw, L.J.; Khosa, F.; Rumberger, J.A.; Berman, D.; Callister, T.; Raggi, P.; Blumenthal, R.S.; Nasir, K. Absence of coronary artery calcification and all-cause mortality. *JACC Cardiovasc. Imaging* **2009**, *2*, 692–700. [CrossRef] [PubMed]

22. Arad, Y.; Goodman, K.J.; Roth, M.; Newstein, D.; Guerci, A.D. Coronary calcification, coronary disease risk factors, C-reactive protein, and atherosclerotic cardiovascular disease events: The St. Francis Heart Study. *J. Am. Coll. Cardiol.* **2005**, *46*, 158–165. [CrossRef] [PubMed]

23. Budoff, M.J.; Shaw, L.J.; Liu, S.T.; Weinstein, S.R.; Mosler, T.P.; Tseng, P.H.; Flores, F.R.; Callister, T.Q.; Raggi, P.; Berman, D.S. Long-term prognosis associated with coronary calcification: Observations from a registry of 25,253 patients. *J. Am. Coll. Cardiol.* **2007**, *49*, 1860–1870. [CrossRef] [PubMed]

24. LaMonte, M.J.; FitzGerald, S.J.; Church, T.S.; Barlow, C.E.; Radford, N.B.; Levine, B.D.; Pippin, J.J.; Gibbons, L.W.; Blair, S.N.; Nichaman, M.Z. Coronary artery calcium score and coronary heart disease events in a large cohort of asymptomatic men and women. *Am. J. Epidemiol.* **2005**, *162*, 421–429. [CrossRef] [PubMed]

25. Hermann, D.M.; Gronewold, J.; Lehmann, N.; Moebus, S.; Jockel, K.H.; Bauer, M.; Erbel, R. Coronary artery calcification is an independent stroke predictor in the general population. *Stroke* **2013**, *44*, 1008–1013. [CrossRef] [PubMed]

26. O'Neal, W.T.; Efird, J.T.; Dawood, F.Z.; Yeboah, J.; Alonso, A.; Heckbert, S.R.; Soliman, E.Z. Coronary artery calcium and risk of atrial fibrillation (from the multi-ethnic study of atherosclerosis). *Am. J. Cardiol.* **2014**, *114*, 1707–1712. [CrossRef] [PubMed]

27. Goff, D.C., Jr.; Lloyd-Jones, D.M.; Bennett, G.; Coady, S.; D'Agostino, R.B.; Gibbons, R.; Greenland, P.; Lackland, D.T.; Levy, D.; O'Donnell, C.J.; *et al.* 2013 ACC/AHA guideline on the assessment of cardiovascular risk: A report of the American College of Cardiology/American Heart Association Task Force on Practice Guidelines. *Circulation* **2014**, *129*, S49–S73. [CrossRef] [PubMed]

28. Ridker, P.M.; Buring, J.E.; Rifai, N.; Cook, N.R. Development and validation of improved algorithms for the assessment of global cardiovascular risk in women: The Reynolds Risk Score. *JAMA* **2007**, *297*, 611–619. [CrossRef] [PubMed]

29. Ridker, P.M.; Paynter, N.P.; Rifai, N.; Gaziano, J.M.; Cook, N.R. C-reactive protein and parental history improve global cardiovascular risk prediction: The Reynolds Risk Score for men. *Circulation* **2008**, *118*, 2243–2251. [CrossRef] [PubMed]

30. DeFilippis, A.P.; Blaha, M.J.; Ndumele, C.E.; Budoff, M.J.; Lloyd-Jones, D.M.; McClelland, R.L.; Lakoski, S.G.; Cushman, M.; Wong, N.D.; Blumenthal, R.S.; *et al.* The association of Framingham and Reynolds Risk Scores with incidence and progression of coronary artery calcification in MESA (Multi-Ethnic Study of Atherosclerosis). *J. Am. Coll. Cardiol.* **2011**, *58*, 2076–2083. [CrossRef] [PubMed]

31. Desai, M.Y; Halliburton, S.; Masri, A.; Kottha, A.; Kuzmiak, S.; Flamm, S.; Schoenhagen, P. Reclassification of cardiovascular risk with coronary calcium scoring in subjects without documented coronary heart disease: Comparison with risk assessment based on Reynolds Risk Score. *J. Am. Coll. Cardiol.* **2012**, *59*, E1186. [CrossRef]

32. Naqvi, T.Z.; Lee, M.S. Carotid Intima-media thickness and plaque in cardiovascular risk assessment. *JACC Cardiovasc. Imaging* **2014**, *7*, 1025–1038. [CrossRef] [PubMed]

33. Lorenz, M.W.; Markus, H.S.; Bots, M.L.; Rosvall, M.; Sitzer, M. Prediction of clinical cardiovascular events with carotid intima-media thickness: A systematic review and meta-analysis. *Circulation* **2007**, *115*, 459–467. [CrossRef] [PubMed]

34. Folsom, A.R.; Kronmal, R.A.; Detrano, R.C.; O'Leary, D.H.; Bild, D.E.; Bluemke, D.A.; Budoff, M.J.; Liu, K.; Shea, S.; Szklo, M.; *et al.* Coronary artery calcification compared with carotid intima-media thickness in the prediction of cardiovascular disease incidence: The Multi-Ethnic Study of Atherosclerosis (MESA). *Arch. Int. Med.* **2008**, *168*, 1333–1339. [CrossRef]

35. Terry, J.G.; Carr, J.J.; Tang, R.; Evans, G.W.; Kouba, E.O.; Shi, R.; Cook, D.R.; Vieira, J.L.; Espeland, M.A.; Mercuri, M.F.; *et al.* Coronary artery calcium outperforms carotid artery intima-media thickness as a noninvasive index of prevalent coronary artery stenosis. *Arterioscler. Thromb. Vasc. Biol.* **2005**, *25*, 1723–1728. [CrossRef] [PubMed]

36. Danad, I.; Raijmakers, P.G.; Kamali, P.; Harms, H.J.; de Haan, S.; Lubberink, M.; van Kuijk, C.; Hoekstra, O.S.; Lammertsma, A.A.; Smulders, Y.M.; *et al.* Carotid artery intima-media thickness, but not coronary artery calcium, predicts coronary vascular resistance in patients evaluated for coronary artery disease. *Eur. Heart J. Cardiovasc. Imaging* **2012**, *13*, 317–323. [CrossRef] [PubMed]

37. Knaapen, P.; Camici, P.G.; Marques, K.M.; Nijveldt, R.; Bax, J.J.; Westerhof, N.; Gotte, M.J.; Jerosch-Herold, M.; Schelbert, H.R.; Lammertsma, A.A.; *et al.* Coronary microvascular resistance: Methods for its quantification in humans. *Basic Res. Cardiol.* **2009**, *104*, 485–498. [CrossRef] [PubMed]

38. Naqvi, T.Z.; Mendoza, F.; Rafii, F.; Gransar, H.; Guerra, M.; Lepor, N.; Berman, D.S.; Shah, P.K. High prevalence of ultrasound detected carotid atherosclerosis in subjects with low Framingham Risk Score: Potential implications for screening for subclinical atherosclerosis. *J. Am. Soc. Echocardiogr.* **2010**, *23*, 809–815. [CrossRef] [PubMed]

39. Davis, P.H.; Dawson, J.D.; Mahoney, L.T.; Lauer, R.M. Increased carotid intimal-medial thickness and coronary calcification are related in young and middle-aged adults: The Muscatine study. *Circulation* **1999**, *100*, 838–842. [CrossRef] [PubMed]

40. Lester, S.J.; Eleid, M.F.; Khandheria, B.K.; Hurst, R.T. Carotid intima-media thickness and coronary artery calcium score as indications of subclinical atherosclerosis. *Mayo Clin. Proc.* **2009**, *84*, 229–233. [CrossRef] [PubMed]

41. Greenland, P.; Alpert, J.S.; Beller, G.A.; Benjamin, E.J.; Budoff, M.J.; Fayad, Z.A.; Foster, E.; Hlatky, M.A.; Hodgson, J.M.; Kushner, F.G.; *et al.* 2010 ACCF/AHA guideline for assessment of cardiovascular risk in asymptomatic adults: Executive summary: A report of the American College of Cardiology Foundation/American Heart Association task force on practice guidelines. *Circulation* **2010**, *122*, 2748–2764. [CrossRef] [PubMed]

42. Ridker, P.M.; Danielson, E.; Fonseca, F.A.; Genest, J.; Gotto, A.M., Jr.; Kastelein, J.J.; Koenig, W.; Libby, P.; Lorenzatti, A.J.; MacFadyen, J.G.; *et al.* Rosuvastatin to prevent vascular events in men and women with elevated C-reactive protein. *N. Engl. J. Med.* **2008**, *359*, 2195–2207. [CrossRef] [PubMed]

43. Ridker, P.M.; Cushman, M.; Stampfer, M.J.; Tracy, R.P.; Hennekens, C.H. Inflammation, aspirin, and the risk of cardiovascular disease in apparently healthy men. *N. Engl. J. Med.* **1997**, *336*, 973–979. [CrossRef] [PubMed]

44. Ridker, P.M.; Hennekens, C.H.; Buring, J.E.; Rifai, N. C-reactive protein and other markers of inflammation in the prediction of cardiovascular disease in women. *N. Engl. J. Med.* **2000**, *342*, 836–843. [CrossRef] [PubMed]

45. Ridker, P.M.; Rifai, N.; Rose, L.; Buring, J.E.; Cook, N.R. Comparison of C-reactive protein and low-density lipoprotein cholesterol levels in the prediction of first cardiovascular events. *N. Engl. J. Med.* **2002**, *347*, 1557–1565. [CrossRef] [PubMed]

46. Koenig, W.; Lowel, H.; Baumert, J.; Meisinger, C. C-reactive protein modulates risk prediction based on the Framingham Score: Implications for future risk assessment: Results from a large cohort study in southern Germany. *Circulation* **2004**, *109*, 1349–1353. [CrossRef] [PubMed]

47. Pai, J.K.; Pischon, T.; Ma, J.; Manson, J.E.; Hankinson, S.E.; Joshipura, K.; Curhan, G.C.; Rifai, N.; Cannuscio, C.C.; Stampfer, M.J.; *et al.* Inflammatory markers and the risk of coronary heart disease in men and women. *New Engl. J. Med.* **2004**, *351*, 2599–2610. [CrossRef] [PubMed]

48. Boekholdt, S.M.; Hack, C.E.; Sandhu, M.S.; Luben, R.; Bingham, S.A.; Wareham, N.J.; Peters, R.J.; Jukema, J.W.; Day, N.E.; Kastelein, J.J.; *et al.* C-reactive protein levels and coronary artery disease incidence and mortality in apparently healthy men and women: The EPIC-Norfolk prospective population study 1993–2003. *Atherosclerosis* **2006**, *187*, 415–422. [CrossRef] [PubMed]

49. Ballantyne, C.M.; Hoogeveen, R.C.; Bang, H.; Coresh, J.; Folsom, A.R.; Heiss, G.; Sharrett, A.R. Lipoprotein-associated phospholipase A2, high-sensitivity C-reactive protein, and risk for incident coronary heart disease in middle-aged men and women in the Atherosclerosis Risk in Communities (ARIC) study. *Circulation* **2004**, *109*, 837–842. [CrossRef] [PubMed]

50. Park, R.; Detrano, R.; Xiang, M.; Fu, P.; Ibrahim, Y.; LaBree, L.; Azen, S. Combined use of computed tomography coronary calcium scores and C-reactive protein levels in predicting cardiovascular events in nondiabetic individuals. *Circulation* **2002**, *106*, 2073–2077. [CrossRef] [PubMed]

51. Lakoski, S.G.; Cushman, M.; Blumenthal, R.S.; Kronmal, R.; Arnett, D.; D'Agostino, R.B., Jr.; Detrano, R.C.; Herrington, D.M. Implications of C-reactive protein or coronary artery calcium score as an adjunct to global risk assessment for primary prevention of CHD. *Atherosclerosis* **2007**, *193*, 401–407. [CrossRef] [PubMed]

52. Blaha, M.J.; Budoff, M.J.; DeFilippis, A.P.; Blankstein, R.; Rivera, J.J.; Agatston, A.; O'Leary, D.H.; Lima, J.; Blumenthal, R.S.; Nasir, K. Associations between C-reactive protein, coronary artery calcium, and cardiovascular events: Implications for the JUPITER population from MESA, a population-based cohort study. *Lancet* **2011**, *378*, 684–692. [CrossRef] [PubMed]

53. Mohlenkamp, S.; Lehmann, N.; Moebus, S.; Schmermund, A.; Dragano, N.; Stang, A.; Siegrist, J.; Mann, K.; Jockel, K.H.; Erbel, R.; *et al.* Quantification of coronary atherosclerosis and inflammation to predict coronary events and all-cause mortality. *J. Am. Coll. Cardiol.* **2011**, *57*, 1455–1464. [CrossRef] [PubMed]

54. Reilly, M.P.; Wolfe, M.L.; Localio, A.R.; Rader, D.J. Study of inherited risk of coronary A. C-reactive protein and coronary artery calcification: The study of Inherited Risk of Coronary Atherosclerosis (SIRCA). *Arterioscler. Thromb. Vasc. Biol.* **2003**, *23*, 1851–1856. [CrossRef] [PubMed]

55. Ding, J.; Hsu, F.C.; Harris, T.B.; Liu, Y.; Kritchevsky, S.B.; Szklo, M.; Ouyang, P.; Espeland, M.A.; Lohman, K.K.; Criqui, M.H.; *et al.* The association of pericardial fat with incident coronary heart disease: The Multi-Ethnic Study of Atherosclerosis (MESA). *Am. J. Clin. Nutr.* **2009**, *90*, 499–504. [CrossRef] [PubMed]

56. Budoff, M.J.; Nasir, K.; Katz, R.; Takasu, J.; Carr, J.J.; Wong, N.D.; Allison, M.; Lima, J.A.; Detrano, R.; Blumenthal, R.S.; *et al.* Thoracic aortic calcification and coronary heart disease events: The Multi-Ethnic Study of Atherosclerosis (MESA). *Atherosclerosis* **2011**, *215*, 196–202. [CrossRef] [PubMed]

57. Zeb, I.; Budoff, M.J.; Katz, R.; Lloyd-Jones, D.; Agatston, A.; Blumenthal, R.S.; Blaha, M.; Blankstein, R.; Carr, J.J.; Nasir, K. Non-alcoholic fatty liver disease is an independent predictor of long-term incident coronary heart disease events—The Multi-Ethnic Study of Atherosclerosis. *Circulation* **2012**, *126*, A13688.

58. Yeboah, J.; Carr, J.J.; Terry, J.G.; Ding, J.; Zeb, I.; Liu, S.; Nasir, K.; Post, W.; Blumenthal, R.S.; Budoff, M.J. Computed tomography-derived cardiovascular risk markers, incident cardiovascular events, and all-cause mortality in nondiabetics: The Multi-Ethnic Study of Atherosclerosis. *Eur. J. Prev. Cardiol.* **2014**, *21*, 1233–1241. [CrossRef] [PubMed]

59. Budoff, M.J.; Nasir, K.; McClelland, R.L.; Detrano, R.; Wong, N.; Blumenthal, R.S.; Kondos, G.; Kronmal, R.A. Coronary calcium predicts events better with absolute calcium scores than age-sex-race/ethnicity percentiles: MESA (Multi-Ethnic Study of Atherosclerosis). *J. Am. Coll. Cardiol.* **2009**, *53*, 345–352. [CrossRef] [PubMed]

60. Kalia, N.K.; Miller, L.G.; Nasir, K.; Blumenthal, R.S.; Agrawal, N.; Budoff, M.J. Visualizing coronary calcium is associated with improvements in adherence to statin therapy. *Atherosclerosis* **2006**, *185*, 394–399. [CrossRef] [PubMed]

61. Taylor, A.J.; Bindeman, J.; Feuerstein, I.; Le, T.; Bauer, K.; Byrd, C.; Wu, H.; O'Malley, P.G. Community-based provision of statin and aspirin after the detection of coronary artery calcium within a community-based screening cohort. *J. Am. Coll. Cardiol.* **2008**, *51*, 1337–1341. [CrossRef] [PubMed]

62. Budoff, M.J.; Hokanson, J.E.; Nasir, K.; Shaw, L.J.; Kinney, G.L.; Chow, D.; Demoss, D.; Nuguri, V.; Nabavi, V.; Ratakonda, R.; *et al.* Progression of coronary artery calcium predicts all-cause mortality. *JACC Cardiovasc. Imaging* **2010**, *3*, 1229–1236. [CrossRef] [PubMed]

63. Berry, J.D.; Liu, K.; Folsom, A.R.; Lewis, C.E.; Carr, J.J.; Polak, J.F.; Shea, S.; Sidney, S.; O'Leary, D.H.; Chan, C.; *et al.* Prevalence and progression of subclinical atherosclerosis in younger adults with low short-term but high lifetime estimated risk for cardiovascular disease: The coronary artery risk development in young adults study and multi-ethnic study of atherosclerosis. *Circulation* **2009**, *119*, 382–389. [CrossRef] [PubMed]

64. Raggi, P.; Cooil, B.; Shaw, L.J.; Aboulhson, J.; Takasu, J.; Budoff, M.; Callister, T.Q. Progression of coronary calcium on serial electron beam tomographic scanning is greater in patients with future myocardial infarction. *Am. J. Cardiol.* **2003**, *92*, 827–829. [CrossRef] [PubMed]

65. Budoff, M.J.; Young, R.; Lopez, V.A.; Kronmal, R.A.; Nasir, K.; Blumenthal, R.S.; Detrano, R.C.; Bild, D.E.; Guerci, A.D.; Liu, K.; *et al.* Progression of coronary calcium and incident coronary heart disease events: MESA (Multi-Ethnic Study of Atherosclerosis). *J. Am. Coll. Cardiol.* **2013**, *61*, 1231–1239. [CrossRef] [PubMed]

66. Budoff, M.J.; Achenbach, S.; Blumenthal, R.S.; Carr, J.J.; Goldin, J.G.; Greenland, P.; Guerci, A.D.; Lima, J.A.; Rader, D.J.; Rubin, G.D.; *et al.* Assessment of coronary artery disease by cardiac computed tomography: A scientific statement from the American Heart Association Committee on Cardiovascular Imaging and Intervention, Council on Cardiovascular Radiology and Intervention, and Committee on Cardiac Imaging, Council on Clinical Cardiology. *Circulation* **2006**, *114*, 1761–1791. [CrossRef] [PubMed]

67. Parker, M.S.; Hui, F.K.; Camacho, M.A.; Chung, J.K.; Broga, D.W.; Sethi, N.N. Female breast radiation exposure during CT pulmonary angiography. *AJR Am. J. Roentgenol.* **2005**, *185*, 1228–1233. [CrossRef] [PubMed]

68. Horton, K.M.; Post, W.S.; Blumenthal, R.S.; Fishman, E.K. Prevalence of significant noncardiac findings on electron-beam computed tomography coronary artery calcium screening examinations. *Circulation* **2002**, *106*, 532–534. [CrossRef] [PubMed]

69. Schragin, J.G.; Weissfeld, J.L.; Edmundowicz, D.; Strollo, D.C.; Fuhrman, C.R. Non-cardiac findings on coronary electron beam computed tomography scanning. *J. Thora. Imaging* **2004**, *19*, 82–86. [CrossRef]

70. Machaalany, J.; Yam, Y.; Ruddy, T.D.; Abraham, A.; Chen, L.; Beanlands, R.S.; Chow, B.J. Potential clinical and economic consequences of noncardiac incidental findings on cardiac computed tomography. *J. Am. Coll. Cardiol.* **2009**, *54*, 1533–1541. [CrossRef] [PubMed]

International Journal of
*Molecular Sciences*

MDPI

Review

# Ultrasound Imaging for Risk Assessment in Atherosclerosis

David C. Steinl [1] and Beat A. Kaufmann [2,*]

[1] Department of Biomedicine, University Hospital Basel, Hebelstrasse 20, Basel 4031, Switzerland;
david.steinl@unibas.ch

[2] Division of Cardiology, University Hospital Basel, Petersgraben 4, Basel 4031, Switzerland

* Correspondence: beat.kaufmann@usb.ch; Tel.: +41-61-328-6712; Fax: +41-61-265-4598

Academic Editor: Michael Henein
Received: 6 February 2015; Accepted: 9 April 2015; Published: 29 April 2015

**Abstract:** Atherosclerosis and its consequences like acute myocardial infarction or stroke are highly prevalent in western countries, and the incidence of atherosclerosis is rapidly rising in developing countries. Atherosclerosis is a disease that progresses silently over several decades before it results in the aforementioned clinical consequences. Therefore, there is a clinical need for imaging methods to detect the early stages of atherosclerosis and to better risk stratify patients. In this review, we will discuss how ultrasound imaging can contribute to the detection and risk stratification of atherosclerosis by (a) detecting advanced and early plaques; (b) evaluating the biomechanical consequences of atherosclerosis in the vessel wall; (c) assessing plaque neovascularization and (d) imaging the expression of disease-relevant molecules using molecular imaging.

**Keywords:** ultrasound; contrast enhanced; molecular imaging; atherosclerosis; micro-bubbles

## 1. Introduction

Atherosclerosis is a systemic, multifactorial disease affecting large arteries throughout the body [1]. The disease process is mainly driven by two underlying processes: a disturbed equilibrium of lipid accumulation and chronic inflammation of the vessel wall [2], which together leads to the buildup of atherosclerotic plaques that, in turn, can lead to a variety of cardiovascular diseases and complications.

Autopsy studies have shown that the pathogenesis of atherosclerosis starts at a young age [3,4] and thus can progress silently over decades before it results in life-threatening vascular complications (acute myocardial infarction, cerebrovascular insult). Death statistics in Europe show that nearly every second death is attributable to cardiovascular diseases [5]. In the last four decades, however, rates of death from cardiovascular disease have declined in the western world, whereas disease prevalence has increased in developing countries, and it is estimated that today, 80% of the global burden of cardiovascular disease is occurring in these countries [6].

Given the natural history of atherosclerosis with a silent progression over several decades, reliable methods for risk estimations in individuals are a clinical need. Current approaches to risk stratification rely on well-established clinical risk scores (Framingham, PROCAM (Prospective Cardiovascular Münster Study), ESC-SCORE (European Society of Cardiology-Systematic Coronary Risk Evaluation)) that incorporate clinical risk factors for atherosclerosis (arterial hypertension, diabetes mellitus, hypercholesterolemia, smoking, and family history for premature coronary artery disease) [7]. However, when applying these risk scores to populations in western countries, approximately 40% of the adult population fall into a medium risk category, in which the benefit of a broad use of preventive strategies is uncertain. Therefore, further risk stratification in this group for the allocation of preventive therapies and/or strategies is desirable. In addition, those methods should optimally allow for the subsequent assessment of therapy effects aimed at inhibiting the progression of atherosclerosis.

*Int. J. Mol. Sci.* **2015**, *16*, 9749–9769

In the last decades, developments in all major non-invasive imaging technologies have been used to assess the atherosclerotic disease process. When envisaging application of such an imaging method in risk prediction, ultrasound has distinct advantages over other imaging modalities in terms of wide availability and low cost.

In this review, we will first give an overview of the vascular biology of the pathogenesis of atherosclerosis. We will then review how conventional ultrasound imaging can be used to assess arteries for the presence of atherosclerosis, and what complex signal processing algorithms can possibly add to the evaluation of conventional ultrasound images. We will then explain the basic principles of molecular imaging using targeted contrast agents, and will review how this novel methodology has been used in relevant animal models of disease to detect the presence and response to therapy of vascular inflammation that drives the pathogenesis of atherosclerosis.

## 2. Atherosclerosis

While incompletely understood, the first step in the development of vascular endothelial inflammation that leads to the development of atherosclerotic plaques, seems to be sub-endothelial deposition of low density lipoprotein-cholesterol (LDL). LDL is subsequently modified to oxidized LDL (oxLDL), a process that is influenced by cardiovascular risk factors such as hypertension, diabetes, or smoking that all increase oxidative stress in the vascular wall [8–10]. Through activation of NF-κB, oxLDL leads to the expression of vascular cell adhesion molecule-1 (VCAM-1) and intercellular adhesion molecule-1 (ICAM-1) on the vascular surface of endothelial cells [11]. By interaction with their counter-ligands $\alpha4{:}\beta1$ and CD11a:CD18 on monocytes [12], the expression of these cell adhesion molecules leads to recruitment of monocytes into the vascular wall [13] (Figure 1).

Once in the vascular wall, the local inflammatory milieu further activates the monocytes, which are converted to macrophages that scavenge oxLDL and act to amplify the inflammatory process via the secretion of cytokines (IL-1β, IL-12, TNF-α) [14]. Continuing recruitment of monocytes leads to intimal thickening and the formation of fatty streaks, which are the earliest lesions of atherosclerosis that can be appreciated macroscopically. Continuing inflammatory cell recruitment leads to plaque growth and recruitment of additional cell types (vascular smooth muscle cells, VSMC). Plaque growth also leads to local tissue hypoxia, which promotes neovascularization of the atherosclerotic lesions [15]. The inflammatory activation of the endothelium not only leads to recruitment of leukocytes, but via the expression of von Willebrand factor (vWF) [16] and tissue factor (TF) [17] also creates a pro-thrombotic environment. Intimal thickening, plaque formation, plaque neovascularization, and the expression of cell adhesion molecules and pro-thrombotic factors are all potential targets for imaging of atherosclerosis with ultrasound.

The structural changes in the arterial wall during the pathogenesis of atherosclerosis can be conceptualized as vascular aging, the pace of which is correlated to the risk for cardiovascular events in individuals [18]. These morphological and functional changes are a consequence of the underlying molecular pathways described above, and include changes of the general vessel morphology with buildup of plaque [19], alterations in the biomechanical properties of the arterial wall [20], and the development of atherosclerotic plaque neo-vascularization [21]. These parameters can be assessed by different non-invasive imaging techniques, usually performed on either the carotid arteries or the aortic arch, the changes of which are accepted surrogate markers for atherosclerotic disease progression in the coronaries and cardiovascular events [22].

*Int. J. Mol. Sci.* **2015**, *16*, 9749–9769

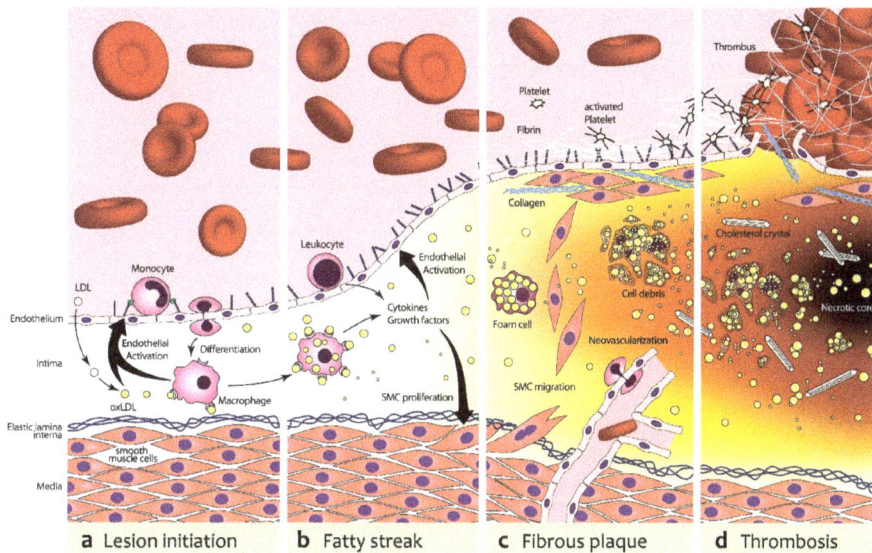

**Figure 1.** Pathogenesis of atherosclerosis. (**a**) In the first stage, low density lipoprotein-cholesterol (LDL) is deposited in the endothelium and undergoes oxidative modification, resulting in oxidized LDL (oxLDL). OxLDL stimulates endothelial cells to express adhesion molecules (vascular cell adhesion molecule-1 (VCAM-1), P-Selectin) and various chemokines (e.g., Monocyte Chemoattractant Protein-1 (MCP-1), Interleukin 8 (IL-8)). This leads to a recruitment of monocytes, which transmigrate into the intima and differentiate to pro-atherogenic macrophages; (**b**) Macrophages harvest residual oxLDL via their scavenger receptors and add to the endothelial activation and, subsequently, leukocyte recruitment with the secretion of Tumor Necrosis Factor α (TNF-α) and IL-6; (**c**) The increasing plaque volume promotes neovascularization. Proliferating smooth muscle cells (SMCs) stabilize the nascent fibrous plaque. With deposition of fibrin and activated platelets on the dysfunctional endothelium that expresses tissue factor (TF) and von Willebrand factor (vWF), a pro-thrombotic milieu is formed; (**d**) Foam cells can undergo apoptosis and release cell-debris and lipids, which will result in the formation of a necrotic core. In addition, proteases secreted from foam cells can destabilize the plaque. This can lead to plaque rupture, in which case extracellular matrix molecules (e.g., collagens, elastin, TF, vWF) catalyze thrombotic events.

## 3. Anatomical Imaging of Atherosclerosis with Ultrasound

Established atherosclerotic lesions can be visualized with anatomical B-mode ultrasound imaging as protrusions of the intima-media (Figure 2) and the number of visualized plaques, as well as total plaque area or total plaque volume was reported to be an independent predictor of future cardiovascular mortality and coronary events [23,24].

**a**

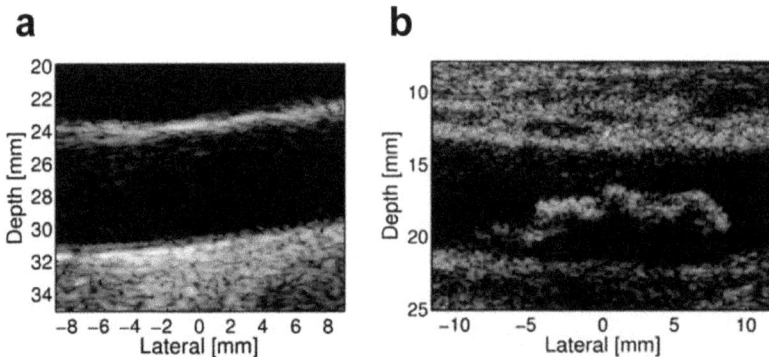

**b**

Figure 2. B-mode imaging of the carotid artery. These images illustrate (**a**) a normal carotid artery; and (**b**) a large atherosclerotic plaque protruding into the lumen of the carotid artery. Reproduced from [25], with permission.

Additional information can be derived from plaque echogenicity. On B-mode ultrasound, plaques that contain a large lipid core appear echolucent, whereas plaque fibrosis and calcification tend to appear echogenic. As proposed by Gray-Weale *et al.* [26], carotid plaques can be classified into four categories as echolucent, predominantly echolucent, predominantly echogenic, or echogenic. In patients with carotid stenosis, the degree of plaque echolucency correlated with increased risk for cerebrovascular events [27]. Similarly, standardized measurements of the decrease in gray scale levels within carotid plaques over time have been correlated to the risk of subsequent cardiovascular events [28].

Plaques that are large enough to be visualized with ultrasound develop relatively late in the pathogenesis of atherosclerosis. However, increases in the carotid intima media thickness (C-IMT), which occur prior to plaque development, can be measured with high-resolution ultrasound. An increase of C-IMT has been found to be associated with risk for cardiovascular events in several large observational studies [29,30]. Thus, C-IMT measurements have a value in population studies, and may even be useful for initial evaluations of the effect of new therapies targeting atherosclerosis before the start of large-scale morbidity and mortality trials [31]. However, a recent meta-analysis has cast doubt on the value of C-IMT for risk prediction in individual patients, showing little added benefit over traditional risk assessment using the Framingham Score with small net reclassification improvements in risk category that are unlikely to be clinically relevant [32]. There are several possible reasons for this lack of additional prognostic information. C-IMT is frequently measured in the common carotid artery, whereas advanced lesions prone to complications develop in the bulb or proximal internal carotid artery. While C-IMT correlates to age and blood pressure, it also offers little advantage over these traditional risk factors for predicting events [33]. Another reason may be that the differences in C-IMT between risk strata is around 200 μm, which is below the axial resolution of ultrasound systems commonly used for these measurements. Algorithms that help measuring C-IMT more precisely [34] or the use of three-dimensional ultrasound that would allow volumetric assessment of plaque burden [35] might potentially help to improve diagnostic value of C-IMT. In this respect, a recent study that used two-dimensional short axis images of the carotid arteries to assess a global, three-dimensional plaque burden of both carotids has shown incremental predictive value over traditional risk factors, which was, in addition, comparable to coronary artery calcium scoring [36].

Catheters in the millimeter size domain offer the possibility for intravascular ultrasound (IVUS) to gain information from inside the vessel in return for sacrificing non-invasiveness. As opposed to angiography, which shows only plaques that lead to a coronary stenosis, IVUS is able to also detect plaques with an eccentric remodeling that do not narrow the vessel lumen. This is important, since eccentric plaque remodeling does not preclude these plaques from causing complications such as plaque rupture, which can lead to vessel occlusion and myocardial infarction [37].

The atheroma burden measured with grayscale IVUS correlates closely with histology findings [38–41]. It has been shown that when patients with known coronary artery disease are followed up after examination with IVUS, the majority of acute coronary syndromes are due to complications at sites in the coronary tree, which had exhibited an eccentric plaque without stenosis [37].

Gray-scale IVUS does reveal some information about the composition of atherosclerotic lesions and plaques can be classified based on their visual appearance as echolucent ("soft"), hypoechogenic (fibrous), hyperechogenic with or without shadowing (calcified), or intermediate form lesions [42]. However, with gray-scale IVUS calcified plaques cannot be discriminated from densely fibromuscular tissue, both of them hyperechogenic, while lipid-rich plaques cannot be discriminated from fibrotic plaques or intraplaque hemorrhages that appear hypoechogenic to echolucent. Thus, it does not come as a surprise that, in clinical studies, IVUS has not been capable to discriminate between plaques found in patients with stable angina and unstable atherosclerotic lesions [43].

## 4. Biomechanical Imaging

In the evaluation of atherosclerosis, valuable information can not only be obtained from wall morphology or plaque composition but can also be derived from the arterial wall mechanical properties, which change along with the progression of atherosclerosis [44]. Fibrosis, calcification, and increased smooth muscle cell proliferation change the biomechanical properties of the arterial walls, which translate into an increase in Young's elastic modulus and changes of other parameters of vessel wall deformability or stiffness.

Because in less elastic arteries the speed of pulse wave propagation is increased due to a diminished "Windkessel" capacity, measurements of arterial pulse wave velocity (PWV) with transoesophageal echo (TOE) or transthoracic echo (TTE) doppler ultrasound imaging have been shown to be an independent predictor of atheroma burden and cardiovascular events [45]. However, while a strong association of changes in PWV with age and blood pressure has consistently been shown, the association of these changes with other risk factors for atherosclerosis is unclear [46].

While non-invasively measured PWV is used to assesses changes in the mechanical properties of long segments of arteries and thus, in overall risk prediction, invasive techniques have also been developed in an effort to match changes in local biomechanical properties to plaque composition and, thus, to the risk for vascular complications. Elastography, which derives strain from high-frequency IVUS for a set intravascular pressure differential has been shown to identify plaque components *in vitro* in excised human arteries [47], with fatty, potentially unstable plaques exhibiting a higher deformability and higher strain values. Similar techniques are used to assess vessel strain with palpography, however, in contrast to elastography, only the first 450 µm starting from the luminal side of the vessel are examined. Because the blood pressure, which is the force leading to deformation of the vessel wall, is primarily acting on the luminal side of the vessel wall, this technique is regarded as being more robust for differentiating between different plaque components compared to elastography. [48]. Palpography has been validated *in vivo* in an animal model of atherosclerosis [49] and was able to differentiated plaques with increased fatty or macrophage content from other, more stable plaques. In a study using three-dimensional palpography in patients either with stable angina, unstable angina or acute myocardial infarction, the number of highly deformable plaques on plapography correlated with the clinical presentation [50].

For the non-invasive assessment of vessel elasticity, high frame rate ultrasound imaging [51], velocity vector imaging [52], or strain imaging of the carotid arteries as a reference vessel for cardiovascular risk have been used. Speckle tracking strain imaging of carotid artery plaques has been validated against sonomicrometry [53] and preliminary studies in humans showed an association of carotid artery circumferential strain derived using speckle tracking with previous strokes [54].

## 5. Contrast Enhanced Ultrasound Imaging of Atherosclerosis

Ultrasound contrast agents were developed in the early 1990s with the goal of contrasting the blood pool to allow for better delineation of the left ventricular endocardial border. The use of microbubbles for this purpose is based on observations made by Gramiak and Shah, who noticed a signal enhancement of blood in the aorta after injection of mechanically agitated saline containing small air bubbles [55]. These air bubbles were very short lived and also not suitable for venous injection. Later, ultrasound contrast agents for intravenous injection were developed. Commercially available and FDA-certified are microbubble-based contrast agents, which have a gaseous core encapsulated with a thin shell. To increase microbubble stability, large-molecule, hydrophobic gases such as sulfur hexafluoride or perfluorocarbons are used. For encapsulation, the microbubble shell is composed of combinations of albumin, galactose, lipids, or polymers. These microbubbles measure on average 1–2 μm in diameter and thus circulate freely throughout the microcirculation [56]. In addition to microbubbles, nanoscale echogenic liposomes have also been used as ultrasound contrast agents [57].

### 5.1. Physical Aspects of Contrast Enhanced Ultrasound Imaging

Microbubbles undergo volumetric oscillations with compression during the pressure peaks, and expansion during the pressure nadirs of an incident ultrasound wave [58,59]. These volumetric oscillations result in a strong backscattered acoustic signal that can be detected by ultrasound systems.

The oscillation of a microbubble depends on the frequency and the intensity of an incident sound wave [60]. With low transmitted powers, microbubbles will oscillate at the same frequency as the incident sound wave and thus will backscatter sound waves at the same frequency (linear backscatter). At medium transmitted powers, microbubbles show non-linear oscillations and will thus backscatter sound waves at both the incident frequency and multiples of the incident frequency (harmonic frequencies). As tissue shows a lesser degree of nonlinear behavior, this phenomenon can be exploited to improve the contrast to tissue signal ratio. At high-transmitted powers, the microbubble shell will be disrupted and the microbubble will be destroyed.

### 5.2. Plaque Neovascularization

Healthy arteries exhibit a vascularization of the outer adventitial wall layers but not of the intima and inner media, which are supplied with nutrients by the luminal blood flow. However, with a thickening of the arterial wall as a consequence of the buildup of atherosclerotic plaque, tissue hypoxia develops and stimulates sprouting of the Vasa vasorum into the medial and intimal layers of plaques [61], resulting in plaque neovascularization. Neovessels that develop in large plaques lack pericytes and are prone to rupture [62], which may lead to acute atherosclerotic complications. Plaque neovessels have been shown to promote plaque progression through deposition of cholesterol and macrophages and enlargement of the plaque necrotic core [63]. Thus, it is thought that the detection of plaque neovascularization could be used as a marker for the risk for subsequent cardiovascular events.

Contrast enhanced ultrasound imaging has been used to detect neovascularization in carotid plaques (Figure 3). Initial studies have shown that qualitative assessments of plaque neovascularization, as detected with contrast-enhanced ultrasound, correlate to histologically determined micro-vascular density and to plaque echolucency, the latter being a marker of plaque vulnerability [64,65]. Similarly, quantitative data on plaque neovascularization obtained using maximum intensity projection methods have been shown to correlate with histologic neovascularization and inflammatory cell infiltration [66]. A retrospective analysis has shown a correlation of the degree of plaque neovascularization as assessed with contrast-enhanced ultrasound with previous cerebrovascular and coronary events [67]. Interestingly, the distribution of neovascularization within carotid plaques offers additional information on plaque vulnerability with a higher frequency of neovascularization predominantly of the shoulder regions in symptomatic patients [68]. However, prospective studies that assess the predictive value of this method to estimate the risk of future cerebrovascular events have not been published as of to date. Additionally, contrast imaging of the carotid arteries suffers from a pseudoenhancement artefact adjacent to the far wall of the artery that could potentially hamper assessment of plaque neovascularisation in this area [69]. However, novel imaging algorithms using detection methods based on counterpropagation of ultrasound pulses have been described that can potentially eliminate this artefact [70].

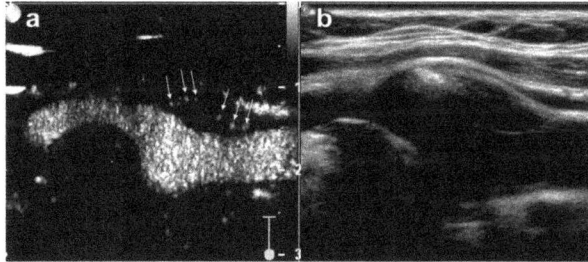

**Figure 3.** Contrast-enhanced ultrasound (CEUS) imaging of plaque neovascularization in a carotid plaque. (**a**) Carotid artery with intraplaque neovascularization on CEUS. The arrows denote microbubbles within neovessels in a plaque at the origin of the internal carotid artery; (**b**) Corresponding B-mode ultrasound image. Reproduced from [71] with permission.

*5.3. Basic Principles of Molecular Imaging with Ultrasound Contrast Agents*

The nonlinear signal generation of freely circulating microbubbles is used in clinical practice for opacification of the blood pool or for the assessment of organ perfusion. However, the goal of molecular imaging is the specific attachment of these microbubbles to a relevant target of interest. An ultrasound image of these retained microbubbles should then represent the level of expression of the target in the tissue. The target-specific retention of microbubbles can be accomplished by either (1) modifications of the microbubble shell components that facilitate attachment of microbubbles to activated leukocytes; or (2) the conjugation of ligands specific for a disease molecule to the microbubble surface. An example for microbubble targeting using modified shell characteristics is the incorporation of negatively charged phospholipids into the microbubble shell. The negative microbubble surface charge leads to complement deposition on its surface, which, in turn, promotes attachment to activated leukocytes [72]. Similarly, microbubbles with an albumin shell have been shown to bind to monocytes and neutrophils through attachment to the leucocyte β2-integrin Mac-1 (CD11b/CD18) [73]. Much more versatile is the strategy of coupling ligands like antibodies, glycoproteins, carbohydrates or peptides to the microbubble surface. Coupling of the ligands is done either directly to the microbubble shell surface or more often to a polyethylene glycol (PEG) spacer arm which projects the ligand further away from the microbubble surface. It has been shown that the length of the PEG spacer arm can influence targeting efficiency of microbubbles [74]. For the conjugation of the ligands to the PEG spacer arms, simple biotin-streptavidin linking has been used, whereas covalent bonding with maleimide is also possible. Using both strategies, several thousand ligands can usually be conjugated per square micron of microbubble shell surface [75].

Protocols for ultrasound molecular imaging usually involve intravenous bolus injections of several million microbubbles. Over several minutes, microbubbles will adhere to a target of interest within tissue, whereas freely circulating microbubbles are being cleared from the circulation due to shell disintegration and gas loss or clearance in the liver. The presence of attached microbubbles can then be derived from imaging sequences before and after destruction of microbubbles by high-power ultrasound impulses at a timepoint when the total number of freely circulating microbubbles remains nearly constant over a short timespan and is not significantly affected by destruction pulses in the scanplane. On such imaging sequences, the signal before destruction represents the signal generated from the sum of attached and remaining freely circulating microbubbles, whereas, after destruction, only signals from freely circulating microbubbles will be recorded (Figure 4).

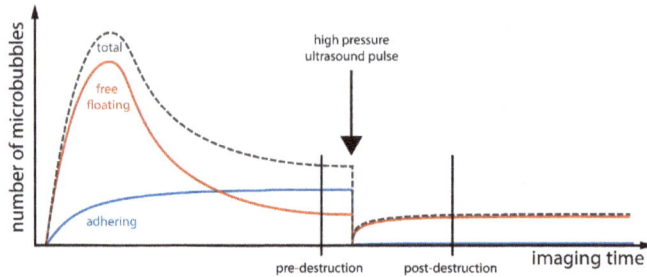

**Figure 4.** Algorithm for ultrasound molecular imaging. After a bolus injection of targeted microbubbles, a percentage of the total number of microbubbles injected will adhere to the molecule of interest. Freely circulating microbubbles will be cleared in the liver over several minutes. The signal intensity before destruction at the site of interest will be the summation of adhering and free floating microbubbles. The signal intensity after destruction and a short period of reperfusion is generated only by the remaining circulating free microbubbles. Digital subtraction of pre- and post-destruction images can then be used to obtain signal from adhered microbubbles only. Adapted from [76].

Digital subtraction of signals before and after destruction will then yield the enhancement from attached microbubbles. As microbubbles can be destroyed *in vivo*, repeated injections and imaging sequences, and thus the assessment of several targets in a short timeframe is possible using this technique. However, several factors have to be taken into account for successful ultrasound molecular imaging (Figure 5). The target molecule should be as specific as possible for a particular disease and should possess a limited constitutive expression. As microbubbles are pure intravascular tracers, the molecular target by necessity has to be present on the endothelial surface. The ligand that is used for targeting should be attached in high density to the microbubble surface. It should be specific for the target molecule with the highest possible affinity (fast on-rate, slow off-rate), in particular when used for targeting in high shear conditions such as in large arteries. While, theoretically, microbubble attachment to cells could lead to signal dampening, this has been shown to be minimal *in vitro* [77]. Given these data from *in vitro* studies, sizeable damping of attached microbubbles is unlikely to occur *in vivo*. In this respect, it is important to realize that the subtraction techniques that are commonly employed in ultrasound molecular imaging do not measure the number of adhering microbubbles, but rather the signal generated from the sum of adhering microbubbles, which also depends on other factors for a given number of particles (e.g., attenuation of ultrasound to and from the target particles, size of the particles). Finally, it should be noted that contrast-enhanced ultrasound molecular imaging is in a preclinical stage of development, and no large clinical trials using this technique have been performed to date.

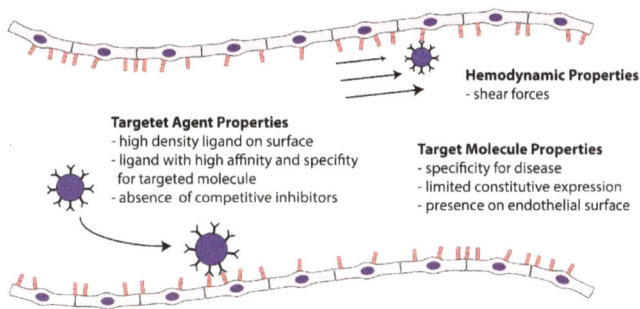

**Figure 5.** Determinants of targeted micobubble retention. The determinants of targeted microbubble retention within a vessel is dependent on various factors, including the targeted molecules and the contrast agent itself as well as hemodynamic properties. Adapted from [78].

*5.4. Ultrasound Molecular Imaging of Vascular Inflammation in Atherosclerosis*

Inflammation plays an important role in the initiation and progression of atherosclerosis, Leukocyte recruitment to the vessel wall and subsequent transmigration into the underlying tissue is one of the most important contributors to atherosclerosis [79]. Invading monocytes and lymphocytes serve as a source for pro-inflammatory cytokines, reactive oxygen species, pro-angiogenic signals and bioactive products that promote smooth muscle cell migration and further inflammatory cell recruitment as well as pro-thrombotic compounds that have been implicated in the progression of atherosclerosis as well as in complications such as total vessel occlusion in the case of plaque rupture.

Perfluorocarbon-exposed sonicated dextrose albumin (PESDA) microbubbles have been shown to attach to inflamed, dysfunctional endothelium after stretch-induced injury of the carotid arteries during a high fat diet in a pig model [80]. Similarly, increased signal from PESDA microbubbles has been shown in a rat model of early aortic atherosclerosis. The attachment of PESDA microbubbles is complement-mediated, as complement depletion reduced targeted signals [81].

Recruitment of leukocytes to the arterial wall is mediated by cell adhesion molecules that are expressed on the endothelial surface in response to inflammatory stimuli. P-Selectin is responsible for the initial tethering of leukocytes to the activated endothelium by interaction with glycosylated P-Selectin glycoprotein ligand-1 (PSGL-1) [82]. In a mouse model of atherosclerosis, P-Selectin-dependent monocyte rolling has been demonstrated on early atherosclerotic lesions [83] and ultrasound molecular imaging of the expression of P-Selectin has also been validated in the detection of very early atherosclerosis [84]. P-Selectin is, therefore, an important effector in the early inflammatory process and was first successfully used with ultrasound molecular imaging in the detection of renal tissue injury [73] and later in models of myocardial ischemia with the use of sialyl Lewis(x) as a ligand for P-Selectin [85] or anti-P-Selectin antibodies [86].

After initial tethering by P-Selectin, rolling and cytokine-triggered firm adhesion of leukocytes is primarily mediated by endothelial cell adhesion molecules of the immunoglobulin gene superfamily (Vascular cell adhesion molecule 1, VCAM-1 and intercellular adhesion molecule 1, ICAM-1) by interaction with their leukocyte counterligands $\alpha4{:}\beta1$ and $\alpha L{:}\beta2/\alpha M{:}\beta2$ [87]. In murine models of atherosclerosis, slow rolling of monocytes on the surface of atherosclerotic plaques appears to be largely mediated by VCAM-1 [88]. With regards to the use as early markers of disease, VCAM-1 and ICAM-1 have been demonstrated to be transcriptionally upregulated on the endothelium of mouse models of atherosclerosis during the very early stages of disease development [89]. Initial *in vitro* studies have confirmed that ICAM-1 expressed on activated human coronary endothelial cells could be targeted using microbubbles [90]. Early *in vivo* studies using echogenic liposomes, targeted to either VCAM-1 or ICAM-1, showed increased signal intensity in yucatan miniswine at sites of injured endothelium, with the drawback of very high intra-arterial doses of liposomes and invasive intravascular ultrasound (IVUS) imaging [91]. VCAM-1 is a prominent target for preclinical contrast-enhanced ultrasound molecular imaging (Figure 6) and was successfully used *in vivo* on ApoE $^{-/-}$ mice to detect vascular inflammation in established atherosclerotic lesions in the thoracic [92] and abdominal aorta [93].

While the aforementioned studies demonstrated signal enhancement in established atherosclerosis, the most important value of ultrasound molecular imaging is thought to be in the screening of individuals during the very early phases of atherosclerotic disease development. Accordingly, it has been shown that microbubble contrast agents targeted to VCAM-1 or P-Selectin can detect vascular inflammation not only in advanced atherosclerosis but also at lesion-prone sites in murine models of early atherosclerosis [84]. In addition, molecular imaging has been shown to be capable of detecting the effects of anti-oxidative [94,95] or anti-inflammatory [96] treatment on the endothelial phenotype.

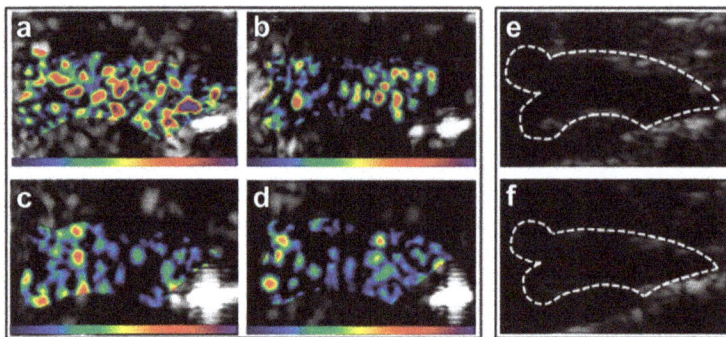

**Figure 6.** Molecular imaging of the effect of statin treatment on VCAM-1 expression in a mouse model of atherosclerosis. The images show (**a**) high signal in a non-treated animal after injection of VCAM-1 targeted microbubbles; (**b**) low signal in the same mouse after injection of microbubbles bearing a control antibody, whereas in a mouse treated with statins; both VCAM-1 targeted (**c**) and control microbubbles (**d**) show low signal. The color scale for the contrast enhanced ultrasound (CEU) images is shown at the bottom of each frame; (**e**,**f**) illustrate the outline of the ascending aorta on B-mode ultrasound images, which was used as a region of interest for acoustic intensity measurements. Reproduced from [96] with permission.

*5.5. Ultrasound Molecular Imaging of Thrombus Formation and Vascular Thrombogenic Potential in Atherosclerosis*

In established atherosclerosis, thrombus formation at vulnerable lesions is responsible for complications of atherosclerotic disease such as myocardial infarction or stroke. While thrombus formation within the heart chambers, or even in the aorta, is possible using conventional ultrasound techniques, the detection of smaller thrombi in vessels, such as the coronary arteries, is not possible, and molecular imaging targeted to individual components of thrombi could be of potential value for imaging in small vessels. Unger *et al.* described *in vitro* enhancement of signal from thrombi by targeting microbubbles to activated glycoprotein (GP) IIb/IIIa (integrin $\alpha_{IIb}:\beta_3$) on platelets by using a hexapeptide mimicking the carboxyterminal fibrinogen sequence [97]. Similarly, others have used RGD peptide sequences to target activated platelets *in vivo* [98]. Monoclonal antibodies targeting activated GP IIb/IIIa have been used to detect clots and assess the effect of thrombolysis in animal models of carotid artery injury [99,100]. Fusion of antibodies directed against GP IIb/IIIa with fibrinolytic molecules has been successfully used for low-dose targeted thrombolysis in mouse models [101].

While thrombotic complications of atherosclerosis lead to direct clinical complications, platelets-endothelial interactions also play a role in vascular inflammation throughout the pathogenesis of atherosclerosis. P-Selectin, vWF, GP Ibα mediated platelet-endothelial interactions accelerate the formation of atherosclerotic plaques via secretion of cytokines, thus potentiating the vascular inflammation [102–104]. Ultrasound molecular imaging has been used in animal models to assess vascular thrombogenicity by targeting activated von Willebrand factor [105]. Coupling of the A1 domain of von Willebrand factor to the microbubble surface has been used in murine atherosclerosis to target GP Ibα on platelets adhering to inflamed endothelium and has been shown to detect responses to short- or long-term therapies aimed at reducing vascular thrombogenicity [94].

## 6. Conclusions

The progress that has been made in the last years in understanding the molecular mechanisms that cause atherosclerosis will lead to novel and improved therapeutic options. Implicit in this is the need for a better *in vivo* assessment of the development of atherosclerosis, and improved risk stratification. Ultrasound imaging is a widely available tool that can assess the morphologic and functional consequences of atherosclerosis. However, it should be noted that while several ultrasound technologies that have recently been developed, such as virtual histology or the contrast based assessment of plaque neovascularization, show promise for risk stratification in specific settings, prospective clinical studies using these techniques are currently lacking. While the aforementioned techniques have already been tested in the clinical arena, ultrasound molecular imaging is still at a preclinical stage, with feasibility proven in a wide array of disease models. However, for translation of these techniques into the clinical field, further technical developments will be necessary. The conjugation strategy using biotin-streptavidin linking of ligands to microbubbles that is widely used in preclinical studies is unlikely to be translatable due to concerns for interference of streptavidin with fatty acid synthesis and gluconeogenesis. As a consequence, biotin-streptavidin conjugation will likely be replaced by covalent bond strategies using amine or sulfhydryl groups. Likewise, antibody-based ligands are likely to be replaced by smaller molecule, lower cost, non-immunogenic ligands. A current limitation of the imaging hardware used for molecular imaging is the inability of ultrasound systems to distinguish attached microbubbles from those that are freely circulating. As a result, this makes the use of post-processing necessary for the detection of signals from targeted microbubbles. However, the steadily increasing computer processing power, which will allow for ultrafast ultrasound imaging in the future, could possibly overcome this limitation and allow for on-line distinction of attached microbubbles from those that are freely circulating [106,107].

Despite these limitations, ultrasound imaging, which is a widely available, relatively easy to use and low-cost non-invasive imaging technique, is well positioned to be of value in risk assessment and diagnoses of atherosclerosis in the future.

**Acknowledgments:** Beat A. Kaufmann is supported by grant 310030_149718 from the Swiss National Science Foundation.

**Author Contributions:** David C. Steinl drafted the manuscript, Beat A. Kaufmann revised the manuscript.

**Conflicts of Interest:** The authors declare no conflict of interest.

## References

1. Ross, R. Atherosclerosis—An inflammatory disease. *N. Engl. J. Med.* **1999**, *340*, 115–126. [CrossRef] [PubMed]
2. Wildgruber, M.; Swirski, F.K.; Zernecke, A. Molecular imaging of inflammation in atherosclerosis. *Theranostics* **2013**, *3*, 865–884. [CrossRef] [PubMed]
3. Enos, W.F.; Holmes, R.H.; Beyer, J. Coronary disease among United-States soldiers killed in action in korea. *JAMA J. Am. Med. Assoc.* **1953**, *152*, 1090–1093. [CrossRef]
4. Strong, J.P.; Malcom, G.T.; McMahan, C.A.; Tracy, R.E.; Newman, W.P., 3rd; Herderick, E.E.; Cornhill, J.F. Prevalence and extent of atherosclerosis in adolescents and young adults: Implications for prevention from the pathobiological determinants of atherosclerosis in youth study. *JAMA* **1999**, *281*, 727–735. [CrossRef] [PubMed]

5. Health Status, OECD Database 2014. Available online: https://stats.oecd.org/index.aspx?DataSetCode= HEALTH_STAT (accessed on 20 April 2015).

6. Murray, C.J.L.; Vos, T.; Lozano, R.; Naghavi, M.; Flaxman, A.D.; Michaud, C.; Ezzati, M.; Shibuya, K.; Salomon, J.A.; Abdalla, S.; *et al.* Disability-adjusted life years (dalys) for 291 diseases and injuries in 21 regions, 1990–2010: A systematic analysis for the global burden of disease study 2010. *Lancet* **2012**, *380*, 2197–2223. [CrossRef] [PubMed]

7. Ginghina, C.; Bejan, I.; Ceck, C.D. Modern risk stratification in coronary heart disease. *J. Med. Life* **2011**, *4*, 377–386. [PubMed]

8. Witztum, J.L.; Steinberg, D. Role of oxidized low-density-lipoprotein in atherogenesis. *J. Clin. Investig.* **1991**, *88*, 1785–1792. [CrossRef] [PubMed]

9. Kunsch, C.; Medford, R.M. Oxidative stress as a regulator of gene expression in the vasculature. *Circ. Res.* **1999**, *85*, 753–766. [CrossRef] [PubMed]

10. Stocker, R.; Keaney, J.F. Role of oxidative modifications in atherosclerosis. *Physiol. Rev.* **2004**, *84*, 1381–1478. [CrossRef] [PubMed]

11. Marui, N.; Offermann, M.K.; Swerlick, R.; Kunsch, C.; Rosen, C.A.; Ahmad, M.; Alexander, R.W.; Medford, R.M. Vascular cell-adhesion molecule-1 (VCAM-1) gene-transcription and expression are regulated through an antioxidant sensitive mechanism in human vascular endothelial-cells. *J. Clin. Investig.* **1993**, *92*, 1866–1874. [CrossRef] [PubMed]

12. Bevilacqua, M.P. Endothelial-leukocyte adhesion molecules. *Annu. Rev. Immunol.* **1993**, *11*, 767–804. [CrossRef] [PubMed]

13. Huo, Y.Q.; Hafezi-Moghadam, A.; Ley, K. Role of vascular cell adhesion molecule-1 and fibronectin connecting segment-1 in monocyte rolling and adhesion on early atherosclerotic lesions. *Circ. Res.* **2000**, *87*, 153–159. [CrossRef] [PubMed]

14. Moore, K.J.; Sheedy, F.J.; Fisher, E.A. Macrophages in atherosclerosis: A dynamic balance. *Nat. Rev. Immunol.* **2013**, *13*, 709–721. [CrossRef] [PubMed]

15. Falk, E.; Shah, P.K.; Fuster, V. Coronary plaque disruption. *Circulation* **1995**, *92*, 657–671. [CrossRef] [PubMed]

16. Theilmeier, G.; Michiels, C.; Spaepen, E.; Vreys, I.; Collen, D.; Vermylen, J.; Hoylaerts, M.F. Endothelial von willebrand factor recruits platelets to atherosclerosis-prone sites in response to hypercholesterolemia. *Blood* **2002**, *99*, 4486–4493. [CrossRef] [PubMed]

17. Steffel, J.; Hermann, M.; Greutert, H.; Gay, S.; Luscher, T.F.; Ruschitzka, F.; Tanner, F.C. Celecoxib decreases endothelial tissue factor expression through inhibition of c-Jun terminal NH2 kinase phosphorylation. *Circulation* **2005**, *111*, 1685–1689. [CrossRef] [PubMed]

18. Nilsson, P.M.; Boutouyrie, P.; Cunha, P.; Kotsis, V.; Narkiewicz, K.; Parati, G.; Rietzschel, E.; Scuteri, A.; Laurent, S. Early vascular ageing in translation: From laboratory investigations to clinical applications in cardiovascular prevention. *J. Hypertens.* **2013**, *31*, 1517–1526. [CrossRef] [PubMed]

19. Glagov, S.; Weisenberg, E.; Zarins, C.K.; Stankunavicius, R.; Kolettis, G.J. Compensatory enlargement of human atherosclerotic coronary arteries. *N. Engl. J. Med.* **1987**, *316*, 1371–1375. [CrossRef] [PubMed]

20. Aggoun, Y.; Bonnet, D.; Sidi, D.; Girardet, J.P.; Brucker, E.; Polak, M.; Safar, M.E.; Levy, B.I. Arterial mechanical changes in children with familial hypercholesterolemia. *Arterioscler. Thromb. Vasc. Biol.* **2000**, *20*, 2070–2075. [CrossRef] [PubMed]

21. Doyle, B.; Caplice, N. Plaque neovascularization and antiangiogenic therapy for atherosclerosis. *J. Am. Coll. Cardiol.* **2007**, *49*, 2073–2080. [CrossRef] [PubMed]

22. Amato, M.; Montorsi, P.; Ravani, A.; Oldani, E.; Galli, S.; Ravagnani, P.M.; Tremoli, E.; Baldassarre, D. Carotid intima-media thickness by B-mode ultrasound as surrogate of coronary atherosclerosis: Correlation with quantitative coronary angiography and coronary intravascular ultrasound findings. *Eur. Heart J.* **2007**, *28*, 2094–2101. [CrossRef] [PubMed]

23. Stork, S.; Feelders, R.A.; van den Beld, A.W.; Steyerberg, E.W.; Savelkoul, H.F.J.; Lamberts, S.W.J.; Grobbee, D.E.; Bots, M.L. Prediction of mortality risk in the elderly. *Am. J. Med.* **2006**, *119*, 519–525. [CrossRef] [PubMed]

24. Mathiesen, E.B.; Johnsen, S.H.; Wilsgaard, T.; Bonaa, K.H.; Lochen, M.L.; Njolstad, I. Carotid plaque area and intima-media thickness in prediction of first-ever ischemic stroke a 10-year follow-up of 6584 men and women: The tromso study. *Stroke* **2011**, *42*, 972–978. [CrossRef] [PubMed]

25. Allen, J.D.; Ham, K.L.; Dumont, D.M.; Sileshi, B.; Trahey, G.E.; Dahl, J.J. The development and potential of acoustic radiation force impulse (ARFI) imaging for carotid artery plaque characterization. *Vasc. Med.* **2011**, *16*, 302–311. [CrossRef] [PubMed]

26. Gray-Weale, A.C.; Graham, J.C.; Burnett, J.R.; Byrne, K.; Lusby, R.J. Carotid artery atheroma: Comparison of preoperative B-mode ultrasound appearance with carotid endarterectomy specimen pathology. *J. Cardiovasc. Surg.* **1988**, *29*, 676–681.

27. Mathiesen, E.B.; Bonaa, K.H.; Joakimsen, O. Echolucent plaques are associated with high risk of ischemic cerebrovascular events in carotid stenosis: The tromso study. *Circulation* **2001**, *103*, 2171–2175. [CrossRef] [PubMed]

28. Reiter, M.; Effenberger, I.; Sabeti, S.; Mlekusch, W.; Schlager, O.; Dick, P.; Puchner, S.; Amighi, J.; Bucek, R.A.; Minar, E.; *et al.* Increasing carotid plaque echolucency is predictive of cardiovascular events in high-risk patients. *Radiology* **2008**, *248*, 1050–1055. [CrossRef] [PubMed]

29. Nambi, V.; Chambless, L.; Folsom, A.R.; He, M.; Hu, Y.J.; Mosley, T.; Volcik, K.; Boerwinkle, E.; Ballantyne, C.M. Carotid intima-media thickness and presence or absence of plaque improves prediction of coronary heart disease risk the aric (atherosclerosis risk in communities) study. *J. Am. Coll. Cardiol.* **2010**, *55*, 1600–1607. [CrossRef] [PubMed]

30. Bots, M.L.; Hoes, A.W.; Koudstaal, P.J.; Hofman, A.; Grobbee, D.E. Common carotid intima-media thickness and risk of stroke and myocardial infarction—The rotterdam study. *Circulation* **1997**, *96*, 1432–1437. [CrossRef] [PubMed]

31. Peters, S.A.; den Ruijter, H.M.; Grobbee, D.E.; Bots, M.L. Results from a carotid intima-media thickness trial as a decision tool for launching a large-scale morbidity and mortality trial. *Circ. Cardiovasc. Imaging* **2013**, *6*, 20–25. [CrossRef] [PubMed]

32. Den Ruijter, H.M.; Peters, S.A.; Anderson, T.J.; Britton, A.R.; Dekker, J.M.; Eijkemans, M.J.; Engstrom, G.; Evans, G.W.; de Graaf, J.; Grobbee, D.E.; *et al.* Common carotid intima-media thickness measurements in cardiovascular risk prediction: A meta-analysis. *JAMA* **2012**, *308*, 796–803. [CrossRef] [PubMed]

33. Finn, A.V.; Kolodgie, F.D.; Virmani, R. Correlation between carotid intimal/medial thickness and atherosclerosis: A point of view from pathology. *Arterioscler. Thromb. Vasc. Biol.* **2010**, *30*, 177–181. [CrossRef] [PubMed]

34. Molinari, F.; Zeng, G.; Suri, J.S. An integrated approach to computer-based automated tracing and its validation for 200 common carotid arterial wall ultrasound images: A new technique. *J. Ultrasound Med.* **2010**, *29*, 399–418. [PubMed]

35. Sillesen, H.; Muntendam, P.; Adourian, A.; Entrekin, R.; Garcia, M.; Falk, E.; Fuster, V. Carotid plaque burden as a measure of subclinical atherosclerosis: Comparison with other tests for subclinical arterial disease in the high risk plaque bioimage study. *JACC Cardiovasc. Imaging* **2012**, *5*, 681–689. [CrossRef] [PubMed]

36. Baber, U.; Mehran, R.; Sartori, S.; Schoos, M.M.; Sillesen, H.; Muntendam, P.; Garcia, M.J.; Gregson, J.; Pocock, S.; Falk, E.; *et al.* Prevalence, impact, and predictive value of detecting subclinical coronary and carotid atherosclerosis in asymptomatic adults: The bioimage study. *J. Am. Coll. Cardiol.* **2015**, *65*, 1065–1074. [CrossRef] [PubMed]

37. Yamagishi, M.; Terashima, M.; Awano, K.; Kijima, M.; Nakatani, S.; Daikoku, S.; Ito, K.; Yasumura, Y.; Miyatake, K. Morphology of vulnerable coronary plaque: Insights from follow-up of patients examined by intravascular ultrasound before an acute coronary syndrome. *J. Am. Coll. Cardiol.* **2000**, *35*, 106–111. [CrossRef] [PubMed]

38. Nishimura, R.A.; Edwards, W.D.; Warnes, C.A.; Reeder, G.S.; Holmes, D.R.; Tajik, A.J.; Yock, P.G. Intravascular ultrasound imaging-*in vitro* validation and pathological correlation. *J. Am. Coll. Cardiol.* **1990**, *16*, 145–154. [CrossRef] [PubMed]

39. Prati, F.; Arbustini, E.; Labellarte, A.; Dal Bello, B.; Sommariva, L.; Mallus, M.T.; Pagano, A.; Boccanelli, A. Correlation between high frequency intravascular ultrasound and histomorphology in human coronary arteries. *Heart (Br. Card. Soc.)* **2001**, *85*, 567–570. [CrossRef]

40. Palmer, N.D.; Northridge, D.; Lessells, A.; McDicken, W.N.; Fox, K.A.A. *In vitro* analysis of coronary atheromatous lesions by intravascular ultrasound-reproducibility and histological correlation of lesion morphology. *Eur. Heart J.* **1999**, *20*, 1701–1706. [CrossRef] [PubMed]

41. Potkin, B.N.; Bartorelli, A.L.; Gessert, J.M.; Neville, R.F.; Almagor, Y.; Roberts, W.C.; Leon, M.B. Coronary-artery imaging with intravascular high-frequency ultrasound. *Circulation* **1990**, *81*, 1575–1585. [CrossRef] [PubMed]

42. Mintz, G.S.; Nissen, S.E.; Anderson, W.D.; Bailey, S.R.; Erbel, R.; Fitzgerald, P.J.; Pinto, F.J.; Rosenfield, K.; Siegel, R.J.; Tuzcu, E.M.; *et al.* American college of cardiology clinical expert consensus document on standards for acquisition, measurement and reporting of intravascular ultrasound studies (ivus). A report of the american college of cardiology task force on clinical expert consensus documents. *J. Am. Coll. Cardiol.* **2001**, *37*, 1478–1492. [CrossRef] [PubMed]

43. Schoenhagen, P.; Stone, G.W.; Nissen, S.E.; Grines, C.L.; Griffin, J.; Clemson, B.S.; Vince, D.G.; Ziada, K.; Crowe, T.; Apperson-Hanson, C.; *et al.* Coronary plaque morphology and frequency of ulceration distant from culprit lesions in patients with unstable and stable presentation. *Arterioscler. Thromb. Vasc. Biol.* **2003**, *23*, 1895–1900. [CrossRef] [PubMed]

44. Wuyts, F.L.; Vanhuyse, V.J.; Langewouters, G.J.; Decraemer, W.F.; Raman, E.R.; Buyle, S. Elastic properties of human aortas in relation to age and atherosclerosis—A structural model. *Phys. Med. Biol.* **1995**, *40*, 1577–1597. [CrossRef] [PubMed]

45. Van Sloten, T.T.; Schram, M.T.; van den Hurk, K.; Dekker, J.M.; Nijpels, G.; Henry, R.M.; Stehouwer, C.D. Local stiffness of the carotid and femoral artery is associated with incident cardiovascular events and all-cause mortality: The hoorn study. *J. Am. Coll. Cardiol.* **2014**, *63*, 1739–1747. [CrossRef] [PubMed]

46. Cecelja, M.; Chowienczyk, P. Dissociation of aortic pulse wave velocity with risk factors for cardiovascular disease other than hypertension: A systematic review. *Hypertension* **2009**, *54*, 1328–1336. [CrossRef] [PubMed]

47. De Korte, C.L.; Pasterkamp, G.; van der Steen, A.F.; Woutman, H.A.; Bom, N. Characterization of plaque components with intravascular ultrasound elastography in human femoral and coronary arteries *in vitro*. *Circulation* **2000**, *102*, 617–623. [CrossRef] [PubMed]

48. Schaar, J.A.; van der Steen, A.F.; Mastik, F.; Baldewsing, R.A.; Serruys, P.W. Intravascular palpography for vulnerable plaque assessment. *J. Am. Coll. Cardiol.* **2006**, *47*, C86–C91. [CrossRef] [PubMed]

49. De Korte, C.L.; Sierevogel, M.J.; Mastik, F.; Strijder, C.; Schaar, J.A.; Velema, E.; Pasterkamp, G.; Serruys, P.W.; van der Steen, A.F. Identification of atherosclerotic plaque components with intravascular ultrasound elastography *in vivo*: A yucatan pig study. *Circulation* **2002**, *105*, 1627–1630. [CrossRef] [PubMed]

50. Schaar, J.A.; Regar, E.; Mastik, F.; McFadden, E.P.; Saia, F.; Disco, C.; de Korte, C.L.; de Feyter, P.J.; van der Steen, A.F.; Serruys, P.W. Incidence of high-strain patterns in human coronary arteries: Assessment with three-dimensional intravascular palpography and correlation with clinical presentation. *Circulation* **2004**, *109*, 2716–2719. [CrossRef] [PubMed]

51. Kruizinga, P.; Mastik, F.; van den Oord, S.C.; Schinkel, A.F.; Bosch, J.G.; de Jong, N.; van Soest, G.; van der Steen, A.F. High-definition imaging of carotid artery wall dynamics. *Ultrasound Med. Biol.* **2014**, *40*, 2392–2403. [CrossRef] [PubMed]

52. Svedlund, S.; Gan, L.M. Longitudinal common carotid artery wall motion is associated with plaque burden in man and mouse. *Atherosclerosis* **2011**, *217*, 120–124. [CrossRef] [PubMed]

53. Widman, E.; Caidahl, K.; Heyde, B.; D'Hooge, J.; Larsson, M. Ultrasound speckle tracking strain estimation of *in vivo* carotid artery plaque with *in vitro* sonomicrometry validation. *Ultrasound Med. Biol.* **2015**, *41*, 77–88. [CrossRef] [PubMed]

54. Tsai, W.C.; Sun, Y.T.; Liu, Y.W.; Ho, C.S.; Chen, J.Y.; Wang, M.C.; Tsai, L.M. Usefulness of vascular wall deformation for assessment of carotid arterial stiffness and association with previous stroke in elderly. *Am. J. Hypertens.* **2013**, *26*, 770–777. [CrossRef] [PubMed]

55. Gramiak, R.; Shah, P.M. Echocardiography of the aortic root. *Investig. Radiol.* **1968**, *3*, 356–366. [CrossRef]

56. Jayaweera, A.R.; Edwards, N.; Glasheen, W.P.; Villanueva, F.S.; Abbott, R.D.; Kaul, S. *In vivo* myocardial kinetics of air-filled albumin microbubbles during myocardial contrast echocardiography. Comparison with radiolabeled red blood cells. *Circ. Res.* **1994**, *74*, 1157–1165. [CrossRef] [PubMed]

57. Demos, S.M.; Onyuksel, H.; Gilbert, J.; Roth, S.I.; Kane, B.; Jungblut, P.; Pinto, J.V.; McPherson, D.D.; Klegerman, M.E. *In vitro* targeting of antibody-conjugated echogenic liposomes for site-specific ultrasonic image enhancement. *J. Pharm. Sci.* **1997**, *86*, 167–171. [CrossRef] [PubMed]

58. de Jong, N.; Frinking, P.J.; Bouakaz, A.; Goorden, M.; Schourmans, T.; Jingping, X.; Mastik, F. Optical imaging of contrast agent microbubbles in an ultrasound field with a 100-MHz camera. *Ultrasound Med. Biol.* **2000**, *26*, 487–492. [CrossRef] [PubMed]

59. Dayton, P.A.; Morgan, K.E.; Klibanov, A.L.; Brandenburger, G.H.; Ferrara, K.W. Optical and acoustical observations of the effects of ultrasound on contrast agents. *IEEE Trans. Ultrasonics Ferroelectr. Freq. Control* **1999**, *46*, 220–232. [CrossRef]

60. Kaufmann, B.A.; Wei, K.; Lindner, J.R. Contrast echocardiography. *Curr. Probl. Cardiol.* **2007**, *32*, 51–96. [CrossRef] [PubMed]

61. Barger, A.C.; Beeuwkes, R., 3rd; Lainey, L.L.; Silverman, K.J. Hypothesis: Vasa vasorum and neovascularization of human coronary arteries. A possible role in the pathophysiology of atherosclerosis. *N. Engl. J. Med.* **1984**, *310*, 175–177. [CrossRef] [PubMed]

62. Sluimer, J.C.; Kolodgie, F.D.; Bijnens, A.; Maxfield, K.; Pacheco, E.; Kutys, B.; Duimel, H.; Frederik, P.M.; van Hinsbergh, V.W.M.; Virmani, R.; *et al.* Thin-walled microvessels in human coronary atherosclerotic plaques show incomplete endothelial junctions relevance of compromised structural integrity for intraplaque microvascular leakage. *J. Am. Coll. Cardiol.* **2009**, *53*, 1517–1527. [CrossRef] [PubMed]

63. Kolodgie, F.D.; Gold, H.K.; Burke, A.P.; Fowler, D.R.; Kruth, H.S.; Weber, D.K.; Farb, A.; Guerrero, L.J.; Hayase, M.; Kutys, R.; *et al.* Intraplaque hemorrhage and progression of coronary atheroma. *N. Engl. J. Med.* **2003**, *349*, 2316–2325. [CrossRef] [PubMed]

64. Coli, S.; Magnoni, M.; Sangiorgi, G.; Marrocco-Trischitta, M.M.; Melisurgo, G.; Mauriello, A.; Spagnoli, L.; Chiesa, R.; Cianflone, D.; Maseri, A. Contrast-enhanced ultrasound imaging of intraplaque neovascularization in carotid arteries: Correlation with histology and plaque echogenicity. *J. Am. Coll. Cardiol.* **2008**, *52*, 223–230. [CrossRef] [PubMed]

65. Staub, D.; Partovi, S.; Schinkel, A.F.; Coll, B.; Uthoff, H.; Aschwanden, M.; Jaeger, K.A.; Feinstein, S.B. Correlation of carotid artery atherosclerotic lesion echogenicity and severity at standard us with intraplaque neovascularization detected at contrast-enhanced us. *Radiology* **2011**, *258*, 618–626. [CrossRef] [PubMed]

66. Hoogi, A.; Adam, D.; Hoffman, A.; Kerner, H.; Reisner, S.; Gaitini, D. Carotid plaque vulnerability: Quantification of neovascularization on contrast-enhanced ultrasound with histopathologic correlation. *AJR Am. J. Roentgenol.* **2011**, *196*, 431–436. [CrossRef] [PubMed]

67. Staub, D.; Patel, M.B.; Tibrewala, A.; Ludden, D.; Johnson, M.; Espinosa, P.; Coll, B.; Jaeger, K.A.; Feinstein, S.B. Vasa vasorum and plaque neovascularization on contrast-enhanced carotid ultrasound imaging correlates with cardiovascular disease and past cardiovascular events. *Stroke* **2010**, *41*, 41–47. [CrossRef] [PubMed]

68. Saito, K.; Nagatsuka, K.; Ishibashi-Ueda, H.; Watanabe, A.; Kannki, H.; Iihara, K. Contrast-enhanced ultrasound for the evaluation of neovascularization in atherosclerotic carotid artery plaques. *Stroke* **2014**, *45*, 3073–3075. [CrossRef] [PubMed]

69. Ten Kate, G.L.; Renaud, G.G.; Akkus, Z.; van den Oord, S.C.; ten Cate, F.J.; Shamdasani, V.; Entrekin, R.R.; Sijbrands, E.J.; de Jong, N.; Bosch, J.G.; *et al.* Far-wall pseudoenhancement during contrast-enhanced ultrasound of the carotid arteries: Clinical description and *in vitro* reproduction. *Ultrasound Med. Biol.* **2012**, *38*, 593–600. [CrossRef] [PubMed]

70. Renaud, G.; Bosch, J.G.; Ten Kate, G.L.; Shamdasani, V.; Entrekin, R.; de Jong, N.; van der Steen, A.F. Counter-propagating wave interaction for contrast-enhanced ultrasound imaging. *Phys. Med. Biol.* **2012**, *57*, L9–L18. [CrossRef] [PubMed]

71. Staub, D.; Schinkel, A.F.L.; Coll, B.; Coli, S.; van der Steen, A.F.W.; Reed, J.D.; Krueger, C.; Thomenius, K.E.; Adam, D.; Sijbrands, E.J.; *et al.* Contrast-enhanced ultrasound imaging of the vasa vasorum from early atherosclerosis to the identification of unstable plaques. *JACC Cardiovasc. Imaging* **2010**, *3*, 761–771. [CrossRef] [PubMed]

72. Lindner, J.R.; Song, J.; Xu, F.; Klibanov, A.L.; Singbartl, K.; Ley, K.; Kaul, S. Noninvasive ultrasound imaging of inflammation using microbubbles targeted to activated leukocytes. *Circulation* **2000**, *102*, 2745–2750. [CrossRef] [PubMed]

73. Lindner, J.R.; Coggins, M.P.; Kaul, S.; Klibanov, A.L.; Brandenburger, G.H.; Ley, K. Microbubble persistence in the microcirculation during ischemia/reperfusion and inflammation is caused by integrin- and complement-mediated adherence to activated leukocytes. *Circulation* **2000**, *101*, 668–675. [CrossRef] [PubMed]

74. Khanicheh, E.; Mitterhuber, M.; Kinslechner, K.; Xu, L.; Lindner, J.R.; Kaufmann, B.A. Factors affecting the endothelial retention of targeted microbubbles: Influence of microbubble shell design and cell surface projection of the endothelial target molecule. *J. Am. Soc. Echocardiogr.* **2012**, *25*, 460–466. [CrossRef] [PubMed]

75. Takalkar, A.M.; Klibanov, A.L.; Rychak, J.J.; Lindner, J.R.; Ley, K. Binding and detachment dynamics of microbubbles targeted to P-selectin under controlled shear flow. *J. Control. Release* **2004**, *96*, 473–482. [CrossRef] [PubMed]

76. Lindner, J.R. Molecular imaging with contrast ultrasound and targeted microbubbles. *J. Nucl. Cardiol.* **2004**, *11*, 215–221. [CrossRef] [PubMed]

77. Lankford, M.; Behm, C.Z.; Yeh, J.; Klibanov, A.L.; Robinson, P.; Lindner, J.R. Effect of microbubble ligation to cells on ultrasound signal enhancement: Implications for targeted imaging. *Investig. Radiol.* **2006**, *41*, 721–728. [CrossRef]

78. Lindner, J.R. Molecular imaging of vascular phenotype in cardiovascular disease: New diagnostic opportunities on the horizon. *J. Am. Soc. Echocardiogr.* **2010**, *23*, 343–350. [CrossRef] [PubMed]

79. Fenyo, I.M.; Gafencu, A.V. The involvement of the monocytes/macrophages in chronic inflammation associated with atherosclerosis. *Immunobiology* **2013**, *218*, 1376–1384. [CrossRef] [PubMed]

80. Tsutsui, J.M.; Xie, F.; Cano, M.; Chomas, J.; Phillips, P.; Radio, S.J.; Lof, J.; Porter, T.R. Detection of retained microbubbles in carotid arteries with real-time low mechanical index imaging in the setting of endothelial dysfunction. *J. Am. Coll. Cardiol.* **2004**, *44*, 1036–1046. [CrossRef] [PubMed]

81. Anderson, D.R.; Tsutsui, J.M.; Xie, F.; Radio, S.J.; Porter, T.R. The role of complement in the adherence of microbubbles to dysfunctional arterial endothelium and atherosclerotic plaque. *Cardiovasc. Res.* **2007**, *73*, 597–606. [CrossRef] [PubMed]

82. McEver, R.P.; Cummings, R.D. Perspectives series: Cell adhesion in vascular biology. Role of PSGL-1 binding to selectins in leukocyte recruitment. *J. Clin. Investig.* **1997**, *100*, 485–491. [CrossRef] [PubMed]

83. Ramos, C.L.; Huo, Y.; Jung, U.; Ghosh, S.; Manka, D.R.; Sarembock, I.J.; Ley, K. Direct demonstration of P-selectin- and VCAM-1-dependent mononuclear cell rolling in early atherosclerotic lesions of apolipoprotein E-deficient mice. *Circ. Res.* **1999**, *84*, 1237–1244. [CrossRef] [PubMed]

84. Kaufmann, B.A.; Carr, C.L.; Belcik, J.T.; Xie, A.; Yue, Q.; Chadderdon, S.; Caplan, E.S.; Khangura, J.; Bullens, S.; Bunting, S.; *et al.* Molecular imaging of the initial inflammatory response in atherosclerosis: Implications for early detection of disease. *Arterioscler. Thromb. Vasc. Biol.* **2010**, *30*, 54–59. [CrossRef] [PubMed]

85. Villanueva, F.S.; Lu, E.; Bowry, S.; Kilic, S.; Tom, E.; Wang, J.; Gretton, J.; Pacella, J.J.; Wagner, W.R. Myocardial ischemic memory imaging with molecular echocardiography. *Circulation* **2007**, *115*, 345–352. [CrossRef] [PubMed]

86. Kaufmann, B.A.; Lewis, C.; Xie, A.; Mirza-Mohd, A.; Lindner, J.R. Detection of recent myocardial ischaemia by molecular imaging of P-selectin with targeted contrast echocardiography. *Eur. Heart J.* **2007**, *28*, 2011–2017. [CrossRef] [PubMed]

87. Walpola, P.L.; Gotlieb, A.I.; Cybulsky, M.I.; Langille, B.L. Expression of ICAM-1 and VCAM-1 and monocyte adherence in arteries exposed to altered shear stress. *Arterioscler. Thromb. Vasc. Biol.* **1995**, *15*, 2–10. [CrossRef] [PubMed]

88. Dansky, H.M.; Barlow, C.B.; Lominska, C.; Sikes, J.L.; Kao, C.; Weinsaft, J.; Cybulsky, M.I.; Smith, J.D. Adhesion of monocytes to arterial endothelium and initiation of atherosclerosis are critically dependent on vascular cell adhesion molecule-1 gene dosage. *Arterioscler. Thromb. Vasc. Biol.* **2001**, *21*, 1662–1667. [CrossRef] [PubMed]

89. Iiyama, K.; Hajra, L.; Iiyama, M.; Li, H.; DiChiara, M.; Medoff, B.D.; Cybulsky, M.I. Patterns of vascular cell adhesion molecule-1 and intercellular adhesion molecule-1 expression in rabbit and mouse atherosclerotic lesions and at sites predisposed to lesion formation. *Circ. Res.* **1999**, *85*, 199–207. [CrossRef] [PubMed]

90. Villanueva, F.S.; Jankowski, R.J.; Klibanov, S.; Pina, M.L.; Alber, S.M.; Watkins, S.C.; Brandenburger, G.H.; Wagner, W.R. Microbubbles targeted to intercellular adhesion molecule-1 bind to activated coronary artery endothelial cells. *Circulation* **1998**, *98*, 1–5. [CrossRef] [PubMed]

91. Hamilton, A.J.; Huang, S.L.; Warnick, D.; Rabbat, M.; Kane, B.; Nagaraj, A.; Klegerman, M.; McPherson, D.D. Intravascular ultrasound molecular imaging of atheroma components *in vivo*. *J. Am. Coll. Cardiol.* **2004**, *43*, 453–460. [CrossRef] [PubMed]

92. Kaufmann, B.A.; Sanders, J.M.; Davis, C.; Xie, A.; Aldred, P.; Sarembock, I.J.; Lindner, J.R. Molecular imaging of inflammation in atherosclerosis with targeted ultrasound detection of vascular cell adhesion molecule-1. *Circulation* **2007**, *116*, 276–284. [CrossRef] [PubMed]

93. Wu, J.; Leong-Poi, H.; Bin, J.; Yang, L.; Liao, Y.; Liu, Y.; Cai, J.; Xie, J.; Liu, Y. Efficacy of contrast-enhanced us and magnetic microbubbles targeted to vascular cell adhesion molecule-1 for molecular imaging of atherosclerosis. *Radiology* **2011**, *260*, 463–471. [CrossRef] [PubMed]

94. Liu, Y.; Davidson, B.P.; Yue, Q.; Belcik, T.; Xie, A.; Inaba, Y.; McCarty, O.J.; Tormoen, G.W.; Zhao, Y.; Ruggeri, Z.M.; *et al.* Molecular imaging of inflammation and platelet adhesion in advanced atherosclerosis effects of antioxidant therapy with nadph oxidase inhibition. *Circ. Cardiovasc. Imaging* **2013**, *6*, 74–82. [CrossRef] [PubMed]

95. Khanicheh, E.; Qi, Y.; Xie, A.; Mitterhuber, M.; Xu, L.; Mochizuki, M.; Daali, Y.; Jaquet, V.; Krause, K.H.; Ruggeri, Z.M.; *et al.* Molecular imaging reveals rapid reduction of endothelial activation in early atherosclerosis with apocynin independent of antioxidative properties. *Arterioscler. Thromb. Vasc. Biol.* **2013**, *33*, 2187–2192. [CrossRef] [PubMed]

96. Khanicheh, E.; Mitterhuber, M.; Xu, L.; Haeuselmann, S.P.; Kuster, G.M.; Kaufmann, B.A. Noninvasive ultrasound molecular imaging of the effect of statins on endothelial inflammatory phenotype in early atherosclerosis. *PLoS ONE* **2013**, *8*, e58761. [CrossRef] [PubMed]

97. Unger, E.C.; McCreery, T.P.; Sweitzer, R.H.; Shen, D.; Wu, G. *In vitro* studies of a new thrombus-specific ultrasound contrast agent. *Am. J. Cardiol.* **1998**, *81*, 58g–61g. [CrossRef] [PubMed]

98. Wu, W.; Wang, Y.; Shen, S.; Wu, J.; Guo, S.; Su, L.; Hou, F.; Wang, Z.; Liao, Y.; Bin, J. *In vivo* ultrasound molecular imaging of inflammatory thrombosis in arteries with cyclic Arg-Gly-Asp-modified microbubbles targeted to glycoprotein IIB/IIIA. *Investig. Radiol.* **2013**, *48*, 803–812. [CrossRef]

99. Wang, X.; Hagemeyer, C.E.; Hohmann, J.D.; Leitner, E.; Armstrong, P.C.; Jia, F.; Olschewski, M.; Needles, A.; Peter, K.; Ahrens, I. Novel single-chain antibody-targeted microbubbles for molecular ultrasound imaging of thrombosis: Validation of a unique noninvasive method for rapid and sensitive detection of thrombi and monitoring of success or failure of thrombolysis in mice. *Circulation* **2012**, *125*, 3117–3126. [CrossRef] [PubMed]

100. Alonso, A.; Della Martina, A.; Stroick, M.; Fatar, M.; Griebe, M.; Pochon, S.; Schneider, M.; Hennerici, M.; Allemann, E.; Meairs, S. Molecular imaging of human thrombus with novel abciximab immunobubbles and ultrasound. *Stroke* **2007**, *38*, 1508–1514. [CrossRef] [PubMed]

101. Wang, X.; Palasubramaniam, J.; Gkanatsas, Y.; Hohmann, J.D.; Westein, E.; Kanojia, R.; Alt, K.; Huang, D.; Jia, F.; Ahrens, I.; *et al.* Towards effective and safe thrombolysis and thromboprophylaxis: Preclinical testing of a novel antibody-targeted recombinant plasminogen activator directed against activated platelets. *Circ. Res.* **2014**, *114*, 1083–1093. [CrossRef] [PubMed]

102. Burger, P.C.; Wagner, D.D. Platelet P-selectin facilitates atherosclerotic lesion development. *Blood* **2003**, *101*, 2661–2666. [CrossRef] [PubMed]

103. Massberg, S.; Brand, K.; Gruner, S.; Page, S.; Muller, E.; Muller, I.; Bergmeier, W.; Richter, T.; Lorenz, M.; Konrad, I.; *et al.* A critical role of platelet adhesion in the initiation of atherosclerotic lesion formation. *J. Exp. Med.* **2002**, *196*, 887–896. [CrossRef] [PubMed]

104. Henn, V.; Slupsky, J.R.; Grafe, M.; Anagnostopoulos, I.; Forster, R.; Muller-Berghaus, G.; Kroczek, R.A. Cd40 ligand on activated platelets triggers an inflammatory reaction of endothelial cells. *Nature* **1998**, *391*, 591–594. [CrossRef] [PubMed]

105. McCarty, O.J.; Conley, R.B.; Shentu, W.; Tormoen, G.W.; Zha, D.; Xie, A.; Qi, Y.; Zhao, Y.; Carr, C.; Belcik, T.; *et al.* Molecular imaging of activated von willebrand factor to detect high-risk atherosclerotic phenotype. *JACC Cardiovasc. Imaging* **2010**, *3*, 947–955. [CrossRef] [PubMed]

106. Couture, O.; Bannouf, S.; Montaldo, G.; Aubry, J.F.; Fink, M.; Tanter, M. Ultrafast imaging of ultrasound contrast agents. *Ultrasound Med. Biol.* **2009**, *35*, 1908–1916. [CrossRef] [PubMed]

107. Mauldin, F.W., Jr.; Dhanaliwala, A.H.; Patil, A.V.; Hossack, J.A. Real-time targeted molecular imaging using singular value spectra properties to isolate the adherent microbubble signal. *Phys. Med. Biol.* **2012**, *57*, 5275–5293. [CrossRef] [PubMed]

International Journal of
*Molecular Sciences*

MDPI

*Review*

# Ultrasound Tissue Characterization of Vulnerable Atherosclerotic Plaque

Eugenio Picano [1,*] and Marco Paterni [2]

1   Biomedicine Department, NU School of Medicine, Astana 010000, Kazakistan
2   CNR (Consiglio Nazionale Ricerche), Institute of Clinical Physiology, 56124 Pisa, Italy;
    marco.paterni@ifc.cnr.it
*   Correspondence: picano@ifc.cnr.it; Tel.: +39-050-3152398; Fax: +39-050-3152166

Academic Editor: Michael Henein
Received: 30 January 2015; Accepted: 24 April 2015; Published: 5 May 2015

**Abstract:** A thrombotic occlusion of the vessel fed by ruptured coronary atherosclerotic plaque may result in unstable angina, myocardial infarction or death, whereas embolization from a plaque in carotid arteries may result in transient ischemic attack or stroke. The atherosclerotic plaque prone to such clinical events is termed high-risk or vulnerable plaque, and its identification in humans before it becomes symptomatic has been elusive to date. Ultrasonic tissue characterization of the atherosclerotic plaque is possible with different techniques—such as vascular, transesophageal, and intravascular ultrasound—on a variety of arterial segments, including carotid, aorta, and coronary districts. The image analysis can be based on visual, video-densitometric or radiofrequency methods and identifies three distinct textural patterns: hypo-echoic (corresponding to lipid- and hemorrhage-rich plaque), iso- or moderately hyper-echoic (fibrotic or fibro-fatty plaque), and markedly hyperechoic with shadowing (calcific plaque). Hypoechoic or dishomogeneous plaques, with spotty microcalcification and large plaque burden, with plaque neovascularization and surface irregularities by contrast-enhanced ultrasound, are more prone to clinical complications than hyperechoic, extensively calcified, homogeneous plaques with limited plaque burden, smooth luminal plaque surface and absence of neovascularization. Plaque ultrasound morphology is important, along with plaque geometry, in determining the atherosclerotic prognostic burden in the individual patient. New quantitative methods beyond backscatter (to include speed of sound, attenuation, strain, temperature, and high order statistics) are under development to evaluate vascular tissues. Although not yet ready for widespread clinical use, tissue characterization is listed by the American Society of Echocardiography roadmap to 2020 as one of the most promising fields of application in cardiovascular ultrasound imaging, offering unique opportunities for the early detection and treatment of atherosclerotic disease.

**Keywords:** atherosclerosis; plaque; tissue characterization; ultrasound

---

## 1. Tissue Characterization of Vulnerable Plaque: From Histology to Ultrasound

The underlying hypothesis in tissue characterization studies is that a different biochemical structure, internal architectural arrangement or physiologic state of normal *vs.* diseased tissue can affect the physical properties of the tissue and can therefore be detected by ultrasound. Tissue characterization can be performed using three main approaches with increasing degrees of complexity and accuracy: visual eyeballing, software-assisted videodensitometry of standard digitized images, and backscatter analysis of native radiofrequency signal (Figure 1). Visual eyeballing is the "first generation" approach (arising in the 1980s) and is still the only clinically viable option for large-scale use, but it can only detect the most obvious changes in tissue structure such as a hypoechoic, hyperechoic, or calcified carotid plaque. Videodensitometry is the "second-generation" approach, implemented

*Int. J. Mol. Sci.* **2015**, *16*, 10121–10133

since the mid-1990s, more objective than visual assessment and based on quantitative analysis of digitized video images. It samples the commercial video signal downstream to the processing chain distorting the linear relationship between received signal and displayed image. Radiofrequency analysis is a more technologically demanding "third generation" approach, commercially developed over the last 15 years and theoretically the most accurate, since the native ultrasonic signal is sampled upstream to the video display, and is not distorted by the post-processing function of the imaging chain. According to recent recommendations, "The long history of the ultrasound tissue characterization technique compared with its rare clinical use tells its own story in relation to its difficulty" [1]. This procedure is complex, subject to artifacts related to image settings, and the exact location of the sample volume. Calibrated backscatter has a value as a marker of fibrosis and calcification, but—the guidelines conclude—this methodology remains more of a research instrument than a clinical tool in echocardiography.

In spite of these recognized difficulties, the clinical yield of ultrasonic tissue characterization remains especially attractive in atherosclerosis, especially for the acoustic identification of vulnerable or high-risk plaques, a challenging but achievable target—as recently outlined by National Heart Lung and Blood Institute (NHLBI) Working Group [2]—for future research in the field. The carotid plaque is defined as "a focal structure that encroaches into the arterial lumen of at least 0.5 mm or 50% of the surrounding intima-media thickness or demonstrated a thickness of greater than or equal to 1.5 mm" [3]. For the clinician, there is a need to characterize "vulnerable plaque", *i.e.*, the plaque susceptible to rupture, which can give rise to clinical complications, from embolization to thrombosis leading to symptoms, myocardial infarction, stroke and death. The vulnerability features are only weakly related to plaque size and stenosis and are also related to plaque morphology and histologic content: plaque size matters, but shape and content of the plaque also matter. Vulnerable, high-risk plaque is histologically different from stable, benign, clinically asymptomatic plaque—not only regarding its larger plaque burden but also for its higher content of lipids, with necrotic cores due to invasion of lipid pools by macrophages and other inflammatory cells with speckled micro-calcification (Table 1). The necrotic core can show hemorrhages due to extravasation of erythrocytes from the intimal neo-vascularization originating from the adventitia. The fibrous cap is usually thin, and the luminal contours may be irregular rather than smooth. All these histologic features can leave their readout on a variety of acoustic parameters, based on acoustic backscatter, attenuation, spatial texture, angular variability, plaque neo-vascularization detected through contrast administration, and acoustic internal homogeneity of spatial gray-level distribution. In order to have a comprehensive evaluation of plaque prognostic potential we need the assessment of plaque hemodynamic severity—as can be optimally provided by Duplex scan including Doppler and conventional B-mode—but also better insight into plaque content and morphology, as potentially provided by the tissue characterization approach [4].

**Table 1.** The vulnerable plaque read-out: from histology to ultrasound.

| Histology | Ultrasound |
| --- | --- |
| Outward remodeling | Stenosis > 70% |
| Decreased Fibrous Tissue | Hypoechoic core |
| Increased Lipid-Hemorrhages | Hypoechoic core |
| More necrotic core | Dishomogeneous texture |
| Macrophages—inflammation | Dishomogeneous texture |
| Micro-calcification | Spotty hyper-dense foci |
| Endothelial rupture | Irregular border by CEUS |
| Intimal neovessel formation | Vascularization by CEUS |

CEUS, contrast-enhanced ultrasound.

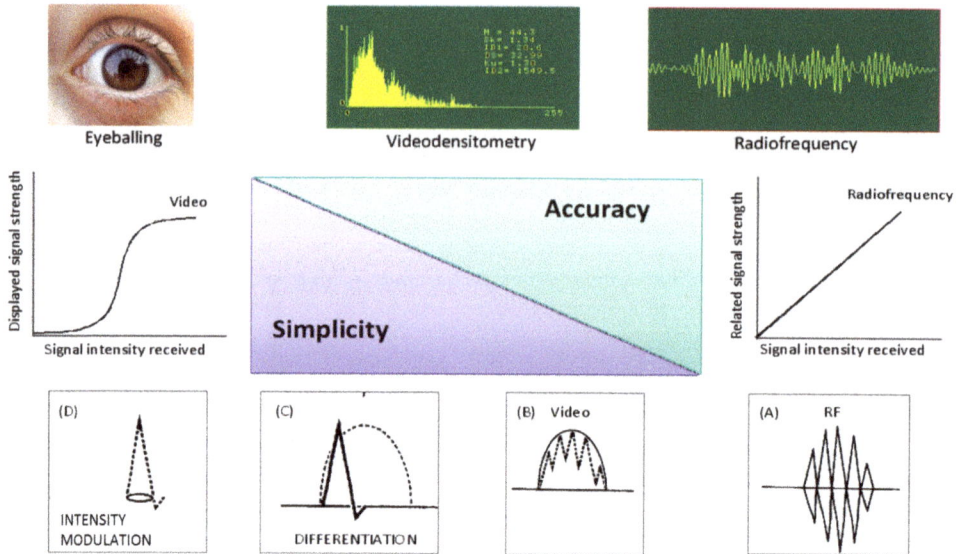

**Figure 1.** Approaches to tissue characterization: visual eyeballing analysis; videodensitometric analysis of digitized image by descriptors of image brightness and gray level spatial distribution; backscatter based sampling of received signal. The latter method is the most technically demanding, available in some but not all commercially available instruments, but it works on a linear relationship between received and displayed signal. This relationship is non-linear for visual and videodensitometric methods, working downstream to the electronic chain of signal processing, shown in the bottom panel, from right (panel **A**, original radiofrequency) to left (panels **B–D**, video, distorted, signal).

## 2. Tissue Characterization of the Atherosclerotic Plaque: *Ex Vivo* Studies

In the 1984 edition of Braunwald's classic Textbook of Cardiology, Wissler stated that "at our present stage of technology and knowledge, it is virtually impossible to evaluate the quantities of the major components in any given plaque in the human subject at any specific time, short of surgical removal or examination at autopsy" [5]. Since then, there has been growing interest in characterization of the acoustic properties of the vascular wall to identify and define the composition of atherosclerotic plaque non-invasively, and several *in vitro* studies have established a solid experimental foundation of ultrasonic tissue characterization of the atherosclerotic plaque. These studies aimed (1) to clarify the biological determinants of vascular acoustic properties; (2) to test, under controlled conditions, ultrasonic parameters of potential diagnostic use; (3) to propose an anatomic-geometrical model of arterial scatterers in different stages of normality and disease; and (4) to orient the technological efforts necessary to translate the most meaningful laboratory information into a clinically feasible tool.

The normal wall can be distinguished from atherosclerotic plaque by a variety of acoustic parameters. The peak amplitude value of reflected signal is low in normal walls and fatty plaques, intermediate in fibrous plaques, and highest in calcific plaques. This index is strongly phase-sensitive and angle-dependent, but also very simple [6]. Plaques can also be identified with parameters based on acoustic attenuation [7]. In particular, calcific plaques show an attenuation 700% greater than that found in normal wall, and 300% greater than that found in atherosclerotic non-calcific plaques. This finding is the experimental counterpart of the clinical echo finding of acoustic shadowing associated with focal calcification. A third parameter of interest for characterizing atherosclerotic plaque as well as in the elaboration of an anatomic-geometrical model of arterial scatterers is the backscatter angular dependence [8]. In contrast to the echoes arising from the myocardium, which are relatively independent of the angle of incidence of the ultrasonic beam to the tissue, those arising from arterial walls are generally said to be of a specular type. This is a major limitation to any application based on a quantitative diagnostic approach *in vivo* because specular reflectors give rise to a signal whose amplitude is highly dependent on the angle of incidence of the ultrasonic beam to the tissue target. The backscatter coefficient, measured at the single frequency of 10 MHz, was evaluated at a normal angle of incidence of the interrogating beam to the tissue sample and over an angular span of 60° (±30° around normal incidence). Angular scattering measurements identified a directive and a non-directive pattern. The directive pattern was characterized by a strongly angle-dependent backscatter that falls abruptly when the beam is moved slightly away from normal incidence. This pattern was typical of calcific, fibrous, and to a lesser extent, fibrofatty and normal samples. The non-directive pattern is characterized by a backscatter that is not significantly angle-dependent and fluctuates throughout the entire angular range. This was typical of fatty samples.

The histological architecture and biochemical composition of the arterial wall might be a reasonable morphological substrate for the recorded difference in angular scattering, which is determined by size and orientation of the scatterers relative to the ultrasonic beam. A directive angular response may be attributable to simple planar organization of the targets within the tissue. Scatterers in the normal wall might be physically identified in the thin elastic membrane present within the normal media layer and oriented perpendicularly to the beam axis. They give rise to directional scattering typical of that in structures in which large plane interfaces exist within the scattering volume. In fibrous and calcified specimens, the scatterers might be physically identified in thick collagen bundles and calcium laminae, which like elastic membranes are oriented perpendicularly to the beam. This might explain the very high directivity of these plaques. In fatty plaques, lipids accumulate in the intima, mainly in the amorphous state but also as cholesterol crystals. Such crystals are comparable in size to the wavelength of the beam, and are spatially arranged in a random fashion. The absence of a spatial orientation and the small size of the scatterers both contribute to the nondirective type of angular scattering in the plaques. In the fibrofatty plaque, the markedly directive response is probably attributable to the fibrous cap; however, the coexistence within the scattering volume of a nondirective structure (the fatty core, absent in the purely fibrous plaques) partially blunts the directivity of the angular response, which is substantially less than in the fibrous samples.

Another potentially useful parameter is the spatial distribution of echo density in an arterial region of interest. In this approach, the information is less dependent on the absolute value of echodensity, and more related to the relative value of different pixels within a region of interest [9]. The shape of the integrated backscatter amplitude distribution is more spread out and flat in the atherosclerotic region.

Another approach is the analysis of the echo signal in time domain (across the depth wall), conceptually similar to the old A-mode representation of ultrasound [10]. If one measures only the first interface aqueous-intimal echo, there is a variable amplitude value for all plaque subsets except for lipidic plaque, which shows a consistently low amplitude value.

These findings were confirmed by different laboratories using qualitative assessment of B-mode images [11] or more quantitative backscatter analysis [12–14] and the overall conclusion is that the atherosclerotic plaque composition leaves several ultrasonic fingerprints which can be fruitfully used for tissue characterization of the vulnerable plaque (Table 1, right side). Even under ideal imaging conditions (*in vitro*, no interposed tissue, controlled angle of insonation, quantitative analysis of reflected signal) lipids and hemorrhages cannot be distinguished by ultrasound, and both appear as low echogenic ("soft") tissue.

### 3. *In Vivo* Ultrasonic Tissue Characterization

Clinical studies have confirmed that the vulnerable, lipid-rich plaque can be identified in the carotid with all three approaches of tissue characterization: visual eyeballing [15–19], videodensitometry [20–22] and backscatter [23–25] (Table 2). Whatever the method, plaque morphology assessment is critically dependent on image quality and the echogenicity is usually normalized for an internal standard, such as—for visual assessment—the flowing blood (black) or far wall media-adventitia surface (white). Eyeballing characterization of plaque texture is subjective and operator-dependent, polluted by technological speckle, but it remains an attractive option since it is simple, straightforward, and still capable of detecting quickly and simply any obvious changes in plaque composition [15–18] with an acceptable reproducibility in controlled conditions when compared to more complex methods [19]. Videodensitometry is quantitative, and still widely applicable in the current era, with most instruments generating a picture describing the image texture through mean gray level and higher order statistics (such as entropy) for spatial distribution of texture.

**Table 2.** Ultrasonic tissue characterization: tools.

| Parameter | B-Mode Ultrasound Imaging | | |
|---|---|---|---|
| | Vascular | Transesophageal | Intravascular |
| Ultrasound frequency | 5–15 | 5–10 | 15–20 |
| Signal-to-noise ratio | ++ | ++ | +++ |
| Accuracy | ++ | ++ | +++ |
| Prognostic value | ++ | ++ | +++ |
| Applicability | Bedside | Echo lab | Cath lab |
| Invasiveness | Non-invasive | Semi-invasive | Invasive |
| Main target artery | Carotid (femoral) | Thoracic Aorta | Coronary |

− = poor; ± = fair; + = good; ++ = very good; +++ = excellent.

Videodensitometry is quantitative, and still widely applicable in the current era, with most instruments generating a picture describing the image texture through mean gray level and higher order statistics (such as entropy) for spatial distribution of texture. The vulnerable plaque is more often hypoechoic, with lower mean gray level and higher entropy values (an index of spatial heterogeneity of gray level distribution) in the region of interest [19–22]. Calibrating the system in a 256 gray level scale with blood equal to 0 (=black), perfect white = 255 and the far wall, strongly reflective adventitial interface = 190, the lipid-hemorrhagic plaques (with blood-like backscatter) remain below 30, with stable plaques showing higher mean gray levels and—with texture analysis—higher entropy values associated with less homogeneous spatial pattern. Radiofrequency analysis is more technologically demanding, and theoretically more accurate, since the "native" ultrasonic signal is sampled, which is not distorted by the post-processing function of the imaging chain [23–25]. It can be applied in the carotid and in the coronary [24,26] districts, with similar findings. Calibrating the system with blood equal to 0 decibels (dB) and the perfect artificial reflector equal to 50 dB, the lipid-hemorrhagic plaques (with blood-like backscatter) remain below 14, fibrous and fibro-fatty plaques in between 14 and 26, and calcific plaques above 27. Although the radiofrequency approach is the most quantitative, the other, simpler approaches can also provide clinically valuable information for *in vivo* characterization of the ultrasonic plaque. The echogenicity is usually expressed with a qualitative score with visual assessment (from black = 1 to white = 4), in grey level units with videodensitometry (from black = 0 to white = 255), in absolute decibel values in radiofrequency analysis (Table 3). Additional echographic features of plaque instability are the neovascularization of the plaque and the irregular contour of plaque surface which are best detected by contrast-enhanced ultrasound (CEUS) [27–29].

### 4. Ultrasound Plaque Morphology as an Index of Clinical Instability

Although the field suffers from lack of standardization, steady changes in image technology, and lack of prospective randomized trials, over the last 30 years a series of observational studies has built up respectable evidence that an unstable plaque morphology by ultrasound identifies a higher risk subset when evaluated by different approaches (visual, videodensitometry, backscatter, CEUS) in different districts (carotid, aorta, coronary arteries) with different methods (vascular, transesophageal, intravascular ultrasound), on different populations (from acute and stable coronary patients to stable and unstable cerebrovascular or peripheral artery disease patients to asymptomatic persons at risk).

Subjects with echo-lucent and/or heterogeneous and/or neovascularized (by CEUS) atherosclerotic plaques in carotid arteries have increased risk of ischemic cerebrovascular and cardiovascular events independently of both degree of stenosis and cardiovascular risk factors [30–37].

The link between echo plaque structure and prognosis do not appear to be limited to the carotid arteries but may apply to virtually all vascular districts where atherosclerotic plaques can be imaged by ultrasound technology including femoral artery [38]. In the ascending thoracic aorta, non-calcified aortic plaques detected by TEE in brain infarction have been associated with a tenfold increased risk of subsequent events when compared to calcified plaque [39].

In coronary arteries evaluated by intracoronary ultrasound, a meta-analysis of 16 studies totaling 1693 patients who underwent PCI showed that the necrotic core (hypoechoic) component derived from virtual histology—IVUS at the minimum lumen sizes were significantly greater in the embolization (no-reflow) group compared with the no-embolization group [40]. Necrotic core was identified in the color-coded analysis as "red" (different from green, fibrotic, yellow-green, fibrofatty, and white, dense calcium). The larger the amount of attenuated coronary plaque by intracoronary ultrasound, the greater the likelihood of no-reflow [41].

Recent studies support the concept that plaque instability is not merely a local vascular incident but rather that plaque instability exists simultaneously at multiple sites of the vascular bed [42]. Similar features can be recognized in unstable patients not only in the symptomatic carotid plaque but also in the contralateral asymptomatic side, suggesting that the vulnerable plaque is also a part of the systemic inflammatory process of the vulnerable patient [43]. Ultrasonic features of instability are potentially a biomarker not only of the vulnerable plaque but—in a certain sense—of the vulnerable patient.

## 5. Clinical Implications

Information on plaque characterization can be obtained by ultrasound and is clinically important for several reasons. First, ultrasonically heterogeneous or soft plaques are associated with lipids and hemorrhages and have a greater tendency to ulceration, embolization, and development of symptoms. Second, anti-atherosclerotic statin treatment may induce a biochemical remodeling of the atherosclerotic plaque, with greater effect on lipidic components than on overall plaque size, which appear more echo-dense (and therefore less vulnerable) after therapy both in the coronary [44,45] and the carotid [46–51] arteries.

**Table 3.** Plaque imaging by ultrasound: criteria of instability.

| Type of Plaque | Unstable | Stable |
|---|---|---|
| Visual assessment | Hypo-, Anechoic<br>Heterogeneous<br>Irregular surface | Iso-, Hyper-echoic<br>Homogeneous<br>Regular surface |
| Videodensitometry | Low median gray level<br>High entropy | High median gray level<br>Low entropy |
| Radiofrequency | <13 dB | 14–33 dB |
| CEUS | Neovessel Present | Neovessel Absent |

Higher values correspond to higher echodensity. Visual assessment of echogenicity refers black as blood and white as far-wall adventitia interface (Gray-Weale, 1988 [16]). Homogeneity is defined according to Joakimsen, 1997 [17] and surface regularity as in Ibrahimi, 2014 [43]. Videodensitometry values are expressed in median grey levels (MGL, 0 black–255 white), with 0 = black as blood and 190 = bright as far-wall adventitia (Ibrahimi, 2014 [43]). Backscatter values are expressed in decibels (dB, calibrated with 0 dB = blood and 50 dB = stainless steel specular interface), according to Kawasaky, 2001 [24]. CEUS binary criteria for intimal neovascularization were proposed by Coli, 2008 [27].

Whatever the method available (visual, videodensitometry or backscatter-based), a simple description of plaque morphology can be helpful to the clinician as a clue to separate stable and unstable plaques (Table 4). Conventional echography and ultrasonic tissue characterization are not mutually exclusive, and any commercial device has—or will soon have—both conventional imaging and quantitative tissue characterization imaging built into the same hybrid basic hardware. Information on tissue characterization of the atherosclerotic plaque can be obtained with other imaging techniques, including MRI, CCTA, PET with FDG and—invasively—by OCT [4].

**Table 4.** Ultrasound appearance and plaque risk.

| Risk | Low-Risk | High-Risk |
| --- | --- | --- |
| Plaque border profile | Smooth | Irregular |
| Echo-density | Iso-, Hyper-echoic | Hypo-, Anechoic |
| Plaque luminal border * | Regular | Irregular |
| Plaque neovascularization * | Absent | Present |
| Spotty calcification | Rare | Frequent |
| Massive calcification | Frequent | Rare |
| Plaque burden | Low (<40% stenosis) | High (>70% stenosis) |

\* By CEUS, contrast-enhanced ultrasound. Carotid plaques are imaged by Duplex scan, coronary plaques by invasive intracoronary ultrasound.

The ultrasound approach has clear advantages over other clinically viable imaging approaches used to detect the vulnerable plaque: non-invasive (differently from OCT—although intracoronary ultrasound can be used, with higher frequencies and better signal-to-noise ratio than vascular ultrasound), radiation-free (differently from CCTA and PET), low cost and high spatial and temporal resolution [4]. Carotid duplex imaging is suitable for evaluation of extracranial cerebral vessels, but cannot image intracranial portion of the carotid artery and is less accurate in presence of dense calcification [3].

## 6. Conclusions

Ultrasonic tissue characterization of atherosclerotic plaque began 30 years ago, and while not yet ready for clinical use it is today regarded as one of the most promising fields of application in cardiovascular ultrasound imaging. In particular, it has the potential to identify—for any given stenosis—vulnerable, lipid-rich, unstable plaques more prone to complications such as embolization and rupture, and also receiving the greatest benefit from pharmacological and mechanical intervention strategies to prevent such events. Although fascinating, the imaging approach has inherent limitations, since not all ruptured plaques have histologic features of vulnerability (and 20% have none of them), and not all vulnerable plaques by histology criteria do eventually rupture in their natural history [3]. With these caveats, the imaging of plaque vulnerability remains a reasonable approach to bridging the current gap in understanding clinical manifestations of atherosclerotic disease, especially regarding the striking clinical mismatch between atherosclerosis extent and severity and its clinical manifestations, and the therapeutic mismatch between reductions of clinical events obtained with statins and the limited—if any—reduction in atheroma size. New quantitative methods beyond backscatter (to include speed of sound, attenuation, strain, temperature, and high order statistics) will be developed to evaluate vascular tissues. These image methods may offer opportunities for the early detection and treatment of the disease [52]. Once the methodology and analysis have been standardized, the stage will be set for future prospective randomized trials to evaluate whether quantitative tissue characterization-based information on plaque vulnerability can be used to tailor risk and treatment in patients with clinically symptomatic and high-risk asymptomatic atherosclerosis.

**Acknowledgments:** The authors thank Letizia Morelli for secretarial assistance.

**Conflicts of Interest:** The authors declare no conflict of interest.

## References

1.  Mor-Avi, V.; Lang, R.M.; Badano, L.P.; Belohlavek, M.; Cardim, N.M.; Derumeaux, G.; Galderisi, M.; Marwick, T.; Nagueh, S.F.; Sengupta, P.P.; *et al.* Current and evolving echocardiographic techniques for the quantitative evaluation of cardiac mechanics: ASE/EAE consensus statement on methodology and indications endorsed by the Japanese Society of Echocardiography. *Eur. J. Echocardiogr.* **2011**, *12*, 167–205. [CrossRef] [PubMed]

2.  Fleg, J.L.; Stone, G.W.; Fayad, Z.A.; Granada, J.F.; Hatsukami, T.S.; Kolodgie, F.D.; Ohayon, J.; Pettigrew, R.; Sabatine, M.S.; Tearney, G.J.; *et al.* Detection of high-risk atherosclerotic plaque: Report of the NHLBI Working Group on current status and future directions. *JACC Cardiovasc. Imaging* **2012**, *5*, 941–955. [CrossRef] [PubMed]

3.  Stein, J.H.; Korcarz, C.E.; Hurst, R.T.; Lonn, E.; Kendall, C.B.; Mohler, E.R.; Najjar, S.S.; Rembold, C.M.; Post, W.S.; American Society of Echocardiography Carotid Intima-Media Thickness Task Force. Use of carotid ultrasound to identify subclinical cardiovascular disease risk: Consensus statement from the American society of echocardiography. *J. Am. Soc. Echocardiogr.* **2008**, *21*, 93–111. [CrossRef] [PubMed]

4.  Gallino, A.; Stuber, M.; Crea, F.; Falk, E.; Corti, R.; Lekakis, J.; Schwitter, J.; Camici, P.; Gaemperli, O.; di Valentino, M.; *et al. In vivo* imaging of atherosclerosis. *Atherosclerosis* **2012**, *224*, 25–36. [CrossRef] [PubMed]

5.  Wissler, R.W. Principles of the pathogenesis of atherosclerosis. In *Heart Disease: A Textbook of Cardiovascular Medicine*; Braunwald, E., Ed.; Saunders: Philadelphia, PA, USA, 1984; pp. 1183–1194.

6.  Picano, E.; Landini, L.; Distante, A.; Sarnelli, R.; Benassi, A.; L'Abbate, A. Different degrees of atherosclerosis detected by backscattered ultrasound: An *in vitro* study on fixed human aortic walls. *J. Clin. Ultrasound* **1983**, *11*, 375–379. [CrossRef] [PubMed]

7.  Picano, E.; Landini, L.; Distante, A.; Benassi, A.; Sarnelli, R.; L'Abbate, A. Fibrosis, lipids, and calcium in human atherosclerotic plaque. *In vitro* differentiation from normal aortic walls by ultrasonic attenuation. *Circ. Res.* **1985**, *56*, 556–562.

8.  Picano, E.; Landini, L.; Distante, A.; Salvadori, M.; Lattanzi, F.; Masini, M.; L'Abbate, A. Angle dependence of ultrasonic backscatter in arterial tissues: A study *in vitro*. *Circulation* **1985**, *72*, 572–576. [CrossRef] [PubMed]

9.  Picano, E.; Landini, L.; Lattanzi, F.; Mazzarisi, A.; Sarnelli, R.; Distante, A.; Benassi, A.; L'Abbate, A. The use of frequency histograms of ultrasonic backscatter amplitudes for detection of atherosclerosis *in vitro*. *Circulation* **1986**, *74*, 1093–1098. [CrossRef] [PubMed]

10. Picano, E.; Landini, L.; Lattanzi, F.; Salvadori, M.; Benassi, A.; L'Abbate, A. Time domain echo pattern evaluations from normal and atherosclerotic arterial walls: A study *in vitro*. *Circulation* **1988**, *77*, 654–659. [CrossRef] [PubMed]

11. Wolverson, M.K.; Bashiti, H.M.; Peterson, G.J. Ultrasonic tissue characterization of atheromatous plaques using a high resolution real time scanner. *Ultrasound Med. Biol.* **1983**, *9*, 599–609. [CrossRef] [PubMed]

12. Barzilai, B.; Saffitz, J.E.; Miller, J.G.; Sobel, B.E. Quantitative ultrasonic characterization of the nature of atherosclerotic plaques in human aorta. *Circ. Res.* **1987**, *60*, 459–463. [CrossRef] [PubMed]

13. Hiro, T.; Leung, C.Y.; Karimi, H.; Farvid, A.R.; Tobis, J.M. Angle dependence of intravascular ultrasound imaging and its feasibility in tissue characterization of human atherosclerotic tissue. *Am. Heart J.* **1999**, *137*, 476–481. [CrossRef] [PubMed]

14. Komiyama, N.; Berry, G.J.; Kolz, M.L.; Oshima, A.; Metz, J.A.; Preuss, P.; Brisken, A.F.; Pauliina Moore, M.; Yock, P.G.; Fitzgerald, P.J. Tissue characterization of atherosclerotic plaques by intravascular radiofrequency signal analysis: An *in vitro* study of human coronary artery. *Am. Heart J.* **2000**, *140*, 565–574. [CrossRef] [PubMed]

15. Reilly, L.M.; Lusby, R.J.; Hughes, L.; Ferrell, L.D.; Stoney, R.J.; Ehrenfeld, W.K. Carotid plaque histology using real time ultrasonography. Clinical and therapeutic implications. *Am. J. Surg.* **1983**, *146*, 188–193. [CrossRef] [PubMed]

16. Gray-Weale, A.C.; Graham, J.C.; Burnett, J.R.; Byrne, K.; Lusby, R.J. Carotid artery atheroma: Comparison of preoperative B-mode ultrasound appearance with carotid endarterectomy specimen pathology. *J. Cardiovasc. Surg.* **1988**, *29*, 676–681.

17. European Carotid Plaque Study Group. Carotid artery composition—Relation to clinical presentation and ultrasound B-mode imaging. *Eur. J. Endovasc. Surg.* **1995**, *10*, 23–32.

18. Joakimsen, O.; Bﬁﬁona, K.H.; Stensland-Bugge, E. Reproducibility of ultrasound assessment of carotid plaque occurrence, thickness, and morphology. The Tromsø study. *Stroke* **1997**, *28*, 2201–2207. [CrossRef] [PubMed]

19. Grønholdt, M.L.; Wiebe, B.M.; Laursen, H.; Nielsen, T.G.; Schroeder, T.V.; Sillesen, H. Lipid-rich carotid artery plaques appear echolucent on ultrasound B-mode images and may be associated with intraplaque hemorrhage. *Eur. J. Vasc. Endovasc. Surg.* **1997**, *14*, 439–445. [CrossRef] [PubMed]

20. Mazzone, A.M.; Urbani, M.P.; Picano, E.; Paterni, M.; Borgatti, E.; de Fabritiis, A.; Landini, L. *In vivo* ultrasonic parametric imaging of carotid atherosclerotic plaque by videodensitometric technique. *Angiology* **1995**, *46*, 663–672. [CrossRef] [PubMed]

21. El-Barghouty, N.M.; Levine, T.; Ladva, S.; Flanagan, A.; Nicolaides, A. Histological verification of computerized carotid plaque characterization. *Eur. J. Vasc. Endovasc. Surg.* **1996**, *11*, 414–416. [CrossRef] [PubMed]

22. Baroncini, L.A.V.; Pazin Filho, A.; Murta, O., Jr.; Martins, A.R.; Ramos, S.O.; Cherri, J.; Piccinato, C.E. Ultrasonic tissue characterization of vulnerable carotid plaque: Correlation between videodensitometric method and histological exam. *Cardiov. Ultrasound* **2006**, *4*. [CrossRef]

23. Urbani, M.P.; Picano, E.; Parenti, G.; Mazzarisi, A.; Fiori, L.; Paterni, M.; Pelosi, G.; Landini, L. *In vivo* radiofrequency-based ultrasonic tissue characterization of the atherosclerotic plaque. *Stroke* **1993**, *24*, 1507–1512. [CrossRef] [PubMed]

24. Kawasaki, M.; Takatsu, H.; Noda, T.; Ito, Y.; Kunishima, A.; Arai, M.; Nishigaki, K.; Takemura, G.; Morita, N.; Minatoguchi, S.; *et al.* Noninvasive quantitative tissue characterization and two-dimensional color-coded map of human atherosclerotic lesions using ultrasound integrated backscatter. Comparison between histology and integrated backscatter images before and after death. *J. Am. Coll. Cardiol.* **2001**, *38*, 486–492. [CrossRef] [PubMed]

25. Waki, H.; Masuyama, T.; Mori, H.; Maeda, T.; Kitade, K.; Moriyasu, K.; Tsujimoto, M.; Fujimoto, K.; Koshimae, N.; Matsuura, N. Ultrasonic tissue characterization of the atherosclerotic carotid plaque: Histologic correlates of carotid integrated backscatter. *Circ. J.* **2003**, *67*, 1013–1016. [CrossRef] [PubMed]

26. Gussenhoven, W.J.; Essed, C.E.; Frietman, P.; Mastik, F.; Lancee, C.; Slager, C.; Serruys, P.; Gerritsen, P.; Pieterman, H.; Bom, N. Intravascular echographic assessment of vessel wall characteristics: A correlation with histology. *Int. J. Cardiovasc. Imaging* **1989**, *4*, 105–116. [CrossRef]

27. Coli, S.; Magnoni, M.; Sangiorgi, G.; Marrocco-Trischitta, M.M.; Melisurgo, G.; Mauriello, A.; Spagnoli, L.; Chiesa, R.; Cianflone, D.; Maseri, A. Contrast-enhanced ultrasound imaging of intraplaque neovascularization in carotid arteries: Correlation with histology and plaque echogenicity. *J. Am. Coll. Cardiol.* **2008**, *52*, 223–230. [CrossRef] [PubMed]

28. Faggioli, G.L.; Pini, R.; Mauro, R.; Pasquinelli, G.; Fittipaldi, S.; Freyrie, A.; Serra, C.; Stella, A. Identification of carotid vulnerable plaque in contrast-enhanced ultrasound: Correlation with plaque histology, symptoms and cerebral computed tomography. *Eur. J. Vasc. Endovasc. Surg.* **2011**, *41*, 238–248. [CrossRef] [PubMed]

29. Partovi, S.; Loebe, M.; Aschwanden, M.; Baldi, T.; Jäger, K.A.; Feinstein, S.B.; Staub, D. Contrast-enhanced ultrasound for assessing carotid atherosclerosis plaque lesions. *Am. J. Radiol.* **2012**, *198*, W13–W19.

30. Gronholdt, M.L.; Nordestgaard, B.G.; Schroeder, T.V.; Vorstrup, S.; Sillesen, H. Ultrasound echolucent carotid plaques predict future strokes. *Circulation* **2001**, *104*, 68–73. [CrossRef] [PubMed]

31. Mathiesen, E.B.; Bonaa, K.H.; Joakimsen, O. Echo-lucent plaques are associated with high risk of ischemic cerebrovascular events in carotid stenosis. *Circulation* **2001**, *103*, 2171–2175. [CrossRef] [PubMed]

32. Honda, O.; Sugiyama, S.; Kugiyama, K.; Fukushima, H.; Nakamura, S.; Koide, S.; Kojima, S.; Hirai, N.; Kawano, H.; Soejima, H.; *et al.* Echolucent carotid plaques predict future coronary events in patients with coronary artery disease. *J. Am. Coll. Cardiol.* **2004**, *43*, 1177–1184. [CrossRef] [PubMed]

33. Biasi, G.M.; Froio, A.F.; Diethrich, E.B.; Deleo, G.; Galimberti, S.; Mingazzini, P.; Nicolaides, A.N.; Griffin, M.; Raithel, D.; Reid, D.B.; *et al.* Carotid plaque echolucency increases the risk of stroke in carotid stenting. The Imaging in Carotid Angioplasty and Risk of Stroke (ICAROS) study. *Circulation* **2004**, *110*, 756–762. [CrossRef] [PubMed]

34. Petersen, C.; Peçanha, P.B.; Venneri, L.; Pasanisi, E.; Pratali, L.; Picano, E. The impact of carotid plaque presence and morphology on mortality outcome in cardiological patients. *Cardiovasc. Ultrasound* **2006**, *4*. [CrossRef]

35. Yamada, K.; Kawasaki, M.; Yoshima, S.; Enomoto, Y.; Asano, T.; Minatocuchi, S.; Iwana, T. Prediction of silent ischemic lesions after carotid stenting using integrated backscatter ultrasouns and magnetic resonance imaging. *Atherosclerosis* **2010**, *208*, 161–166. [CrossRef] [PubMed]

36. Irie, Y.; Katakami, N.; Kaneto, H.; Takahara, M.; Nishio, M.; Kasami, R.; Sakamoto, K.; Umayahara, Y.; Sumitsuji, S.; Ueda, Y.; *et al.* The utility of ultrasound tissue characterization in the prediction of cardiovascular events in diabetic patients. *Atherosclerosis* **2013**, *230*, 399–403. [CrossRef] [PubMed]

37. Zhu, Y.; Deng, Y.B.; Liu, Y.N.; Bi, X.J.; Sun, J.; Tang, Q.Y.; Deng, Q. Use of carotid plaque neovascularization of contrast-enhanced ultrasound to predict coronary events in patients with coronary artery disease. *Radiology* **2013**, *268*, 54–61. [CrossRef] [PubMed]

38. Schmidt, C.; Fagerberg, B.; Hulthe, J. Non-stenotic echolucent ultrasound-assessed femoral artery plaques are predictive for future cardiovascular events in middle-aged men. *Atherosclerosis* **2005**, *181*, 125–130. [CrossRef] [PubMed]

39. Cohen, A.; Tzourio, C.; Bertrand, B.; Chauvel, C.; Bousser, M.G.; Amarenco, P. Aortic plaque morphology and vascular events: A follow-up study in patients with ischemic stroke. FAPS Investigators. French Study of Aortic Plaques in Stroke. *Circulation* **1997**, *96*, 3838–3841. [CrossRef] [PubMed]

40. Jang, J.S.; Jin, H.Y.; Seo, J.S.; Yang, T.H.; Kim, D.K.; Park, Y.A.; Cho, K.I.; Park, Y.H.; Kim, D.S. Meta-analysis of plaque composition by intravascular ultrasound and its relation to distal emboli after percutaneous coronary interventions. *Am. J. Cardiol.* **2013**, *111*, 968–972. [CrossRef] [PubMed]

41. Xu, Y.; Mintz, G.S.; Tam, A.; McPherson, J.A.; Iñiguez, A.; Fajadet, J.; Fahy, M.; Weisz, G.; de Bruyne, B.; Serruys, P.W.; *et al.* Prevalence, distribution, predictors and outcomes of patients with calcified nodules in native coronary arteries: A 3-vessel intravascular ultrasound anaysis from Providing Regional Observations to Study predictors of Events in the Coronary Tree (PROSPECT). *Circulation* **2012**, *126*, 537–545. [CrossRef] [PubMed]

42. Lombardo, A.; Biasucci, L.M.; Lanza, G.A.; Coli, S.; Silvestri, P.; Cianflone, D.; Liuzzo, G.; Burzotta, F.; Crea, F.; Maseri, A. Inflammation as a possible link between coronary and carotid plaque instability. *Circulation* **2004**, *109*, 3158–3163. [CrossRef] [PubMed]

43. Ibrahimi, P.; Jashari, F.; Johansson, E.; Gronlund, C.; Bajraktari, G.; Wester, P.; Henein, M.Y. Vulnerable plaques in the contralateral carotid arteries in symptomatic patients: A detailed ultrasonic analysis. *Atherosclerosis* **2014**, *235*, 526–531. [CrossRef] [PubMed]

44. Kawasaki, M.; Sano, K.; Okubo, M.; Yokoyama, H.; Ito, Y.; Murata, I.; Tsuchiya, K.; Minatoguchi, S.; Zhou, X.; Fujita, H.; *et al.* Volumetric quantitative analysis of tissue characteristics of coronary plaques after statin therapy using three-dimensional integrated backscatter intravascular ultrasound. *J. Am. Coll. Cardiol.* **2005**, *45*, 1946–1953. [CrossRef] [PubMed]

45. Yokoyama, M.; Komiyama, N.; Courtney, B.K.; Nakayama, T.; Namikawa, S.; Kuriyama, N.; Koizumi, T.; Nameki, M.; Fitzgerald, P.J.; Komuro, I. Plasma low-density lipoprotein reduction and structural effects on coronary atherosclerotic plaques by atorvastatin as clinically assessed with intravascular ultrasound radio-frequency signal analysis: A randomized prospective study. *Am. Heart J.* **2005**, *150*, 287. [CrossRef] [PubMed]

46. Yamada, K.; Yoshimura, S.; Kawasaki, M.; Enomoto, Y.; Asano, T.; Minatoguchi, S.; Iwama, T. Effects of atorvastatin on carotid atherosclerotic plaques: A randomized trial with quantitative tissue characterization of carotid atherosclerotic plaques with integrated backscatter. *Cerebrovasc. Dis.* **2009**, *28*, 417–424. [CrossRef] [PubMed]

47. Watanabe, K.; Sugiyama, S.; Kugiyama, K.; Honda, O.; Fukushima, H.; Koga, H.; Horibata, Y.; Hirai, T.; Sakamoto, T.; Yoshimura, M.; *et al.* Stabilization of carotid atheroma assessed by quantitative ultrasound analysis in non-hypercholesterolemic patients with coronary artery disease. *J. Am. Coll. Cardiol.* **2005**, *46*, 2022–2030. [CrossRef] [PubMed]

48. Kadoglou, N.P.; Gerasimidis, T.; Moumtzouoglou, A.; Kapelouzou, A.; Sailer, N.; Fotiadis, G.; Vitta, I.; Katinios, A.; Kougias, P.; Bandios, S.; *et al.* Intensive lipid-lowering therapy ameliorates novel calcification markers and GSM score in patients with carotid stenosis. *Eur. J. Vasc. Endovasc. Surg.* **2008**, *35*, 661–668. [CrossRef] [PubMed]

49. Yamagami, H.; Sakaguchi, M.; Furukado, S.; Hoshi, T.; Abe, Y.; Hougaku, H.; Hori, M.; Kitagawa, K. Statin therapy increases carotid plaque echogenicity in hypercholesterolemic patients. *Ultrasound Med. Biol.* **2008**, *34*, 1353–1359. [CrossRef] [PubMed]

50. Della-Morte, D.; Moussa, I.; Elkind, M.S.; Sacco, R.L.; Rundek, T. The short-term effects of atorvastatin on carotid plaque morphology assessed by computer-assisted gray scale densitometry: A pilot study. *Neurol. Res.* **2011**, *33*, 991–994. [CrossRef] [PubMed]

51. Ostling, G.; Gonçalves, I.; Wikstrand, J.; Berglund, G.; Nilsson, J.; Hedblad, B. Long-term treatment with low dose metoprolol is associated with increased plaque echogenicity: The β-blocker cholesterol lowering asymptomatic plaque study. *Atherosclerosis* **2011**, *215*, 440–445. [CrossRef] [PubMed]

52. Pellikka, P.A.; Douglas, P.S.; Miller, J.G.; Abraham, T.P.; Baumann, R.; Buxton, D.B.; Byrd, B.F., III; Chen, P.; Cook, N.L.; Gardin, J.M.; *et al.* American Society of Echocardiography Cardiovascular technology and research summit: A roadmap for 2020. *J. Am. Soc. Echocardiogr.* **2013**, *26*, 325–338. [CrossRef] [PubMed]

International Journal of
*Molecular Sciences*

MDPI

*Article*

# Ultrasound Assessment of Carotid Plaque Echogenicity Response to Statin Therapy: A Systematic Review and Meta-Analysis

Pranvera Ibrahimi, Fisnik Jashari *, Gani Bajraktari, Per Wester and Michael Y. Henein

Department of Public Health and Clinical Medicine, Umeå University, Umeå 901 87, Sweden;
pranvera.ibrahimi@medicin.umu.se (P.I.); ganibajraktari@yahoo.co.uk (G.B.); per.wester@medicin.umu.se (P.W.);
michael.henein@medicin.umu.se (M.Y.H.)
* Correspondence: fisnik.jashari@medicin.umu.se; Tel.: +46-90-785-1431

Academic Editor: William Chi-shing Cho
Received: 16 February 2015; Accepted: 5 May 2015; Published: 12 May 2015

**Abstract:** Objective: To evaluate in a systematic review and meta-analysis model the effect of statin therapy on carotid plaque echogenicity assessed by ultrasound. Methods: We have systematically searched electronic databases (PubMed, MEDLINE, EMBASE and Cochrane Center Register) up to April, 2015, for studies evaluating the effect of statins on plaque echogenicity. Two researchers independently determined the eligibility of studies evaluating the effect of statin therapy on carotid plaque echogenicity that used ultrasound and grey scale median (GSM) or integrated back scatter (IBS). Results: Nine out of 580 identified studies including 566 patients' carotid artery data were meta-analyzed for a mean follow up of 7.2 months. A consistent increase in the echogenicity of carotid artery plaques, after statin therapy, was reported. Pooled weighted mean difference % (WMD) on plaque echogenicity after statin therapy was 29% (95% CI 22%–36%), $p < 0.001$, $I^2 = 92.1\%$. In a meta-regression analysis using % mean changes of LDL, HDL and hsCRP as moderators, it was shown that the effects of statins on plaque echogenicity were related to changes in hsCRP, but not to LDL and HDL changes from the baseline. The effect of statins on the plaque was progressive; it showed significance after the first month of treatment, and the echogenicity continued to increase in the following six and 12 months. Conclusions: Statin therapy is associated with a favorable increase of carotid plaque echogenicity. This effect seems to be dependent on the period of treatment and hsCRP change from the baseline, independent of changes in LDL and HDL.

**Keywords:** carotid atherosclerosis; plaque echogenicity; ultrasound; statins

## 1. Introduction

Carotid atherosclerosis is an important cause of ischemic stroke, the risk of which is mainly related to the degree of stenosis [1]. Adding the evaluation of carotid plaque echogenicity features, on the other hand, it was found to better risk stratify patients beyond the degree of stenosis. Treatment of carotid atherosclerosis with statins has proven effective in reducing such stroke risk, universally considered to be caused by vulnerable plaques [2].

Many imaging techniques are currently used to identify vulnerable plaque features. The most feasible one with less radiation remains carotid ultrasound, which can accurately identify the presence of the plaque and determine its echogenicity, as well as the degree of stenosis. Vulnerable plaques are known for their high lipid and hemorrhage content, in contrast to stable plaques, which are predominately rich in fibrous tissue and calcification [3,4]. Furthermore, such detailed plaque composition has been found to correlate with the textural features (echogenicity) obtained by ultrasound imaging. This can easily be assessed using off-line plaque image analysis techniques, such as grey scale median (GSM) and integrated backscatter (IBS), with plaques rich in lipid and hemorrhagic content appearing echolucent (low GSM or IBS) and those with fibrous or calcific content appearing echogenic (high GSM or IBS) [5,6].

The effect of statins treatment on plaque regression and change in its features is well documented in the literature, but a consensus analysis is lacking. Such an effect has been reported using various imaging modalities, other than US [7], not only in carotid disease, but also in coronary and aortic disease [8]. The aim of this study was to determine, in a systematic and meta-analysis model, the response of plaque features' "echogenicity" to statin therapy in patients with carotid artery disease.

## 2. Results

### 2.1. Study Selection

We identified 576 studies in total after systematic searching in PubMed (Figure 1). No additional studies were found in MEDLINE, EMBASE or in the Cochrane Center Register. After reading the titles and abstracts of the papers, we first depicted 12 studies that evaluated the effects of statins on plaque echogenicity. Of these studies, three were excluded, two measured the effect of statins on the carotid artery wall (intima-media) echogenicity [9,10] and one was a case report [11]. The remaining nine studies [12–20] were included in the final qualitative analysis (Figure 1, Tables 1 and 2). Out of the nine studies, two studies analyzed two groups of patients separately, and we have included both groups in the meta-analysis. In addition, we have used meta-regression and subgroup analysis to determine the effect of % changes in LDL, HDL, hsCRP, the period of treatment and baseline patients' characteristics on echogenicity change. Statins effect on LDL, HDL and hsCRP were also evaluated (Supplementary Material).

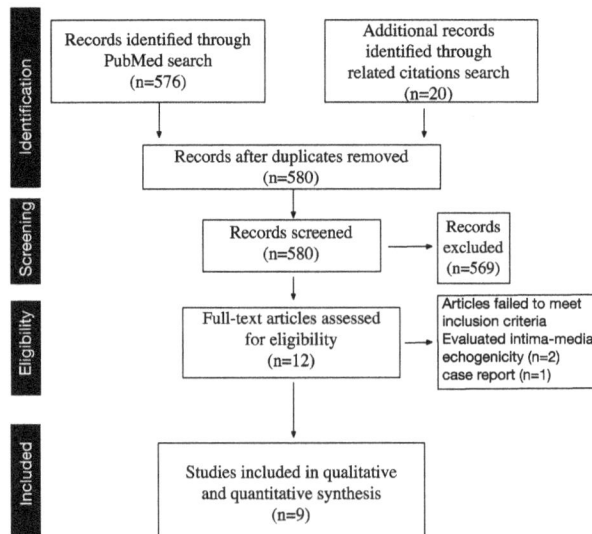

**Figure 1.** Study selection.

### 2.2. Qualitative Assessment and Study Characteristics

#### 2.2.1. Effect of Statins on Plaque Features

All nine studies included were prospective; five were prospective open-label, and four of them were randomized controlled trials [8–11]. Atorvastatin was the most commonly used [13,15,17,18], followed by simvastatin [13], pravastatin and pitavastatin [12,14] and rosuvastatin [20]. In three studies, the statin dosage was fixed [14,16,19], and in the remaining six, it was ranged [12,13,15,17,18,20]. The mean follow up of patients was 7.2 months (1–12). Studies included in this review used different methods to quantify plaque echogenicity. The grey scale median (GSM) was used in five studies to evaluate plaque echogenicity, and integrated backscatter (IBS) analysis was used in the remaining four. In one study, patients were divided into two groups based on the statin dose (intensive *vs.* moderate) [18] and, in the other one, based on treatment strategy (statins + carotid artery stenting (CAS) *vs.* statin alone) [17]. In seven studies [12–14,16–18,20], controls were commenced on a set diet, and in one study [15], patients on statins were compared with those on placebo. The diet type applied in the controls was mentioned in two studies [12,19], which used the Adult Treatment Panel-III lipid-lowering diet. There was only one study that used more than one type of statin in the treated group [13].

In all nine studies presented in this meta-analysis, there was a significant increase of plaque echogenicity after statin therapy. In the compared high (atorvastatin 80 mg/d) and low (atorvastatin 20 mg/d) statin therapy, the GSM was significantly increased more in the group receiving aggressive statin therapy.

In six studies, other ultrasound-derived measurements were evaluated, including: intima-media thickness [12,16,20], plaque thickness [13,19], plaque volume [16] and degree of stenosis [12]. Except for one study that found a decrease in plaque thickness after statin therapy [13], there was no other change, of the above-mentioned measures, observed after statin therapy.

### 2.2.2. Effect of Statins on Blood Lipids and Inflammatory Markers

All studies measured LDL and HDL cholesterol levels, and in all, there was a significant decrease in LDL with statin therapy; in only two [12,14] did the HDL cholesterol significantly increase. However, only one study has evaluated the change of plaque echogenicity after adjusting for LDL cholesterol and its changes from baseline, which concluded that statins' effect on plaque echogenicity was independent of the fall in LDL cholesterol [19].

The included studies used different blood-derived markers of atherosclerosis, such as: high sensitivity CRP (hsCRP) [12–18,20], vasogenic endothelial growth factor (VEGF) [14], interleukins (IL) IL-6 and IL-18 [13], osteopontin (OPN) [15,17,18], osteoprotegerin (OPG) [15,17,18] and tumor necrosis factor-alpha (TNF-α) [14]. Except for IL-6, which was not affected [13], the other markers, hsCRP, VEGF, IL-18, OPN, OPG and TNF-α, were significantly decreased in patients treated with statins compared to controls. Even at only one month after statin therapy, the levels of hsCRP, VEGF and TNF-α decreased significantly [14]. In the study that compared aggressive (atorvastatin 80 mg/d) and modest (atorvastatin 10 mg/d) statin therapy, OPN and OPG levels were lower in patients receiving aggressive statin therapy; however, hsCRP was not different between groups [18].

**Table 1.** Studies included in meta-analysis.

| Author/Year | Population (*n*) | Mean Age ± SD | Gender (Male) | Hypercholesterolemic | Carotid Stenosis | Echogenicity Measured | Minor Score |
|---|---|---|---|---|---|---|---|
| 1. Watanabe *et al.*, 2005 [12] | 30 | 69.9 ± 8.8 | 63% | No | Moderate | IBS | RT |
| 2. Yamagami *et al.*, 2008 [13] | 41 | 63.4 ± 8.3 | 24% | Yes | Moderate | IBS | RT |
| 3. Nakamura *et al.*, 2008 [14] | 33 | 60 ± 9 | 25% | Yes | Moderate | IBS | RT |
| 4. Kadoglou *et al.*, 2008 [15] | 113 | 63.6 ± 9.9 | 67% | Yes | Moderate symptomatic | GSM | 20 |
| 5. Yamada *et al.*, 2009 [16] | 40 | 71 ± 8 | 90% | No | 30%–60% | IBS | RT |
| 6. Kadoglou *et al.*, 2009 [17] | 67 + 46 | 66.7 ± 7.3 | 40% | No | >40% | GSM | 20 |
| 7. Kadoglou *et al.*, 2010 [18] | 66 + 65 | 64.9 ± 10 | 46% | Yes | 30%–60% | GSM | 24 |
| 8. Della-Morte *et al.*, 2011 [19] | 40 | >45 | NA | Yes | NA | GSM | 15 |
| 9. Nohara *et al.*, 2013 [20] | 25 | 63.9 ± 8.1 | 50% | Yes | NA | GSM | 20 |

GSM, grey scale median; IBS, integrated back scatter; RT, randomized trial; SD, standard deviation.

**Table 2.** Studies characteristics and statins effect on plaque echogenicity and LDL, HDL and hsCRP level on the blood.

| Author/Year | Study Design | Statin/Dose | Follow-up (Months) | % Change Echogenicity | % Change LDL | % Change HDL | % Change hsCRP |
|---|---|---|---|---|---|---|---|
| 1. Watanabe et al., 2005 [12] | Randomized case-control trial | Pravastatin | 6 | 14.1 ± 3.3 | 24.5 ± 6.4 | 10.2 ± 6.0 | 45.0 ± 58.3 |
| 3. Yamagami et al., 2008 [13] | Randomized case-control trial | Simvastatin 10 mg | 1 | 10.6 ± 4.3 | 34.2 ± 18.4 | 0 | 43.0 ± 119.8 |
| 2. Nakamura et al., 2008 [14] | Randomized case-control trial | Pitavastatin 4 mg | 12 | 32.1 ± 5.9 | 37.8 ± 12.4 | 9.3 ± 2.0 | 43.7 ± 51.5 |
| 4. Kadoglou et al., 2008 [15] | Open-label prospective trial | Atorvastatin | 6 | 36.0 ± 15.2 | 41.7 ± 19.9 | 4.5 ± 2.4 | 58.9 ± 34.0 |
| 5. Yamada et al., 2009 [16] | Randomized case-control trial | Simvastatin | 6 | 17.0 ± 5.9 | 44.0 ± 23.9 | 0 | 42.1 ± 94.6 |
| 6a. Kadoglou et al., 2009 [17] | Open-label prospective trial | Atorvastatin | 6 | 36.8 ± 9.8 | 38.6 ± 20.0 | 13.4 ± 6.6 | 78.3 ± 74.9 |
| 6b. Kadoglou et al., 2009 [17] | Open-label prospective trial | Atorvastatin + CAS | 6 | 48.4 ± 18.6 | 33.3 ± 15.0 | 4.4 ± 2.2 | 52.1 ± 39.5 |
| 7a. Kadoglou et al., 2010 [18] | Randomized case-control trial | Atorvastatin 10–20 mg | 12 | 32.6 ± 11.7 | 64.5 ± 23.6 | 5.5 ± 2.6 | 52.9 ± 55.2 |
| 7b. Kadoglou et al., 2010 [18] | Randomized case-control trial | Atorvastatin 80 mg | 12 | 51.4 ± 18.4 | 54.2 ± 37.2 | 10.3 ± 6.0 | 65.0 ± 80.0 |
| 8. Della-Morte et al., 2011 [19] | Prospective pilot study | NA | 1 | 21.9 ± 4.8 | 51.4 ± 31.0 | 2.0 ± 1.1 | NA |
| 9. Nohara et al., 2013 [20] | Prospective open label, blinded-endpoint | Rosuvastatin | 12 | 16.9 ± 33.1 | 50.1 ± 22.9 | 8.1 ± 3.6 | NA |

CAS, carotid artery stenting.

## 2.3. Meta-Analysis Results

All nine studies met the inclusion criteria to be included in the meta-analysis. Two of the studies have divided patients into two groups. The first one [18] had two groups on different statin dosage (atorvastatin 10 *vs.* 80 mg), and the other study [17] had two groups on statin therapy; one of them underwent CAS in addition. In total, 566 patients' carotid artery data were meta-analyzed for a mean follow up of 7.2 months. A consistent increase in the echogenicity of carotid artery plaques, after statin therapy, was found. Pooled weighted mean difference % (WMD) on plaque echogenicity after statin therapy was 29% (95% CI: 22%–36%), $p < 0.001$, $I^2 = 92.1\%$ (Figure 2). In these studies, evaluating the effect of statins on plaque echogenicity was the main objective; in addition, their effect on LDL, HDL and hsCRP level was also evaluated. LDL was significantly decreased after statins therapy; the pooled weighted mean difference % (WMD) was −40.8% (95% CI, −48.9%–−32.8%), $p < 0.001$, $I^2 = 81.2\%$ (Supplemental Figure S1). Furthermore, HDL and hsCRP were increased and decreased after statin therapy, respectively (Supplemental Figures S2 and S3).

In a meta-regression analysis, mean changes % of LDL, HDL and hsCRP from baseline were used as moderators to evaluate their association with changes in plaque echogenicity. The increase in plaque echogenicity with statins therapy was independent of LDL ($\beta = 0.32$ (−0.28–0.94), $p < 0.29$) (Figure 3a) and HDL cholesterol ($\beta = 0.91$ (−0.01–3.55), $p = 0.051$) (Figure 3b), but it was related to hsCRP changes from the baseline ($\beta = 1.01$ (0.49–1.52), $p < 0.001$) (Figure 3c).

Patients were divided into subgroups and analyzed based on treatment period. Although the increase of plaque echogenicity was significant even after the first month of treatment, this increase was more evident in the following six and 12 months (Figure 4). Mean difference % was 16.2% (5.2%–27.2%) *vs.* 30.4% (18.2%–42.3%) *vs.* 35.4% (26.3%–44.4%), $p = 0.03$, between 1, 6 and 12 months of treatment, respectively. In addition, we have analyzed separately studies based on baseline cholesterolemia status (hypercholesterolemic *vs.* non-hypercholesterolemic), and it seems that the effects of statins on carotid plaque echogenicity is independent of cholesterol levels at baseline, since it was similarly increased in both groups (Figure 5).

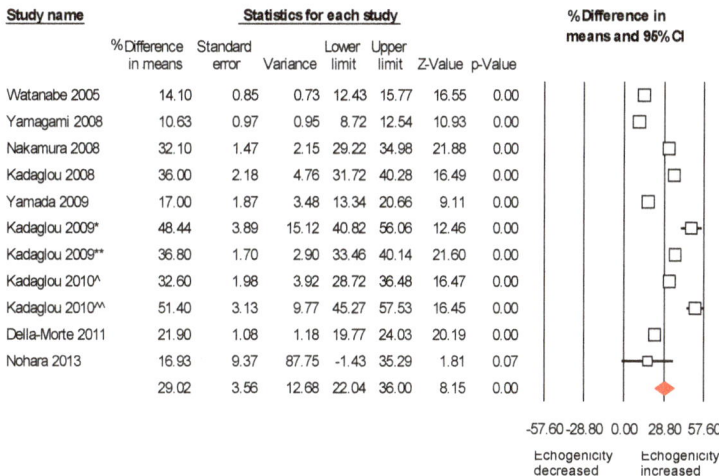

| Study name | % Difference in means | Standard error | Variance | Lower limit | Upper limit | Z-Value | p-Value | % Difference in means and 95% CI |
|---|---|---|---|---|---|---|---|---|
| Watanabe 2005 | 14.10 | 0.85 | 0.73 | 12.43 | 15.77 | 16.55 | 0.00 | |
| Yamagami 2008 | 10.63 | 0.97 | 0.95 | 8.72 | 12.54 | 10.93 | 0.00 | |
| Nakamura 2008 | 32.10 | 1.47 | 2.15 | 29.22 | 34.98 | 21.88 | 0.00 | |
| Kadaglou 2008 | 36.00 | 2.18 | 4.76 | 31.72 | 40.28 | 16.49 | 0.00 | |
| Yamada 2009 | 17.00 | 1.87 | 3.48 | 13.34 | 20.66 | 9.11 | 0.00 | |
| Kadaglou 2009* | 48.44 | 3.89 | 15.12 | 40.82 | 56.06 | 12.46 | 0.00 | |
| Kadaglou 2009** | 36.80 | 1.70 | 2.90 | 33.46 | 40.14 | 21.60 | 0.00 | |
| Kadaglou 2010^ | 32.60 | 1.98 | 3.92 | 28.72 | 36.48 | 16.47 | 0.00 | |
| Kadaglou 2010^^ | 51.40 | 3.13 | 9.77 | 45.27 | 57.53 | 16.45 | 0.00 | |
| Della-Morte 2011 | 21.90 | 1.08 | 1.18 | 19.77 | 24.03 | 20.19 | 0.00 | |
| Nohara 2013 | 16.93 | 9.37 | 87.75 | −1.43 | 35.29 | 1.81 | 0.07 | |
| | 29.02 | 3.56 | 12.68 | 22.04 | 36.00 | 8.15 | 0.00 | |

−57.60 −28.80 0.00 28.80 57.60

Echogenicity decreased     Echogenicity increased

**Figure 2.** Effects of statin on plaque echogenicity. Note: Kadaglou 2009 [17] has divided and analyzed patients in two groups: the first group (*) was on statin, but underwent contralateral carotid artery stenting (CAS), and the second group (**) was treated only with statins. Kadaglou 2010 [18] has divided and analyzed patients into two groups: the first group (^) received atorvastatin 10–20 mg, and the second group (^^) atorvastatin 80 mg.

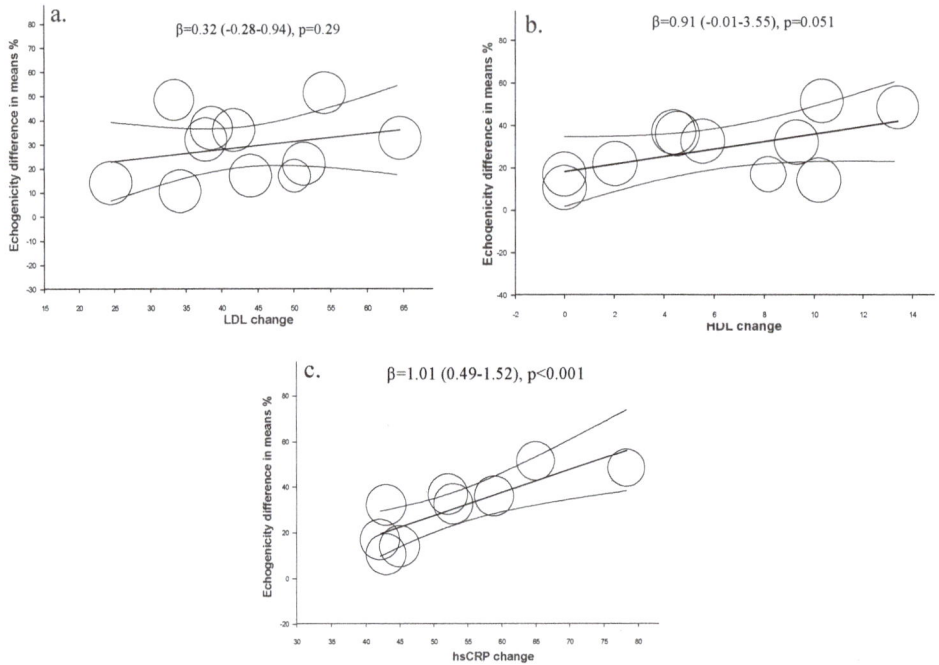

**Figure 3.** Meta-regression. Regression of LDL (**a**), HDL (**b**) and hsCRP (**c**) changes on plaque echogenicity after statin therapy.

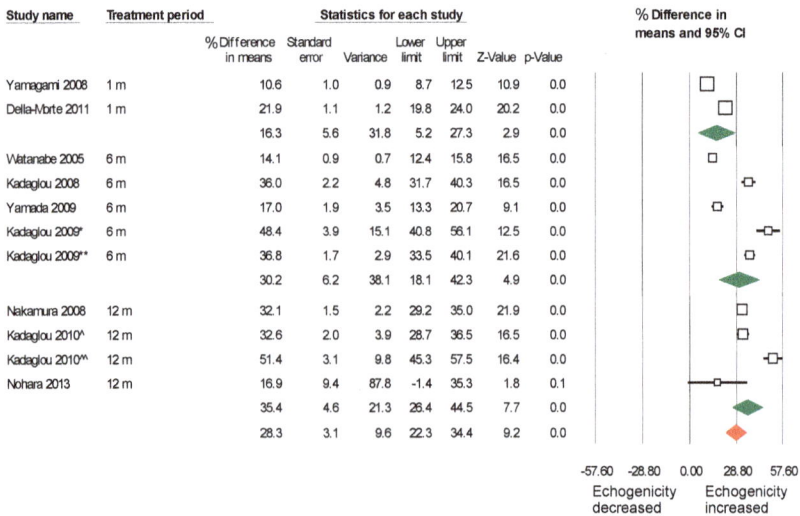

| Study name | Treatment period | Statistics for each study | | | | | | | % Difference in means and 95% CI |
|---|---|---|---|---|---|---|---|---|---|
| | | % Difference in means | Standard error | Variance | Lower limit | Upper limit | Z-Value | p-Value | |
| Yamagami 2008 | 1 m | 10.6 | 1.0 | 0.9 | 8.7 | 12.5 | 10.9 | 0.0 | |
| Della-Morte 2011 | 1 m | 21.9 | 1.1 | 1.2 | 19.8 | 24.0 | 20.2 | 0.0 | |
| | | 16.3 | 5.6 | 31.8 | 5.2 | 27.3 | 2.9 | 0.0 | |
| Watanabe 2005 | 6 m | 14.1 | 0.9 | 0.7 | 12.4 | 15.8 | 16.5 | 0.0 | |
| Kadaglou 2008 | 6 m | 36.0 | 2.2 | 4.8 | 31.7 | 40.3 | 16.5 | 0.0 | |
| Yamada 2009 | 6 m | 17.0 | 1.9 | 3.5 | 13.3 | 20.7 | 9.1 | 0.0 | |
| Kadaglou 2009* | 6 m | 48.4 | 3.9 | 15.1 | 40.8 | 56.1 | 12.5 | 0.0 | |
| Kadaglou 2009** | 6 m | 36.8 | 1.7 | 2.9 | 33.5 | 40.1 | 21.6 | 0.0 | |
| | | 30.2 | 6.2 | 38.1 | 18.1 | 42.3 | 4.9 | 0.0 | |
| Nakamura 2008 | 12 m | 32.1 | 1.5 | 2.2 | 29.2 | 35.0 | 21.9 | 0.0 | |
| Kadaglou 2010^ | 12 m | 32.6 | 2.0 | 3.9 | 28.7 | 36.5 | 16.5 | 0.0 | |
| Kadaglou 2010^^ | 12 m | 51.4 | 3.1 | 9.8 | 45.3 | 57.5 | 16.4 | 0.0 | |
| Nohara 2013 | 12 m | 16.9 | 9.4 | 87.8 | -1.4 | 35.3 | 1.8 | 0.1 | |
| | | 35.4 | 4.6 | 21.3 | 26.4 | 44.5 | 7.7 | 0.0 | |
| | | 28.3 | 3.1 | 9.6 | 22.3 | 34.4 | 9.2 | 0.0 | |

-57.60  -28.80  0.00  28.80  57.60
Echogenicity decreased    Echogenicity increased

**Figure 4.** Analysis of studies based on treatment period in months. The effect of statins on plaque echogenicity was obvious from the first month after treatment, and the effect was progressive on the following six and 12 months (m). (*) was on statin, but underwent contralateral carotid artery stenting (CAS), and (**) was treated only with statins. (^) received atorvastatin 10–20 mg, and (^^) atorvastatin 80 mg.

| Study name | | Statistics for each study | | | | | | | %Difference in means and 95% CI |
|---|---|---|---|---|---|---|---|---|---|
| | H–C | %Difference in means | Standard error | Variance | Lower limit | Upper limit | Z-Value | p-Value | |
| Watanabe 2005 | No | 14.1 | 0.9 | 0.7 | 12.4 | 15.8 | 16.5 | 0.0 | |
| Yamada 2009 | No | 17.0 | 1.9 | 3.5 | 13.3 | 20.7 | 9.1 | 0.0 | |
| Kadaglou 2009* | No | 48.4 | 3.9 | 15.1 | 40.8 | 56.1 | 12.5 | 0.0 | |
| Kadaglou 2009** | No | 36.8 | 1.7 | 2.9 | 33.5 | 40.1 | 21.6 | 0.0 | |
| | | 28.8 | 6.9 | 47.8 | 15.2 | 42.3 | 4.2 | 0.0 | |
| Yamagami 2008 | Yes | 10.6 | 1.0 | 0.9 | 8.7 | 12.5 | 10.9 | 0.0 | |
| Nakamura 2008 | Yes | 32.1 | 1.5 | 2.2 | 29.2 | 35.0 | 21.9 | 0.0 | |
| Kadaglou 2008 | Yes | 36.0 | 2.2 | 4.8 | 31.7 | 40.3 | 16.5 | 0.0 | |
| Kadaglou 2010^ | Yes | 32.6 | 2.0 | 3.9 | 28.7 | 36.5 | 16.5 | 0.0 | |
| Kadaglou 2010^^ | Yes | 51.4 | 3.1 | 9.8 | 45.3 | 57.5 | 16.4 | 0.0 | |
| Della-Morte 2011 | Yes | 21.9 | 1.1 | 1.2 | 19.8 | 24.0 | 20.2 | 0.0 | |
| Nohara 2013 | Yes | 16.9 | 9.4 | 87.8 | −1.4 | 35.3 | 1.8 | 0.1 | |
| | | 29.2 | 4.8 | 23.4 | 19.7 | 38.7 | 6.0 | 0.0 | |
| | | 29.1 | 4.0 | 15.7 | 21.3 | 36.8 | 7.3 | 0.0 | |

-57.60  -28.80  0.00  28.80  57.60

Echogenicity decreased     Echogenicity increased

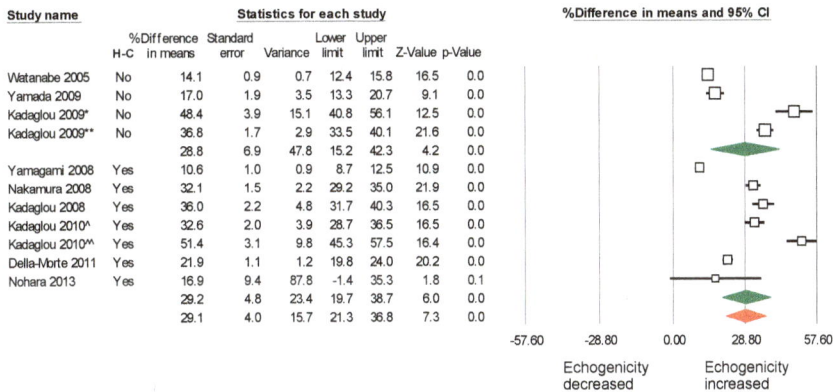

**Figure 5.** The effect of statins on plaque echogenicity was similar in hypercholesterolemic and non-hypercholesterolemic patients. H–C = hypercholesterolemia at baseline. (*) was on statin, but underwent contralateral carotid artery stenting (CAS), and (**) was treated only with statins. (^) received atorvastatin 10–20 mg, and (^^) atorvastatin 80 mg.

## 2.4. Assessment of Potential Publication Bias

No publication bias was noted in our analyses (Table 3).

**Table 3.** Heterogeneity and publication bias measures.

| | Test of Heterogeneity | Publication Bias (Begg and Mazumdar Rank Correlation) | | |
|---|---|---|---|---|
| | $I^2$ | Kendell Tau | Test Statistic Z | *p*-Value |
| Echogenicity | 92.1 | 0.34 | 1.41 | 0.16 |
| LDL | 81.2 | −0.33 | 1.25 | 0.12 |
| HDL | 98.1 | 0.05 | 0.20 | 0.41 |
| hsCRP | 0 | 0 | 0 | 1.0 |

## 3. Discussion

Atherosclerosis is a long-lasting pathology with well-established stages starting with mild wall thickness and ending with complete fibrosis and calcification [1]. Along the course of the disease, plaques are formed mainly based on a lipid core, and they too are subject to structural and functional changes over time [21]. Increasing plaque area and volume while parts of it might be healing and developing fibrosis and spotty calcification characterize active plaque pathology. Soft plaques can easily be identified by various imaging techniques, including ultrasound, which has shown that echolucent plaques are the ones associated with potential complications [22,23], including even micro-emboli [24] and strokes [25].

Statins are well-established treatment for atherosclerosis and its complications. Their beneficial clinical effect, in the form of reduced events, e.g., stroke and coronary syndromes, is through lowering LDL-cholesterol levels [2] and their anti-inflammatory effect [26]; the two mechanisms result in volume reduction and plaque stabilization, as shown by increased plaque echogenicity. Our results support that pathway; however, in addition, they show that the increase in plaque echogenicity seem to be independent of the changes in intima-media thickness, plaque area or volume. These findings suggest that the statins-related increase in plaque echogenicity represents an early effect that could be used for monitoring individual patient's response to therapy. Indeed, evidence exists that the effect of statins on plaque volume appears later, after changes in echogenicity. This is not a unique feature of just carotid disease, but also coronary plaques, which have been shown to demonstrate quantitative regression in volume after 19 months of statins therapy [8]. In our meta-regression analysis, the effect of statins on plaque echogenicity was also independent of changes in LDL and HDL levels, but was related to changes of hsCRP levels. In addition, the increased echogenicity was higher in patients treated for a longer period, again irrespective of the cholesterol level at baseline.

Current data indicate that higher statins doses (atorvastatin 80 mg) have a more potent effect on increasing plaque echogenicity compared to smaller doses (atorvastatin 20 mg). These findings mirror those we previously showed in coronary artery disease, with higher statins doses resulting in a faster rate of coronary calcification compared to smaller doses [27]. Furthermore, the effect of statins duration on plaque features mirrors what we recently reported in the coronary circulation [28].

In addition to the beneficial clinical effect of statins on plaque features and potential stability, our analysis shows that ultrasound carotid imaging plays a pivotal role in early and potential continuous monitoring of such an effect. Atherosclerotic plaques can be detected and their features studied by CT and MRI scanning; however, the two techniques are known for their significant limitations, particularly radiation in the former and claustrophobia in the latter, adding to their higher cost compared with ultrasound. With carotid ultrasound free of those limitations, its accuracy in studying plaque echogenicity makes it emerge as a unique non-invasive image modality ideal for early identification of the disease and accurate monitoring of its progress and response to treatment. Ultrasound, on the other hand, has the limitation of being more time consuming with a potential inter-observer variability, particularly so when using plaque characterization methods that are, to some extent, superior to the conventionally-used intima-media thickness for early disease detection [29]. Finally, statins are usually well-tolerated medications with few adverse effects, including myopathy and elevation of liver enzymes. Rhabdomyolysis is a rare related complication [30]. Likewise, liver dysfunction is very rare, which was a reason for the FDA to remove the old recommendation of routine monitoring of liver enzymes [31]. As for diabetes mellitus, two comprehensive meta-analyses [32,33] have shown a slightly increased risk of diabetes development in subjects on statin therapy; however, the risk is low both in absolute terms and when compared with the reduction in cardiovascular events.

Study limitations: A systematic review based on relevant key words might have missed some relevant publications, but the search was checked by two investigators blinded to each other's means of search. We used only publications in the English language; other relevant ones in different languages might have been missed. Another limitation was the low number of patients included in studies and that different studies have used different statins of variable dosages. Patients were followed up for different periods of time, and in most studies, statins dosage was ranged, thus limiting us to assessing the dose effect on plaque echogenicity using the meta-regression analysis. This is the nature of searching various studies of different designs.

Clinical implications: Our results support the use of carotid ultrasound analysis of plaque features and echogenicity as a marker of plaque stability in response to statins therapy. These changes are independent of plaque area or volume, suggesting that they might reflect an early effect before anatomical response and plaque shrinking is detected. In addition, the effects of statins on the plaque were progressive and independent of baseline cholesterol levels. Applying this method in monitoring individuals at high risk for vascular events might support treatment adjustment for targeting better clinical outcome.

## 4. Experimental Section

### 4.1. Methods

The methodology for this study was based on the Preferred Reporting Items for Systematic Reviews and Meta-Analyses statement [34].

### 4.2. Information Search and Data Collection

Up to April, 2015, we systematically searched electronic databases (PubMed, MEDLINE, EMBASE and Cochrane Center Register) for studies evaluating the effect of statins on carotid plaque echogenicity. The search terms used were: "carotid atherosclerosis", "carotid plaque", "ultrasound" "statins", "HMG-CoA reductase inhibitors" and "lipid-lowering drugs", in various combinations. Two researchers (Pranvera Ibrahimi and Fisnik Jashari), independent of each other, performed the literature search, study selection and data extraction. There was no time, language or publication limit in the literature search. The selected reports were manually searched, and relevant publications, obtained from the reference lists, were retrieved.

### 4.3. Study Eligibility Criteria

Clinical studies that reported results on the effect of statin therapy on the plaque echogenicity (GSM, IBS) evaluated by duplex ultrasound were eligible. Specific inclusion criteria were: (1) observational, non-randomized or randomized studies that explored the effect on statin treatment either as primary or secondary cardiovascular disease prevention; (2) ultrasound of the carotid arteries before and at least once at a follow-up of at least one month; (3) English language articles; (4) studies with $\geq$15 subjects; and (5) ultrasound-based characterization of carotid artery plaque composition. All other studies that used different imaging techniques (e.g., MRI, CT, IVUS, PET) and those that used plaque features other than echogenicity (volume, degree of stenosis, ulceration, neovascularization) as a target for monitoring statin therapy were excluded. We have performed a quality score of the retrieved studies utilizing the methodological index for the non-randomized studies (MINORS) [35]. Studies that scored over 20 out of 24 (or 14 out of the 16 for those non-comparative, but rather solely observational) were considered of adequate quality. In the meta-analysis, we included studies that specified duration of the study and presented plaque echogenicity means and standard deviations prior to and during (or at the completion of) the intervention or the percent change in plaque echogenicity before and during intervention.

### 4.4. Statistical Analyses

For the plaque echogenicity analysis, the treatment effects of interest were the differences in the extent of changes in echogenicity (GSM or IBS), low-density lipoprotein cholesterol (LDL), high-density lipoprotein (HDL) and high sensitivity C-reactive protein (hsCRP) before and after treatment. Because of the significant variation in study size, length and follow-up, as well as patient's characteristics, we have used random-effects. Heterogeneity was measured using $I^2$ statistics. We performed analyses within each imaging group stratified by pre- *vs.* post-treatment. All analyses were conducted using Comprehensive Meta Analysis Version 3 software (Biostat inc., Englewood, NJ, USA).

### 5. Conclusions

Statins therapy is associated with a favorable increase of carotid plaque echogenicity, even after one month of treatment. This effect is independent of changes in plaque morphology and, furthermore, is more profound using higher doses of statins. Finally, the effect of statins seems to be related to the decrease of hsCRP from the baseline rather than dyslipidemia.

**Supplementary Materials:** Supplementary materials can be found at http://www.mdpi.com/1422-0067/16/05/10734/s1.

**Author Contributions:** Pranvera Ibrahimi conceived the study and wrote the manuscript. Pranvera Ibrahimi and Fisnik Jashari performed the literature search, study selection and data extraction and analysis. Gani Bajraktari, Per Wester and Michael Y. Henein critically revised the manuscript.

**Conflicts of Interest:** The authors declare no conflicts of interest.

### References

1. Ibrahimi, P.; Jashari, F.; Nicoll, R.; Bajraktari, G.; Wester, P.; Henein, M.Y. Coronary and carotid atherosclerosis: How useful is the imaging? *Atherosclerosis* **2013**, *231*, 323–333. [CrossRef] [PubMed]

2. Amarenco, P.; Labreuche, J.; Lavallee, P.; Touboul, P.J. Statins in stroke prevention and carotid atherosclerosis: Systematic review and up-to-date meta-analysis. *Stroke J. Cereb. Circ.* **2004**, *35*, 2902–2909. [CrossRef]
3. Martinez-Sanchez, P.; Serena, J.; Alexandrov, A.V.; Fuentes, B.; Fernandez-Dominguez, J.; Diez-Tejedor, E. Update on ultrasound techniques for the diagnosis of cerebral ischemia. *Cerebrovasc. Dis.* **2009**, *27*, 9–18. [PubMed]
4. Seeger, J.M.; Barratt, E.; Lawson, G.A.; Klingman, N. The relationship between carotid plaque composition and neurologic symptoms. *J. Surg. Res.* **1987**, *43*, 78–85. [CrossRef] [PubMed]
5. Gronholdt, M.L.; Nordestgaard, B.G.; Bentzon, J.; Wiebe, B.M.; Zhou, J.; Falk, E.; Sillesen, H. Macrophages are associated with lipid-rich carotid artery plaques, echolucency on B-mode imaging, and elevated plasma lipid levels. *J. Vasc. Surg.* **2002**, *35*, 137–145. [PubMed]
6. El-Barghouty, N.M.; Levine, T.; Ladva, S.; Flanagan, A.; Nicolaides, A. Histological verification of computerised carotid plaque characterisation. *Eur. J. Vasc. Endovasc. Surg.* **1996**, *11*, 414–416. [CrossRef] [PubMed]
7. Makris, G.C.; Lavida, A.; Nicolaides, A.N.; Geroulakos, G. The effect of statins on carotid plaque morphology: A LDL-associated action or one more pleiotropic effect of statins? *Atherosclerosis* **2010**, *213*, 8–20. [CrossRef] [PubMed]
8. Noyes, A.M.; Thompson, P.D. A systematic review of the time course of atherosclerotic plaque regression. *Atherosclerosis* **2014**, *234*, 75–84. [CrossRef] [PubMed]
9. Lind, L.P.S.; den Ruijter, H.M.; Palmer, M.K.; Grobbee, D.E.; Crouse, J.R., 3rd; O'Leary, D.H.; Evans, G.W.; Raichlen, J.S.; Bots, M.L. Effect of rosuvastatin on the echolucency of the common carotid intima-media in low-risk individuals: The METEOR trial. *J. Am. Soc. Echocardiogr.* **2012**, *25*, 1120–1127. [CrossRef] [PubMed]
10. Yamagishi, T.; Kato, M.; Koiwa, Y.; Omata, K.; Hasegawa, H.; Kanai, H. Evaluation of plaque stabilization by fluvastatin with carotid intima-medial elasticity measured by a transcutaneous ultrasonic-based tissue characterization system. *J. Atheroscler. Thromb.* **2009**, *16*, 662–673. [CrossRef] [PubMed]
11. Stivali, G.; Cerroni, F.; Bianco, P.; Fiaschetti, P.; Cianci, R. Images in cardiovascular medicine. Carotid plaque reduction after medical treatment. *Circulation* **2005**, *112*, e276–e277. [CrossRef] [PubMed]
12. Watanabe, K.; Sugiyama, S.; Kugiyama, K.; Honda, O.; Fukushima, H.; Koga, H.; Horibata, Y.; Hirai, T.; Sakamoto, T.; Yoshimura, M.; *et al.* Stabilization of carotid atheroma assessed by quantitative ultrasound analysis in nonhypercholesterolemic patients with coronary artery disease. *J. Am. Coll. Cardiol.* **2005**, *46*, 2022–2030. [CrossRef] [PubMed]
13. Yamagami, H.; Sakaguchi, M.; Furukado, S.; Hoshi, T.; Abe, Y.; Hougaku, H.; Hori, M.; Kitagawa, K. Statin therapy increases carotid plaque echogenicity in hypercholesterolemic patients. *Ultrasound Med. Biol.* **2008**, *34*, 1353–1359. [CrossRef] [PubMed]
14. Nakamura, T.; Obata, J.E.; Kitta, Y.; Takano, H.; Kobayashi, T.; Fujioka, D.; Saito, Y.; Kodama, Y.; Kawabata, K.; Mende, A.; *et al.* Rapid stabilization of vulnerable carotid plaque within 1 month of pitavastatin treatment in patients with acute coronary syndrome. *J. Cardiovasc. Pharmacol.* **2008**, *51*, 365–371. [CrossRef] [PubMed]
15. Kadoglou, N.P.; Gerasimidis, T.; Moumtzouoglou, A.; Kapelouzou, A.; Sailer, N.; Fotiadis, G.; Vitta, I.; Katinios, A.; Kougias, P.; Bandios, S.; *et al.* Intensive lipid-lowering therapy ameliorates novel calcification markers and GSM score in patients with carotid stenosis. *Eur. J. Vasc. Endovasc. Surg.* **2008**, *35*, 661–668. [CrossRef] [PubMed]
16. Yamada, K.; Yoshimura, S.; Kawasaki, M.; Enomoto, Y.; Asano, T.; Minatoguchi, S.; Iwama, T. Effects of atorvastatin on carotid atherosclerotic plaques: A randomized trial for quantitative tissue characterization of carotid atherosclerotic plaques with integrated backscatter ultrasound. *Cerebrovasc. Dis.* **2009**, *28*, 417–424. [CrossRef] [PubMed]
17. Kadoglou, N.P.; Gerasimidis, T.; Kapelouzou, A.; Moumtzouoglou, A.; Avgerinos, E.D.; Kakisis, J.D.; Karayannacos, P.E.; Liapis, C.D. Beneficial changes of serum calcification markers and contralateral carotid plaques echogenicity after combined carotid artery stenting plus intensive lipid-lowering therapy in patients with bilateral carotid stenosis. *Eur. J. Vasc. Endovasc. Surg.* **2010**, *39*, 258–265. [CrossRef] [PubMed]
18. Kadoglou, N.P.; Moumtzouoglou, A.; Kapelouzou, A.; Gerasimidis, T.; Liapis, C.D. Aggressive lipid-lowering is more effective than moderate lipid-lowering treatment in carotid plaque stabilization. *J. Vasc. Surg.* **2010**, *55*, 114–121. [CrossRef]
19. Della-Morte, D.; Moussa, I.; Elkind, M.S.; Sacco, R.L.; Rundek, T. The short-term effect of atorvastatin on carotid plaque morphology assessed by computer-assisted gray-scale densitometry: A pilot study. *Neurol. Res.* **2011**, *33*, 991–994. [CrossRef] [PubMed]

20. Nohara, R.; Daida, H.; Hata, M.; Kaku, K.; Kawamori, R.; Kishimoto, J.; Kurabayashi, M.; Masuda, I.; Sakuma, I.; Yamazaki, T.; *et al.* Effect of long-term intensive lipid-lowering therapy with rosuvastatin on progression of carotid intima-media thickness—Justification for atherosclerosis regression treatment (JART) extension study. *Circ. J.* **2013**, *77*, 1526–1533. [CrossRef] [PubMed]

21. Jashari, F.; Ibrahimi, P.; Nicoll, R.; Bajraktari, G.; Wester, P.; Henein, M.Y. Coronary and carotid atherosclerosis: Similarities and differences. *Atherosclerosis* **2013**, *227*, 193–200. [CrossRef] [PubMed]

22. Nicolaides, A.N.; Kyriacou, E.; Griffin, M.; Sabetai, M.; Thomas, D.J.; Tegos, T.; Geroulakos, G.; Labropoulos, N.; Doré, C.J.; Morris, T.P.; *et al.* Asymptomatic internal carotid artery stenosis and cerebrovascular risk stratification. *J. Vasc. Surg.* **2010**, *52*, 1486–1496. [CrossRef] [PubMed]

23. Ibrahimi, P.; Jashari, F.; Johansson, E.; Gronlund, C.; Bajraktari, G.; Wester, P.; Henein, M.Y. Vulnerable plaques in the contralateral carotid arteries in symptomatic patients: A detailed ultrasound analysis. *Atherosclerosis* **2014**, *235*, 526–531. [CrossRef] [PubMed]

24. Sztajzel, R.; Momjian-Mayor, I.; Comelli, M.; Momjian, S. Correlation of cerebrovascular symptoms and microembolic signals with the stratified gray-scale median analysis and color mapping of the carotid plaque. *Stroke J. Cereb. Circ.* **2006**, *37*, 824–829. [CrossRef]

25. Aburahma, A.F.; Thiele, S.P.; Wulu, J.T., Jr. Prospective controlled study of the natural history of asymptomatic 60% to 69% carotid stenosis according to ultrasonic plaque morphology. *J. Vasc. Surg.* **2002**, *36*, 437–442. [CrossRef] [PubMed]

26. Takemoto, M.; Liao, J.K. Pleiotropic effects of 3-hydroxy-3-methylglutaryl coenzyme a reductase inhibitors. *Arterioscler. Thromb. Vasc. Biol.* **2001**, *21*, 1712–1719. [CrossRef] [PubMed]

27. Schmermund, A.A.S.; Budde, T.; Buziashvili, Y.; Förster, A.; Friedrich, G.; Henein, M.; Kerkhoff, G.; Knollmann, F.; Kukharchuk, V.; Lahiri, A.; *et al.* Effect of intensive *versus* standard lipid-lowering treatment with atorvastatin on the progression of calcified coronary atherosclerosis over 12 months: A multicenter, randomized, double-blind trial. *Circulation* **2006**, *113*, 427–437. [CrossRef] [PubMed]

28. Henein, M.; Granåsen, G.; Wiklund, U.; Schmermund, A.; Guerci, A.; Erbel, R.; Raggi, P. High dose and long-term statin therapy accelerate coronary artery calcification. *Int. J. Cardiol.* **2015**, *184*, 581–586. [CrossRef] [PubMed]

29. Coll, B.; Feinstein, S.B. Carotid intima-media thickness measurements: Techniques and clinical relevance. *Curr. Atheroscler. Rep.* **2008**, *10*, 444–450. [CrossRef] [PubMed]

30. Silva, M.A.; Swanson, A.C.; Gandhi, P.J.; Tataronis, G.R. Statin-related adverse events: A meta-analysis. *Clin. Ther.* **2006**, *28*, 26–35. [CrossRef] [PubMed]

31. Jukema, J.W.; Cannon, C.P.; de Craen, A.J.; Westendorp, R.G.; Trompet, S. The controversies of statin therapy: Weighing the evidence. *J. Am. Coll. Cardiol.* **2012**, *60*, 875–881. [CrossRef] [PubMed]

32. Sattar, N.; Preiss, D.; Murray, H.M.; Welsh, P.; Buckley, B.M.; de Craen, A.J.; Seshasai, S.R.; McMurray, J.J.; Freeman, D.J.; Jukema, J.W.; *et al.* Statins and risk of incident diabetes: A collaborative meta-analysis of randomised statin trials. *Lancet* **2010**, *375*, 735–742. [CrossRef] [PubMed]

33. Preiss, D.; Seshasai, S.R.; Welsh, P.; Murphy, S.A.; Ho, J.E.; Waters, D.D.; DeMicco, D.A.; Barter, P.; Cannon, C.P.; Sabatine, M.S.; *et al.* Risk of incident diabetes with intensive-dose compared with moderate-dose statin therapy: A meta-analysis. *JAMA* **2011**, *305*, 2556–2564. [CrossRef] [PubMed]

34. Liberati, A.A.D.; Tetzlaff, J.; Mulrow, C.; Gøtzsche, P.C.; Ioannidis, J.P.; Clarke, M.; Devereaux, P.J.; Kleijnen, J.; Moher, D. The PRISMA statement for reporting systematic reviews and meta-analyses of studies that evaluate health care interventions: Explanation and elaboration. *PLoS Med.* **2009**, *6*, 1–28. [CrossRef]

35. Slim, K.; Nini, E.; Forestier, D.; Kwiatkowski, F.; Panis, Y.; Chipponi, J. Methodological index for non-randomized studies (*MINORS*): Development and validation of a new instrument. *ANZ J. Surg.* **2003**, *73*, 712–716. [CrossRef] [PubMed]

International Journal of
*Molecular Sciences*

MDPI

*Review*

# Vasa Vasorum in Atherosclerosis and Clinical Significance

Junyan Xu [1,†], Xiaotong Lu [1,†] and Guo-Ping Shi [1,2,*]

1   Second Clinical Medical College, Zhujiang Hospital and Southern Medical University, Guangzhou 510280, China; junyanxu_zj@126.com (J.X.); lxt0115@126.com (X.L.)
2   Department of Medicine, Brigham and Women's Hospital and Harvard Medical School, Boston, MA 02115, USA
*   Correspondence: gshi@rics.bwh.harvard.edu; Tel.: +1-617-525-4358; Fax: +1-617-525-4380
†   These authors contributed equally to this work.

Academic Editor: Michael Henein
Received: 9 April 2015; Accepted: 11 May 2015; Published: 20 May 2015

**Abstract:** Atherosclerosis is a chronic inflammatory disease that leads to several acute cardiovascular complications with poor prognosis. For decades, the role of the adventitial vasa vasorum (VV) in the initiation and progression of atherosclerosis has received broad attention. The presence of VV neovascularization precedes the apparent symptoms of clinical atherosclerosis. VV also mediates inflammatory cell infiltration, intimal thickening, intraplaque hemorrhage, and subsequent atherothrombosis that results in stroke or myocardial infarction. Intraplaque neovessels originating from VV can be immature and hence susceptible to leakage, and are thus regarded as the leading cause of intraplaque hemorrhage. Evidence supports VV as a new surrogate target of atherosclerosis evaluation and treatment. This review provides an overview into the relationship between VV and atherosclerosis, including the anatomy and function of VV, the stimuli of VV neovascularization, and the available underlying mechanisms that lead to poor prognosis. We also summarize translational researches on VV imaging modalities and potential therapies that target VV neovascularization or its stimuli.

**Keywords:** atherosclerosis; vasa vasorum; neovascularization; mechanism; imaging modality; angiogenic therapy

## 1. Introduction

Atherosclerosis is a systemic inflammatory disease that associates with several acute cardiovascular complications triggered by atherosclerotic plaque rupture, which primarily manifests as stroke and myocardial infarction. It remains the leading cause of morbidity and mortality worldwide [1]. Considering its poor prognosis, understanding the pathophysiology of atherosclerosis and exploring potential means to discover populations at risk as well as preventing its progression remain of significant importance. For decades, postmortem evaluations have concluded that the main characteristics of rupture-prone vulnerable plaques include a thin fibrous cap, high lipid content, increased numbers of inflammatory cells, and extensive adventitial and intimal neovascularization [2–5]. Though much emphasis has been placed on intimal accumulation of lipids and inflammatory cells, recent research suggests that the adventitia vasa vasorum (VV) also plays a critical role in transforming advanced but stable lesions into vulnerable plaques at risk for rupture.

VV are defined as small blood vessels that supply or drain the walls of larger arteries and veins, delivering nutrients and oxygen as well as removing systemic "waste" products [6]. The association between VV and atherosclerotic plaque formation was first reported in 1876 by Koster [7]. Later in the 1930s, the rich vascular channels surrounding and penetrating atherosclerotic lesions, namely

VV, were suspected as the source of the plaque hemorrhages for the first time [8,9]. However, the role of VV in atherosclerosis did not attract sufficient attention until half a century later when it was hypothesized that the adventitial VV of coronary arteries allowed atherosclerotic plaques to develop beyond a critical thickness by supplying oxygen and nutrients to the core of the lesions [10]. Since then, numerous studies have demonstrated the relationship between VV neovascularization and atherogenic processes. Retrospective studies on autopsies derived from humans noted that VV density was significantly increased in plaques categorized as vulnerable and prone to rupture as was those in hypercholesterolemic animal models [11,12]. In humans, it was reported that more than 80% of VV neovascularization in coronary atherosclerotic plaques had weak integrity, resulting in leakage and subsequent plaque hemorrhage [13,14]. Increased VV also associated with intimal thickening and endothelial dysfunction in animal models, and these effects could be blocked with angiogenic inhibitors [15,16]. Growing evidence supports that vascular inflammation, a crucial factor in the process of atherosclerosis, is initiated in the adventitia, and extensive inflammatory cell infiltration has been observed in the adventitial neovascular network [17]. The critical role of VV in the process of atherosclerosis has been established and identified as an independent predictor of intraplaque hemorrhage and plaque rupture. As a consequence, imaging modalities were developed to visualize VV neovascularization in the early stage of atherosclerosis. Therapies targeted to VV have emerged as a new approach for the treatment of atherosclerosis.

## 2. Structure and Function of Vasa Vasorum

### 2.1. Structure

Human VV occurs as early as the first week of gestation under an X-ray microscope, and along with the lumen blood supplement, nourishes the vessel wall [18]. VV exists mainly in the adventitial and outer medial layers of the blood vessels with more than 29 medial lamellar units or 0.5 mm lumen diameters [19]. Normal vessels in mice and intramyocardial vessels in humans do not contain VV [6].

Three different types of VV have been identified in bovine aortic walls: the VV externae (VVE), the VV internae (VVI), and the venous VV (VVV). VVE originate from major branches and the VVI originate from the main lumen of the aorta. The VVV drain the arterial wall into companion veins [20]. Direct visualizations of VV in porcine coronary arteries confirmed the coexistence of VVE, VVI, and VVV [21]. VV are often used to refer to VVE in literature because more than 96% of the newly formed microvessels in atherosclerosis sprout from VV in the adventitia and only a small part extend from the vessel lumen [22,23].

High-resolution micro-computed tomography (micro-CT) displayed a more precise VV structure. This technique demonstrated that VV originate from the coronary artery branch and run longitudinally along the vessel wall (first-order VV) and branch to form a circumferential plexus around the main coronary lumen (second-order VV) (Figure 1). Normal hearts had significantly greater first-order than second-order VV density (ratio 3:2), while the second-order vessel density was twofold greater than the first in hypercholesterolemic hearts [12]. Furthermore, while VV branching architecture in non-diseased porcine vasculature showed a dichotomous tree structure similar to the vasculature of systemic circulation structure, VV in diseased arteries presented many more disorganized images [21].

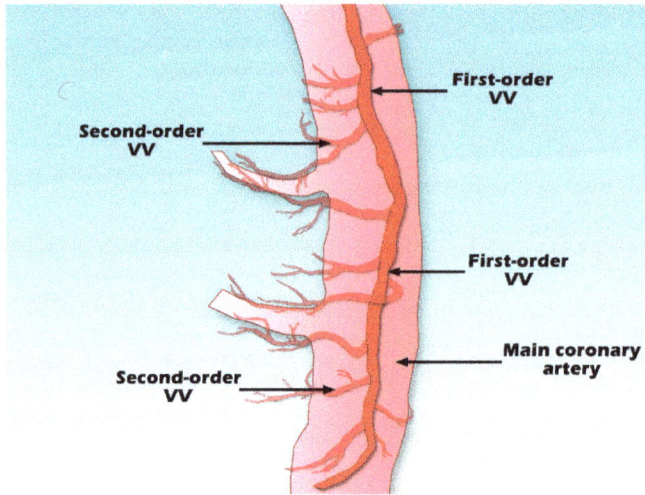

**Figure 1.** Scheme of first-order VV, second-order VV, and main coronary artery.

In atherosclerosis, previous work demonstrated that VV neovessels were immature, irregular, fragile, and prone to extravasation, particularly among those close to the atherosclerotic plaques [24,25]. Plaque neovessels showed poor coverage with mural cells and compromised structural integrity under electron microscopy, including abnormal endothelial cells (ECs), membrane blebs, intracytoplasmic vacuoles, open EC-EC junctions, and basement membrane (BM) detachment. When the mural cells are absent in either normal or atherosclerotic arteries, compromised structural integrity may cause leakage of the intraplaque microvessels [14]. Indeed, insufficient smooth muscle cell (SMC) coating in aberrant intraplaque vessels was observed in symptomatic patients when compared with asymptomatic patients [13].

*2.2. Function*

Rather than merely existing as a structural network, VV are recognized as functional end arteries that exist throughout the body [26,27]. Previous experiments demonstrate their ability to provide oxygen and nourishment to the outer third of the vascular media. Using probes to detect the diffusion of oxygen from the luminal or abluminal side of the canine femoral artery wall, the oxygen level was highest in the outer layers of the vessel wall and decreased as the probe gradually approached the lumen [28]. A more precise experiment by direct measurement of the oxygenation of the arterial wall showed the lowest level of oxygen tension of approximately 10 mmHg at about 300 μm from the lumen [29]. The varying oxygen levels between the inner and outer layers indicate that VV are the primary source of oxygen supply to the adventitia and outer media.

Along with their blood perfusion, the active exchange between VV and parent vessels meets the nutritional demands of the vessel wall and removes the "waste" products, either produced by intramural cells or introduced by diffusional transport through the luminal endothelium [6]. It is worth mentioning that VV also undertakes lipid transportation into the parent vessel wall in rabbit aorta [30], underlying a role of VV in progression of lipid core enlargement. It was also speculated that VV can regulate their own tone and vascular perfusion because the proximal VV form a regularly layered vascular structure of ECs, vascular smooth muscle cells (VSMCs), and surrounding connective tissue [31]. VV dissected from porcine and canine aorta are able to contract and dilate when exposed to endothelin-1 (ET-1) and several other vasodilators [32,33]. VV also participates in restoring the injured vessels [34] and thus are involved in several pathological conditions, including atherosclerosis, abdominal aortic aneurysm (AAA), and pulmonary artery hypertension. In fact, VV were described as "rich vascular channels surrounding and penetrating sclerotic lesions" via injecting India ink into the coronary artery wall, suggesting a role of VV in promoting atherogenesis [35]. Prior evidence suggests that VV neovascularization is a simple reaction that helps meet the demands of the intima and inner layers of the media in response to decreased oxygen concentration and malnourishment. This process gravitates towards preserving the integrity of atherosclerotic plaques. In advanced stages of atherosclerosis, however, the newly formed microvessels become important channels for various inflammatory cells and enable cellular migration to the intima, therefore becoming detrimental to the plaque integrity [36]. Plaques with a high density of neovessels are at a higher risk of hemorrhage, expansion, atherothrombosis, and rupture. As the channels conveying erythrocytes, lipids and inflammatory cells, VV neovessels seem to be multifaceted in atherosclerosis.

## 3. Stimuli of Vasa Vasorum Neovascularization

VV neovascularization is essential to atherogenesis. Metal cannulae-tied transparent visualization of human heart coronary arteries established an association of VV in atherosclerosis progression and associated sequelae [10]. High-fat diet-fed monkeys with atherosclerosis experienced blood flow from VV to the intima-media that was 10 times greater than monkeys on a normal diet. Reduced blood flow in VV directly associated with atherosclerosic lesion regression in monkeys [37]. Since these earlier pioneering studies, accumulating evidence has proven the role of VV neovascularization in both the initiation and progression of atherosclerosis, although many observations in VV neovascularization initiation and stimulation still remain incompletely understood. The extent and distribution of the ectopic neovascularization within the arterial wall depend on a number of physiological and pathological factors.

### 3.1. Hypoxia

Hypoxia and its role in the progression of VV neovascularization have been broadly studied in both cardiovascular and pulmonary arteries. Atherosclerosis, AAA, pulmonary artery hypertension, and many other systemic/pulmonary vascular diseases closely relate to hypoxia and its secondary complications [38,39]. Many factors contribute to the generation of a hypoxia environment.

Intimal thickening is an immediate element that causes hypoxia and the most prominent feature of atherosclerosis throughout the lesion initiation, progression, and ultimate rupture [40,41]. Insufficient arterial oxygen supply due to intimal thickening may shorten oxygen and nutrient diffusion distance between the deep layer of the intima and the luminal surface, resulting in regional hypoxia, ischemic injury of the inner arterial wall, and the eventual induction of VV neovascularization [42]. Increased intimal thickening and plaque growth will enhance the area and degree of hypoxia. Meanwhile, decreased oxygen supply and nourishment of the vessel wall can also originate from the changes VV undergo. As functional end arteries, VV are especially vulnerable to hypoxia [43]. An important consequence of the architecture of VV is that the blood supply cannot reach far enough from the adventitia into the media due to the pressure within the arterial wall, according to Lamé's Law [44]. Several risk factors affect VV blood flow and lesion hypoxia. Aging, compression, or hypertension inevitably increase the arterial vessel tensile force and interfere with VV blood circulation to the inner layers, leading to low oxygen concentration in VV capillaries [6]. Smoking or nicotine inhalation is another known risk factor that contracts the peripheral arteries, thereby reducing the peripheral blood flow as well as blood flow in VV. All these risk factors are common among patients with atherosclerosis. Therefore, poor blood supply from VV induces a hypoxia environment in the intima and part of the media. An increase of lesion oxygen consumption is another risk factor of hypoxia. An active metabolic process within the cholesterol-containing macrophages and foam cells in the lesions contribute to hypoxia by increasing oxygen consumption. Björnrheden *et al.* [45] demonstrated that oxygen consumption was increased in foam cells isolated from the aortic intima-media in atherosclerotic rabbits. Further experiments confirmed that hypoxia correlated with the presence of macrophages and angiogenesis in advanced human carotid plaques, suggesting that hypoxia depended more on the high metabolic demand of lesion inflammatory cells than the vessel wall thickness [46]. Therefore, impaired oxygen diffusion capacity due to intimal thickness, reduced VV blood circulation, and increased oxygen consumption in atherosclerosis together generate an oxygen-insufficient microenvironment.

As a compensatory reaction to the hypoxia, VV tend to sprout across the arterial wall toward the vessel lumen to support the inner layers, called lumenward, according to Zemplenyi *et al.* [42]. In balloon-injured arteries, in which the oxygen supply in the arterial wall is impaired, newly formed VV may compensate the shortage of oxygen supply [46,47]. Spatial VV contents were increased from the dense microvessel network in the adventitia and extended to plaques in the presence of atherosclerosis [48,49]. There was an inverse correlation between low VV contents and decreased oxygenation (*i.e.*, increased expression of hypoxia-inducible factor (HIF)-1α) and increased oxidative stress (*i.e.*, increased superoxide production) within the coronary vessels in atherosclerotic pigs [50]. Additionally, prior studies suggest the signaling pathway of hypoxia-induced angiogenesis is mediated partially via regulating the production of HIF and its downstream factors. HIF is a key regulator of atherosclerosis. It affects multiple pathological events in atherogenesis, including foam cell formation, cellular proliferation, plaque ulceration, lesion hemorrhage, and rupture [51]. HIF-1 is considered the most interrelated factor of the angiogenic process in atherosclerosis. It is a basic helix-loop-helix heterodimer containing HIF-1α and HIF-1β that activates the transcription of hypoxia-inducible genes, such as erythropoietin, vascular endothelial growth factor (VEGF), E26 transformation-specific-1 (Ets-1), heme oxygenase-1 (HO1), inducible nitric oxide synthase (iNOS), and the glycolytic enzyme aldolase A [52,53]. Among them, VEGF and Ets-1 are important regulators of hypoxia-induced angiogenesis via regulating the biology of ECs.

The VEGF family has potent mitogenic and promigratory actions specific for ECs, leading to the conversion to angiogenic phenotypes, which link tightly to neovessel development in both physiological and pathophysiological conditions [54]. VEGF-A is the major subtype of the VEGF family and plays a pivotal role in the induction of neovessels through binding and primarily activating the VEGF receptor type-2 (VEGFR-2, also called KDR or Flk-1) [55]. VEGF acts as a hypoxia-inducible factor [56]. In hypertensive rat aorta, the expressions of VEGF and HIF-1α change concurrently and associate with arterial VV formation. Such a relationship was also reported in atherosclerotic lesions [57]. VEGFR-3 is a receptor for VEGF-C and VEGF-D [58,59]. Although the current consensus is that VEGFR-3 is expressed restrictedly in lymphatic vessels and can induce lymphatic EC proliferation [60,61], an underlying relationship between VEGFR-3 and VV has also been proposed. Immunostaining demonstrated the existence of VEGFR-3 in the VV of the adult aorta and other fenestrated blood vessels in human tissues, such as bone marrow, splenic and hepatic sinusoids, kidney glomeruli, and endocrine glands [62]. In human aortas, VEGF-D is constitutively expressed in normal, fatty streak, and atherosclerotic lesions, as confirmed by both immunostainings and *in situ* hybridization. In atherosclerotic lesions, VEGFR-2 is expressed in SMCs and ECs from the intima, media, and adventitia, whereas VEGFR-3 mainly exists in ECs from the adventitia, which is rich in neovascularization. The VEGF-D/VEGFR-2 cascade was probably the prominent trigger in promoting atherosclerotic plaque neovascularization [63]. From a different study, immunostaining located VEGFR-2 to the luminal endothelium in human atherosclerotic lesions, whereas VEGFR-3 was expressed in SMCs from the media and adventitia, but not in the luminal endothelium [64]. Two studies showed different expression patterns of VEGFR-3 in human atherosclerosis lesions, suggesting that VEGFR-3 has functions other than neovascularization. In a mouse atherosclerosis model, transgenic expression of soluble VEGFR-3 or its mutant did not affect atherosclerotic lesion adventitial VV density, but nearly completely blocked the lymphoid vessel growth, leading to increased plasma cholesterol and triglyceride levels and enhanced atherosclerosis [65]. These studies suggest that VEGFR-3 contributes to atherosclerosis by interacting with VV in addition to lymphoid vessels.

Evidence shows that the Ets transcription factor regulates the expression of matrix metalloproteinase (*MMP*) genes, which are related to the essential steps in angiogenesis and tumor invasion (extracellular matrix degradation and vascular EC migration) [66]. Hypoxia induces Ets-1 expression via the activity of HIF-1 [53], likely via the hypoxia-responsive element (HRE) located at the Ets promoter. In addition to inducing protease expression, Ets-1 also influences angiogenesis by enhancing the transcription of hepatocyte growth factor (HGF) and VEGF, and forming an auto-loop of their upregulation [67]. It is speculated that the role of Ets-1 in atherosclerosis may have the same activities. Indeed, HIF-1, VEGF, and Ets-1 were all expressed in 29 human carotid plaques obtained from carotid endarterectomy [68]. Hypoxia-induced HIF-1a/VEGF/Ets-1 cascade was suggested as important for angiogenesis in human atherosclerosis. Nevertheless, the connections among these hypoxia-induced angiogenesic pathways still remain unclear. For example, in contrast to the studies discussed above, acidic fibroblast growth factor (FGF), basic fibroblast growth factor (bFGF), VEGF, and epidermal growth factor all induce the expression of Ets-1 mRNA in ECs [69,70], suggesting that multiple mechanisms can contribute to hypoxia-induced angiogenesis. Therefore, whether VEGF and Ets play an independent role in the process of VV neovascularization, or act synergistically with other growth factors, merits further investigation (Figure 2).

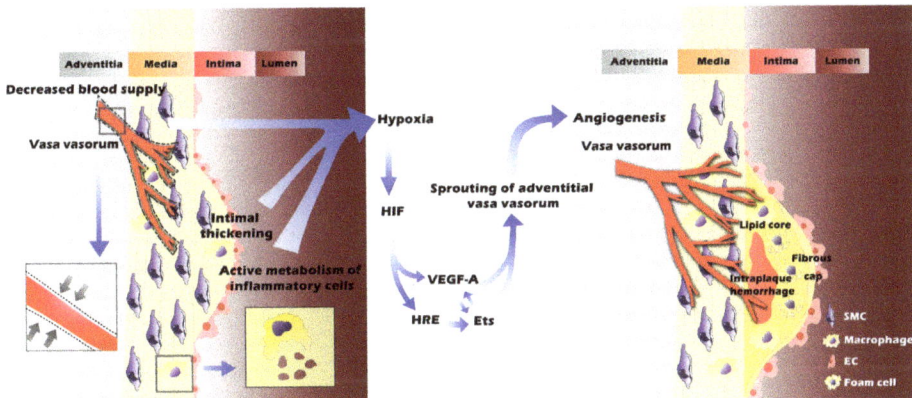

**Figure 2.** Intimal thickening, decreased blood supply (due to high pressure in parent vessel and the stimulated constriction of VV, represented by grey arrows in the left closed box) and active metabolism of inflammatory cells together contribute to hypoxia in atherosclerotic vessels (**left** panel). The oxygen-insufficient microenvironment in inner layers of vessel wall further induces angiogenesis through activating HIF, VEGF-A and Ets signaling pathways. As a result, the formation of intraplaque neovessels originating from VV leads to the progression of atherosclerotic plaques, including intraplaque hemorrhage, lipid core enlargement, inflammatory cell infiltration, and ultimate rupture (**right** panel).

### 3.2. Inflammation

Systemic atherosclerosis, parenchymal inflammation and VV neovascularization are inseparably linked. There are two main hypotheses for the initiation of vascular inflammation. The traditional concept of vascular inflammation includes "inside-out" responses centered on the monocyte adhesion and lipid oxidation hypotheses. However, growing evidence supports a new paradigm of an "outside-in" hypothesis, in which vascular inflammation is initiated in the adventitia and progresses inward toward the intima [17]. Although the initiation of vascular inflammation is still up for debate, VV neovascularization is no doubt triggered and perpetuated by inflammatory reactions within the vascular wall [71]. In fact, the role of inflammation in neovessel formation was proposed decades ago [72]. VV neovascularization occurs most prevalently at the sites of the intima that contain chronic inflammatory cell infiltration, especially the macrophages and lymphocytes. Furthermore, inflammatory cytokines, growth factors, and angiogenic stimuli, which are released by activated inflammatory cells (e.g., macrophages), can enhance not only the inflammation itself but also the development of VV neovascularization [73]. Some studies suggest that vascular inflammation in atherosclerotic lesions closely associates with cell metabolism-created hypoxia, microvascularization, hemorrhage formation, and plaque rupture [74]. However, the exact mechanisms of the inflammation-induced angiogenesis remain unknown.

*3.3. Lipids*

The role of lipids in atherosclerosis is an old topic. Several types of lipid complex or lipid-containing substances have been reported in atheromatous lesions, including modified oxidized-LDL, 7-ketocholesterol (7KCh), cholesterol (such as low-density lipoprotein LDL, high-density lipoprotein HDL, and triglyceride), soluble phospholipids, and eicosanoids [75,76]. These lipid complexes or lipid-containing products can be present in the circulation, released from dead foam cells, or exist on the blood cell membrane. High levels of circulating LDL remain a profound risk factor in predicting cardiovascular events. Circulating LDL is an important source of atherosclerotic plaque lipid content and contributes to atherosclerosis progression, such as lipid deposition/foam cell formation and associated inflammatory process [77]. As the disease progresses, lipid-laden foam cells undergo apoptosis and release free cholesterols, leading to the formation of necrotic cores full of lipids [78]. Studies of red blood cell membrane components present another conceivable origin of lipids in atherosclerosis [79]. Taken together, atherosclerotic plaque is recognized as a reservoir of lipid complexes.

VV neovascularization begins at the radial projection at the site of lipid retention [80]. It provides atherosclerotic lesions with blood as well as lipids [81]. Hypoxia may not be the only factor responsible for medial VSMC proliferation or activation due to the restricted hypoxia in intimal lesions [40]. Therefore, experts hypothesize that VV neovascularization may be lipid-dependent. Indeed, intima-borne lipid promotes angiogenesis by activating VSMC PPAR-$\gamma$ receptors [76]. Conditioned medium from early atheromatous lesions was enriched in oxidized lipid 15-Deoxy-$\delta$-12, 14-prostaglandin J2, and their derivatives. These naturally occurring compounds activate the PPAR-$\gamma$ pathway in subjacent medial VSMCs [82]. In pace with PPAR-$\gamma$ activation, the expression of VEGF-A was upregulated in VSMCs, leading to neovascularization. Cholesterol efflux also regulates angiogenesis via the modulation of lipid rafts and VEGFR-2 signaling in ECs. The decline of the lipid rafts results in lower VEGFR-2 contents on the cell membrane, which leads to the down-regulation of VEGFR signaling and culminates in the inhibition of VEGF-stimulated angiogenesis [83]. This might be an entirely new hypothesis to interpret the modulation of the lipid-mediated angiogenic process. These results were all based on the model of dyslipidemia zebrafish, which might be a better model for the study and evaluation of the early development of atherosclerosis. Besides the aforementioned VEGF-dependent mechanism of angiogenesis, the VEGF-independent pathway also contributes to the angiogenesis in vascular disease. Polyunsaturated fatty acid (PUFA), which is prone to oxidation, is capable of activating the Toll-like receptor 2 (TLR2)/MyD88 pathway after oxidation. This leads to the activation of Rac1 to promote NF-$\kappa$B signaling, causing cell migration and neovascularization [84,85]. Nevertheless, more mechanisms may come to explain VV neovascularization and each mechanism may not be independent but instead may regulate each other.

Growing evidence indicates that the perivascular adipose tissue (PVAT) associates with the inflammatory process in atherosclerosis. PVAT interacts directly with the outer adventitia without fascia or elastic lamina and is capable of conveying signaling molecules (adipokines and cytokines) to the adjacent blood vessels [86]. VV, which penetrate the PVAT, are highly prone to change. Conditioned medium from differentiated murine 3T3-L1 adipocytes concentration-dependently stimulates human saphenous vein and aortic SMC proliferation. There was an about 206% $\pm$ 21% increase of SMC proliferation in the vein and 145% $\pm$ 9% SMC proliferation increase in the aorta at the highest concentration (100 $\mu$L/mL) used, while such an effect was not observed in conditioned medium from premature or undifferentiated adipocytes [87]. Experts concur that adipokines such as visfatin and leptin are the stimuli in SMC proliferation and migration [88]. Compared with conditioned medium from differentiated human subcutaneous and perirenal adipocytes, conditioned medium from differentiated human perivascular adipocytes showed a much stronger ability to induce angiogenesis (elongation and branching) when applied to human coronary artery ECs, consistent with the elevated (two-fold) expression of VEGF in perivascular adipocytes [89]. In advanced stages of atherosclerosis, neovessels were suspected as the conduit for transporting pro/anti-inflammatory mediators into the vascular wall from PVAT. However, since adipocytes are heterogenous in different tissues and even PVAT are biologically and functionally diverse surrounding different blood vessels [90], a detailed relationship between different PVAT locations and properties remain to be tested.

### 4. Factors Leading to Immature and Fragile Vasa Vasorum

Although VV neovascularization is generally recognized as a compensatory reaction to meet the oxygen and nutritional demand of the inner layer of the vascular wall, intraplaque neovascularization meanwhile gives rise to ensuring plaque destabilization, intraplaque hemorrhage (IPH), atherothrombosis and even ultimate plaque rupture. As mentioned, intraplaque VV are immature, irregular, fragile, and prone to extravasation due to the compromised structural integrity. Different hypotheses have been proposed to explain this immaturity and leakage in varying stages and timescales in atherosclerosis.

*4.1. Imbalance of Angiogenic Factors in Proteolytic Environment*

Atherosclerotic plaque contains a wide spectrum of proteolytic proteases, including metalloproteinases (*i.e.*, MMPs), serine proteases (e.g., elastase, coagulation factors, plasmin, tissue-type plasminogen activator and urokinase-type plasminogen activator), and cysteine proteases (e.g., cathepsins) [91]. Previous studies have demonstrated their tight correlation to the pathophysiology of atherosclerosis, in particular, concerning the plaque destabilization. The MMP family is involved in intraplaque angiogenesis and plaque instability. Increased expression of MMP-1, -2, -3, and -9 was detected in atherosclerotic plaques [92]. Urokinase-type plasminogen activator receptor (UPAR) expression was 1.4-fold higher in macrophages and 1.5-fold higher in carotid endarterectomies from Caucasian patients with symptomatic carotid stenosis, compared to the control group [93]. In unstable plaques, increased legumain was detected and converted cathepsin L to its mature 25 kDa form, leading to the intraplaque angiogenesis, macrophages apoptosis, and necrotic core formation [94,95].

In human carotid endarterectomy samples, the levels of placental growth factor (PlGF), VEGF, and angiopoietin-1 (Ang-1) were significantly decreased in culprit plaques/hemorrhagic when compared with culprit plaques/non-hemorrhagic, but both were higher than those from the normal control group. Soluble Tie-2 (receptor of Ang-1 and Ang-2) levels were also increased in the hemorrhagic lesions, although Ang-2 levels were similar between hemorrhagic and non-hemorrhagic lesions [96]. These results suggested an angiogenic/anti-angiogenic imbalance in hemorrhagic plaques. Since the normal formation of neovessels requires a precise regulation to maintain the balance between angiogenic and anti-angiogenic factors, the disturbed balance in hemorrhagic plaques may impede the maturity and structural integrity of intraplaque neovessels.

Previous studies revealed a significant increase in plasmin and leucocyte elastase activities in hemorrhagic plaques [97]. These proteases may degrade these angiogenic factors in hemorrhagic plaques, resulting in angiogenic/anti-angiogenic imbalance and potential neovessel immaturity. VEGF promotes the initiation of immature vessels by vasculogenesis or angiogenic sprouting. Paracrine Ang-1 stabilizes the interactions between ECs and their surrounding support cells (SMCs and pericytes) and extracellular matrix (ECM) via binding to the Tie 2 receptor on the EC surface. Autocrine Ang-2, considered as an antagonist for Tie 2, has adverse effects, leading to vascular regression or angiogenic sensitivity (more plastic and destabilized state) [98]. Alteration of Ang-1/Ang-2 influences the normal process of neovessel formation. Although a shift of Ang-1/Ang-2 ratio was observed in several pathological processes, such as brain arteriovenous malformations [99] and tumor microvessel development [100], its role in atherosclerosis remains largely unknown. The imbalance between Ang-1 and Ang-2 in atherosclerosis may be a major deterrent to neovessel maturation. Decreased activity of Ang-1 and the ratio between Ang-1 and Ang-2 levels biased towards Ang-2 were observed in atherosclerotic plaques along with high microvessel content [101]. This is consistent with the observation that plaques with high microvessel density are at a high risk of intraplaque hemorrhage [102]. In addition, the Ang-1/Ang-2 ratio in favor of Ang-2 was also observed in hemorrhagic plaques, indicating an underlying role of angiopoietin/Tie system in microvessel immaturity [96].

Therefore, angiogenesis may depend on a precise balance of positive and negative regulations. The angiogenic and anti-angiogenic factors act in coordination to form well-structured and functional vessels. A disorder of homeostasis in VV neovascularization influences the proliferation and migration of ECs and their surrounding support cells, thereby leading to compromised structural integrity and aberrant neovessel formation. However, angiogenic factors can be regulated at the levels of both expression and proteolytic degradation. It remains unknown how these angiogenic factors are regulated in VV neovascularization during atherogenesis, a possible focus of future studies.

*4.2. Further Exacerbation of Neovessel Damage: Iron, Cholesterol Crystal and Proteases*

Angiogenic/anti-angiogenic imbalance inherently leads to the extravasation of vessel wall. Iron, cholesterol crystal, and protease activity may cause further exacerbation of neovessel damage, which also facilitates the permeability of intraplaque microvessels. Iron is abundant in the human body, especially in erythrocytes. Differing from normal erythrocytes, erythrocytes in atherosclerotic plaques are prone to undergo rapid lysis to release a large quantity of hemoglobin (Hb) [103]. Extracellular Hb remains susceptible to morphing into ferrihemoglobin via oxidation, and releases heme that contains abundant redox active iron. Redox active iron plays a detrimental role in oxidation reactions *in vivo*, including lipid oxidation. Oxidized LDL from atherosclerosis is cytotoxic to ECs [104,105]. High expression of ferritin and heme oxygenase-1 that control the redox active iron and degraded heme demonstrate a protective compensatory reaction from iron-derived vascular injury [106,107]. Diaspirin cross-linked Hb (DBBF-Hb) and polyethylene glycol (PEG)-conjugated Hb, proposed as blood substitutes, proved to increase arterial microvascular permeability [108]. Therefore, erythrocyte-derived iron may induce vascular damages, establishing a role for heme iron-dependent oxidation in damaging intraplaque immature microvessels. Apart from erythrocyte-derived iron, several other mechanisms, which are not mutually exclusive, have been proposed as underlying causes of vessel injury within plaques, including perforation of microvessels by cholesterol crystals [109] and direct damage induced by protease activity [110].

## 5. Vasa Vasorum Imaging in Atherosclerosis Plaques

Atherothrombosis resulted from plaque rupture is the direct cause of several acute cardiovascular complications (e.g., stroke, myocardial infarction). Therefore, practical imaging systems for early detection of unstable or even asymptomatic atherosclerotic plaques are urgently needed. Angiographic studies showed that non-obstructive plaques caused approximately 75% of cases with acute coronary occlusion [111]. Therefore, traditional focus on the stenosis of atherosclerotic plaque may be not sufficient to predict vulnerable plaques, thereby forcing us to explore more detailed features of possible culprit plaques, such as the fibrous cap, a lipid-rich necrotic core, signs of inflammation, and VV neovascularization. Visualization of arterial VV and intraplaque neovessels has recently emerged as a new surrogate marker for the early detection of atherosclerotic lesions. Coronary VV neovascularization occurs within the first week of experimental hypercholesterolemia, prior to the development of endothelial dysfunction of the host vessel, suggesting the significance of VV visualization in the identification of atherosclerotic vascular disease in the early stage [16]. Thus, a safe, non-invasive, and affordable imaging technique for the detection of VV has important clinical significance.

## 5.1. Anatomic Imaging of Vasa Vasorum in Atherosclerosis

Differences in VV structure between normal and atherosclerotic arteries were first detected by autopsy cinematography. The 3D anatomy and tree-like branching architecture of VV are shown in Figure 1. Efforts have been focused on identifying not only diseased and non-diseased arteries, but also stable and unstable plaques among risky populations through anatomic imaging.

Growing evidence confirming the role of VV neovascularization in atherosclerosis is provided first by micro-CT. With micro-CT, increased VV contents in the proximal left anterior descending coronary artery were detected in hypercholesterolemic pigs [48]. VV density was found about two times higher in nonstenotic and noncalcified stenotic plaques when compared with normal and calcified segments [11]. With the limitation of the subject imaging volume, micro-CT is mostly used to scan specimens from autopsy or small animals such as mice. Thus, micro-CT usually applies to retrospective studies or animal experiments, and is not available for humans.

High-speed whole body scanning capability of multi-slice CT demonstrated its capacity of accurate visualization of atherosclerotic plaques in the coronary arteries [112]. Several attempts have been made to improve the resolution of CT in VV visualization. For example, iodinated nanoparticles were used as contrast agents in CT. CT imaging using this compound increased the X-ray absorption at the targeted sites, hence improving the quality of CT images. CT angiography (CTA) is another technique with high spatial and temporal resolution, allowing detailed anatomical delineation of atherosclerotic arteries. CTA scanning of patients with moderate (50%–70%) stenosis of the internal carotid artery showed that plaques derived from patients with neurological symptoms had a higher proportion of VV enhancement than of that in total patients (34% *vs.* 24.1%), which indicated that atherosclerotic patients with enhancing VV were more likely to be symptomatic [113]. This result indicates that CTA imaging of VV may aid in the identification of patients at an increased risk for ischemic stroke within populations with the same degree of stenosis. However, because of the risks associated with radiation exposure, current American Heart Association and American College of Cardiology guidelines do not recommend CTA as a general screening tool in low-risk, asymptomatic patients [114].

Intravascular ultrasound (IVUS) is broadly used to provide high-resolution tomographic images of the lumen and acquire precise measurements of atherosclerotic plaques *in vivo*. However, IVUS imaging systems, which are developed to examine blood flow within the lumen of large arteries, are not designed initially to detect VV morphology [115]. Recently, a porcine experiment demonstrated that the change of IVUS-based vessel wall flow assessment signals paralleled VV density detected by micro-CT, indicating the potential of IVUS estimation of blood flow to quantify VV density [116]. Contrast-enhanced IVUS with contrast enhancement agents is another prominent method used in VV visualization. Contrast agents can increase IVUS echogenicity enhancement in the adventitia of coronary arteries, which is consistent with the enhancement of VV [117]. O'Malley *et al.* [118] presented analyses of human coronary arteries *in vivo*, and demonstrated the feasibility of contrast-enhanced IVUS imaging of VV density and perfusion in atherosclerotic plaques. Further, IVUS with contrast microbubbles tracing neovascularization in non-culprit coronary atherosclerotic plaques demonstrated a significant mean enhancement after intracoronary injection of microbubbles (from 7.1% $\pm$ 2.2% to 7.6% $\pm$ 2.5%) in the adventitia, which represented the high density of VV in patients with acute coronary syndrome [119]. Other modified techniques, such as contrast-harmonic IVUS and subharmonic contrast IVUS, can visualize contrast agents in adventitial VV [120,121]. Compared with harmonic contrast IVUS, subharmonic (SH20) imaging was even superior to harmonic (H40) imaging in terms of contrast-to-noise and contrast-to-tissue ratio improvement [122]. However, quantitative comparison of harmonic and subharmonic imaging has not been available. Although contrast-enhanced IVUS can provide clear and direct insight into VV in the adventitia, experiments that quantify the neovessels are not available, leading to the limitation of IVUS application in clinical practice. A more accurate index is needed to visualize adventitial VV.

Contrast-enhanced ultrasonography (CEUS), another modality for vascular imaging together with ultrasonographic contrast microbubbles, has developed during the last decade. In a preliminary feasibility study, CEUS enabled the visualization of the adventitial network of VV in human carotid arteries [123]. The enhanced signal was five times higher on average after stimulating atherosclerosis [124,125]. These enhancements correlated with the histological density of intraplaque neovessels. Visualization of VV density by CEUS in an atherosclerotic population also revealed a positive relationship between the abundance of VV and plaque echolucency, a well-accepted marker of high risk lesions [126]. The capability of CEUS in visualizing adventitial VV and intraplaque neovascularization makes it attractive for plaque risk stratification and assessment of anti-atherosclerotic therapy efficacy. Significant linear correlations between CEUS peak video-intensity and histologic VV counting, as well as the cross-sectional area of neovessels were recently reported [127,128]. Video intensity has become a quantitative parameter of CEUS to detect VV and assess the effects of anti-atherosclerotic therapy. Normalized maximal-video intensity enhancement (MVE) in CEUS, which represented the density of VV, demonstrated a positive relationship with plaque volume. A much lower MVE enhancement was observed after four weeks in atorvastatin-treated rabbits (from $0.18 \pm 0.08$ to $0.16 \pm 0.07$, $p = 0.11$) than in untreated rabbits (from $0.18 \pm 0.08$ to $0.25 \pm 0.08$, $p = 0.001$) [129]. In humans, quantitative CEUS was applied to investigate coronary artery disease patients undergoing lipid-lowering therapy with statins, which parelleled the LDL reduction [130]. All these results inspire the development of a standard diagnostic index for CEUS imaging in quantifying VV density, with which we could identify the populations with unstable plaques and plan the medication. To achieve this goal, large multicenter clinical trials on quantitative CEUS scanning of VV in normal and diseased populations are needed. However, a controversial study indicated that the enhancement of CEUS in carotid atherosclerotic plaques might not always reflect the presence of VV, as verified by the immunochemistry results [131]. Thus, more studies are needed to improve imaging quality and certain clinical standards of CEUS in VV visualization and vulnerable plaques prediction.

Optical coherence tomography (OCT) is an intravascular imaging modality using near-infrared light to generate cross-sectional intravascular images [132]. With its high resolution (10–20 μm), which is 10 times higher than that of IVUS and comparable to that of micro-CT, OCT is widely used for the assessment of coronary atherosclerotic plaques [133]. Under OCT, VV are visualized as a no-signal microchannel within the plaque or the adventitia [134]. The greatest challenge for VV detection under intravascular modalities is the extensive motion artifacts inherently associated with arterial pulsations in addition to other physiological movements. These limitations have been minimized in a recent intensity kurtosis OCT technique, which was developed to visualize VV from carotid arteries *in vivo* [135]. Both the blood flow into VV and dynamic motions of the arterial wall were clearly displayed using this OCT technique. In addition to the earlier time-domain OCT, optical frequency domain imaging (OFDI) was developed as a new-generation OCT that was capable of obtaining A-lines at much higher imaging speeds, facilitating rapid, 3D-pullback imaging during the administration of a non-occlusive flush of an optically transparent media such as Lactated Ringer's or radiocontrast [136]. The laboratory results showed that adventitial VV on OFDI *ex vivo* were clearly displayed and appeared to communicate with the coronary adventitia and media. The results were compatible with histological findings and showed much better resolution when compared with other generations of OCT [137]. More recently, quantification of VV with the 3D OCT method has been applied in both animal and clinical studies. Animal studies conducted on swine suggest significant correlations between the microchannel volume (MCV) count by OCT and the amount of VV by micro-CT [138], which was consistent with the human study [137]. A positive correlation between MCV and plaque volume was also detected from this study [138]. In 2012, international guidelines were formed by the International Working Group for Intravascular OCT Standardization and Validation. OCT imaging modality was recommended as a standard reference in clinical practice [136].

## 5.2. Molecular Imaging of Vasa Vasorum in Atherosclerosis

The recent recognition that plaque biological features influence the prognosis of atherosclerosis has led to the transition from the sole anatomical assessment to combined anatomic and functional imaging modalities, enabling the application of molecular imaging in VV detection. Molecular imaging, which originates from cancer imaging, is defined as *in vivo* characterization and the measurement of biological processes at the cellular, molecular, whole organ, and whole body levels [139]. Radionuclide tracers or targeted agents with specific binding capacity to molecular targets were used as markers of biological functions in molecular imaging.

Nuclear positron emission tomography (PET) and single-photon emission computed tomography (SPECT) were the first two methods used for molecular imaging. These techniques mostly detected the hypoxia and inflammation levels of atherosclerotic plaques through fluorodeoxyglucose (FDG) uptake or other targets [140,141]. As for the angiogenesis scanning, PET/SPECT with $^{64}$Cu-labeled VEGF$_{121}$, $^{111}$In-labeled $\alpha v \beta_3$-targeted agent, or other radionuclide tracers have been applied in cancer and post-myocardial infarction neovessel imaging [142–144]. As MMPs participate in intraplaque angiogenesis, MMP inhibitor-based radiotracers bind to activated MMPs and detect angiogenic processes *in vivo*. Several inhibitors have been successfully labeled with $^{123}$I, $^{11}$C, or $^{18}$F for PET imaging, but no animal experiment currently exists [145,146]. Specific uptake of $^{123}$I-labelled MMP inhibitors into the plaques was shown in animals on a high-cholesterol diet by planar gamma camera, the radioactivity of which was 2.72-fold of the common artery [147]. Yet, direct visualization of VV neovascularization through PET or SPECT still remains unavailable. Autoradiography, another nuclear technique, can identify angiogenesis in atherosclerotic plaque *ex vivo* with $^{125}$I-labeled monoclonal antibodies against fibronectin extra-domain B (ED-B) [148]. More recently, $^{99m}$Tc-labeled membrane type 1 MMP monoclonal antibody ($^{99m}$Tc-MT1-MMP mAb) accumulation, which detects active MMP-2 and MMP-13, was found in atheromatous lesions (4.8 ± 1.9, % injected dose × body weight/mm$^2$ × 10$^2$) and positively correlated with membrane type 1 MMP expression [149]. In conclusion, nuclear neovessel imaging could provide diagnostic imaging capability of vulnerable plaques, and further investigations to improve the modalities are strongly required.

Magnetic Resonance imaging (MRI), by virtue of its ability to characterize various pathological components of atherosclerosis plaques with advantages of high-resolution and radiation avoiding, is another promising modality for studying VV. On traditional anatomical dynamic contrast-enhanced MRI (DCE-MRI), the enhancement of the outer rim of the internal carotid arteries represents VV [150]. With the development of $\alpha v \beta_3$ integrin-targeted gadolinium chelates, MRI molecular imaging emerged. With paramagnetic gadolinium-based nanoparticles, increased angiogenesis was detected as a 26% ± 4% and 47% ± 5% signal increase over baseline at 15 and 120 min in cholesterolemic rabbits, while only half of the signal augmentation was detected in cholesterol-fed rabbits that received non-targeted nanoparticles. A heterogeneous spatial distribution of neovessels was also observed through molecular imaging [151]. MRI with $\alpha v \beta 3$-targeted nanoparticles also showed the ability to assess the therapeutic effect of anti-atherosclerotic agents. After 16 weeks of an appetite suppressant treatment with benfluorex, MR enhancement decreased in treated animals when a steady increase was seen in the untreated group [152]. Gadolinium quantitative MRI of VV reported that the total neovessel area in the matched sections from DCE-MRI correlated with the histological measurement with a high correlation coefficient of 0.80, thereby serving as a quantitative parameter in vascular imaging [153]. In humans, *in vivo* DCE-MRI showed that the transfer constant (K$^{trans}$) of gadolinium enhancement in the carotid adventitia is a quantitative measurement of the VV extent, a method confirmed by histological measurements on a carotid endarterectomy specimen [154]. K$^{trans}$ represents a kinetic parameter that characterizes the transfer of the contrast agent from plasma to the extravascular space (e.g., adventitia). The transfer therefore depends on VV density. By estimating K$^{trans}$ via DCE-MRI, the transfer of gadolinium into the extravascular space, calculated from dynamic kinetics of tissue enhancement, correlates strongly with macrophage content, neovascularization, and loose matrix areas measured by histology analysis [155]. These studies indicated that adventitial enhancement seen on DCE-MRI with gadolinium chelates can be used to detect VV neovascularization in the process of atherosclerosis, therefore assessing the risk for plaque rupture. However, previous studies show gadolinium associates with nephrogenic systemic fibrosis in patients with reduced renal function [156]. Thus, exploration of more effective contrast enhancement agents with lower toxicity and higher resolution is needed in the future.

CEUS is another tool in molecular imaging. As discussed, CEUS has been employed not only to improve the evaluation of intima-media thickness, but also to highlight wall irregularities, ulcerations, adventitial VV and neovasculature of the atherosclerotic plaque that are often not visible by standard non-invasive ultrasound imaging [157]. Further development of conjugated microbubbles that bind to specific ligands in thrombotic material or neovessels has led to the term "molecular imaging" in CEUS scanning. Microbubbles coupled to the VEGFR may allow for a detection of neovascularization. Using dual ET-1/VEGFsp receptor (DEspR)-targeted microbubbles, CEUS molecular imaging has detected an increased DEspR-expression in carotid artery lesions and expanded VV neovessels in transgenic rats with carotid artery disease [158]. CEUS with VEGFR-2-targeted microbubbles was used to evaluate the response to sorafenib (a drug that inhibits cell proliferation and neovascularization in several tumors) in a mouse model of hepatocellular carcinoma [159]. The amount of bound microbubbles in the tumor quantified by dedicated software was lower in the treatment group through CEUS molecular imaging.

Together, as summarized in Table 1, molecular imaging shows an exciting potential in VV imaging by increasing the quality of resolution. Further studies are needed to modify the existing molecular imaging modalities and targeted agents that are safe, accurate, and easy to detect.

Table 1. Molecular imaging modalities and targets of vasa vasorum in atherosclerosis models.

| Imaging Modality | Spatial Resolution | Temporal Resolution | Targets (Reference) | Species | In Vivo Imaging | Histological Validation | Results |
|---|---|---|---|---|---|---|---|
| Planar gamma camera | $cm^3$ | Hours | 123I-labelled MMP inhibitors [147] | Apoe$^{-/-}$ mice | + | + | Signal of 123I-labelled inhibitors into plaques in high cholesterol animals was 2.72-folds of the control |
| Autoradiograph | $\mu m^3$ | Milliseconds | ED-B [148] | Apoe$^{-/-}$ mice | − | − | 125I-labeled monoclonal antibodies against ED-B identified the angiogenesis in atherosclerotic plaques ex vivo |
| | | | 99mTc-MT1-MMP mAb [149] | WHHLMI rabbits | + | + | The highest accumulation of 99mTc-MT1-MMP mAb was found in atheromatous lesions in comparison with stable lesions |
| MRI | $mm^3$ | Seconds | Integrin $\alpha v \beta_3$ | Male New Zealand White (NZW) rabbits [151] | + | + | Paramagnetic gadolinium-based nanoparticles showed strong enhancement in atheroscleotic lesions that was twice of the non-targeted nanoparticles |
| | | | | JCR:LA-cp rats [152] | + | + | The enhancement of $\alpha v \beta_3$-targeted nanoparticles was preserved in benfluorex treating group |
| | | | | Humans [153–155] | + | + | Targeted gadolinium compounds detected VV, total area and $K^{trans}$ of the enhancement could be quantitative parameters |
| CEUS | $\mu m^3$ | Milliseconds | ET-1/VEGFsp receptor [158] | Tg25 (hCETP) Dahl-S rats | + | + | Expanded VV in transgenic rats with carotid artery disease were detected by targeted microbubbles |
| | | | VEGFR-2 [159] | Female nude mice | + | − | VEGFR-2 targeted microbubbles were able to evaluate anti-angiogenic effect of sorafenib |

## 6. Anti- and Pro-Angiogenic Therapies on Vasa Vasorum

### 6.1. Anti-Angiogenic Therapies

The rupture of neovessels originating from VV may be the main cause of intraplaque hemorrhage in advanced atherosclerosis. It seems logical to speculate that anti-angiogenic strategies can be used to inhibit plaque growth and stabilize existing plaques, although lipid-controlling, anti-inflammation, and invasive angioplasty are among the current main treatments for atherosclerosis. As discussed above, angiogenic factors control neovascularization. Expression regulation of these factors (e.g., Ets-1, Ang-1 receptor, HIF-1$\alpha$, MMPs) by microRNAs influences EC activation and SMC phenotype switch, thereby decreasing VV neovascularization [160]. In contrast, the effects of angiogenesis inhibitors on atherosclerotic lesions are difficult to classify because the various agents that block angiogenesis do not have uniformed mechanisms of action. Each agent may also have unique effects on different cell types and biochemical pathways that are also present in atherosclerotic lesions. To simplify the understanding of different angiogenic inhibitors, we grouped angiogenic inhibitors into three categories: direct anti-angiogenic molecules (*i.e.*, angiostatics) that are derived from protein proteolysis; inhibitors that target directly or indirectly the angiogenic factors (e.g., VEGF); and others with incompletely characterized mechanisms.

Direct anti-angiogenic compounds, also known as angiostatics, are substances that target ECs and SMCs without affecting endogenous angiogenic factors. Angiostatics were initially identified as tumor-derived factors that inhibit neovascularization of remote metastases of Lewis lung carcinoma (*i.e.*, angiostatin) [161] and hemangioendothelioma (*i.e.*, endostatin) [162]. Endostatin is a fragment from collagen-XVIII proteolysis. In apolipoprotein E-deficient ($Apoe^{-/-}$) mice, chronic treatments with endostatin reduced intimal neovascularization and inhibited plaque growth by 85% in atherosclerosis [163]. However, as the plaque intimal SMC contents were similar between control and treated mice in this research, a mechanism for the anti-angiogenic property of endostatin remains incompletely understood. Angiostatin, a proteolytic fragment of plasminogen, blocks the angiogenic potential of atherosclerotic aortas with a parallel reduction of macrophages in the plaques and around VV [164]. Activated macrophages stimulate angiogenesis by recruiting more inflammatory cells to increase angiogenesis further. Angiostatin inhibition interrupts this positive feedback from inflammatory cells and hinders angiogenesis in atherosclerotic plaques. The mechanisms by which endostatin and angiostatin inhibit angiogenesis are not fully understood. Inhibition of EC and SMC proliferation and migration, without affecting endothelial intracellular signaling pathways, seems to play a vital role in this process [165].

Angiogenic factor inhibitors offer another target choice of anti-angiogenic therapy in atherosclerosis. The inhibition of the VEGF signaling pathway by targeting VEGF and VEGFR-1 has been well studied. Bevacizumab is a VEGF-specific antibody that has the capacity to inhibit VV after local delivery by stent in a rabbit atherosclerosis model [166]. Exogenous application of antibodies against VEGFR-1 reduced the size of early and intermediate plaques by 50% and the growth of advanced lesions by ~25% in $Apoe^{-/-}$ mice [167]. PlGF is a VEGFR-1 ligand. Deficiency of PlGF inhibited early-stage atherosclerosis. On the other hand, increased adventitial expression of PlGF promoted intimal hyperplasia and VV proliferation [168]. Importantly, the anti-angiogenic property of PlGF antibodies acted only on diseased arteries without affecting healthy vessels [169], offering a great advantage for patient treatment. Other angiogenic factor inhibitors include anti-MMP-2 and anti-MMP-9 antibodies that also have the ability to block endothelial cell tubule formation [170]. MicroRNAs that interfere with the expression of angiogenic factors (e.g., Ets-1, Ang-1 receptor, HIF-1$\alpha$, MMPs) also decrease VV density in atherosclerotic plaques [160]. Apart from antibodies and microRNAs that act directly on angiogenic factors, indirect agents also play an anti-angiogenic role in VV angiogenesis. Thalidomide is an anti-angiogenic drug that exerts anti-inflammatory and immune-modulatory effects. Treatment with thalidomide preserved VV spatial density by 45% in high-cholesterol, diet-fed atherosclerosic pigs and was only 1.3-fold higher than that of normal pigs when compared with the 2.4-fold increase in the untreated group [171]. Reduced adventitial VV neovascularization and plaque progression after thalidomide treatment were also seen in *Apoe$^{-/-}$/Ldlr$^{-/-}$* double knockout mice [172]. The anti-angiogenic effect of thalidomide in atherosclerosis was accompanied by the inhibition of VEGF expression.

Another group of molecules remain difficult to classify because of their limited study and complicated mechanisms. ET-1 is a vasoconstrictor that has mitogenic activity on SMCs and participates in VV neovascularization during atherogenesis [173]. ET receptor antagonism (ET-A) application in hypercholesterolemic pigs showed that elevated VV density in the hypercholesterolemia group was greatly preserved by ~32% [174]. Fumagillin nanoparticles are also anti-angiogenic agents that target ECs. Treatment of atherosclerosic rabbits with $\alpha v \beta_3$ integrin-targeted paramagnetic nanoparticles together with fumagillin decreased MRI enhancement and reduced the numbers of microvessels [175]. Anti-angiogenic rPAI-1$_{23}$, a truncated isoform of plasminogen activator inhibitor-1 (PAI-1), greatly reduced VV, especially the second order VV, with a 37% reduction in total vessel area and a 43% reduction in vessel length in atherogenic female *ApoB-48$^{-/-}$/Ldlr$^{-/-}$* mice through inhibition of bFGF, suggesting a significant therapeutic potential of this anti-angiogenic protease inhibitor peptide in atherosclerosis [176]. Further study concluded that rPAI-1$_{23}$ caused the regression of adventitial VV in hypercholesterolemic mice by increasing plasmin and MMP activities that degrade perlecan, nidogen, and fibrin in the extracellular milieu. Without a supportive scaffold from these extracellular matrix proteins, ECs may undergo apoptosis, leading to VV regression [177]. The anti-proliferative and anti-inflammatory drug, 3-deazaadenosine (c3Ado), dose-dependently prevents the proliferation and migration of human coronary artery ECs. It also inhibited VV neovascularization along the descending aortas in *Apoe$^{-/-}$/Ldl$^{-/-}$* double knockout mice [178]. Here we summarize this last category of molecules in Table 2.

Table 2. Anti-angiogenic molecules with incompletely characterized mechanisms.

| Compounds | Year | Functions | Species | Possible Mechanisms | Results |
|---|---|---|---|---|---|
| ET-A | 1993 [173], 2002 [174] | Inhibiting ET-1 receptor | Female domestic pigs | Inhibiting mitogenic activity of SMCs | Elevated VV density in hypercholesterolemia pigs were greatly preserved by ~32% after ET-A treatment |
| Fumagillin nanoparticle | 2006 [175] | An anti-angiogenic agent that targets $\alpha v \beta_3$ integrin | Male NZW rabbits | Not investigate | MRI enhancement and the numbers of microvessels are decreased in fumagillin-treated cholesterol-fed rabbits |
| rPAI-1$_{23}$ | 2009 [176], 2011 [177] | A truncated isoform of plasminogen activator inhibitor-1 (PAI-1) | Female ApoB-48$^{-/-}$/Ldlr$^{-/-}$ mice | Reducing FGF-2 expression. Increasing plasmin and MMP activities on degrading compounds (i.e., perlecan, nidogen, fibrin) that produce supportive scaffold in the extracellular milieu, leading to apoptosis of ECs | A 37% reduction in total vessel area and a 43% reduction in vessel length of the second order VV are observed as a result of apoptosis of ECs |
| c3Ado | 2009 [178] | An anti-proliferative and anti-inflammatory drug | Apoe$^{-/-}$/Ldl$^{-/-}$ double knockout mice | Preventing the proliferation and migration of ECs | VV neovascularization is inhibited dose dependently |

*6.2. Pro-Angiogenic Therapies*

Based on what has been discussed, anti-angiogenic therapies seem to be effective and significant in atherosclerosis medication, at least in animal models. However, a recent report from colorectal cancer patients noted that the anti-VEGF monoclonal antibody bevacizumab showed a higher risk (3.5 additional cases/1000 person-years) of arterial thromboembolic events (e.g., stroke, myocardial infarction, arterial embolism and thrombosis, and angina) despite its anti-angiogenic properties [179]. This observation put into question the risk of angiogenesis inhibitors to atherothrombosis and associated complications. It seems that the role of neovascularization is far more complicated than imagined. A case report from four patients suggested that VV are a source of collateral circulation after carotid artery occlusion secondary to atherosclerotic disease [180]. It has been demonstrated that VSMC-rich lesions are stable because of their high cellular content, whereas acellular lesions with a higher degree of calcification, fibrosis, and lipids are more prone to fracture or rupture [181]. It could be argued that, by enriching the supply of nutrients to the plaque core, neovascularization may increase plaque cellularity, thereby acting as an underlying cause of plaque stabilization.

Studies on HMG-CoA reductase inhibitors support the feasibility of pro-angiogenic therapies in atherosclerosis. Statins, such as atorvastatin and simvastatin, demonstrated beneficial effects in reducing atherosclerotic VV neovascularization independent of lipid lowering [129,182]. While high doses of cerivastatin (2.5 mg/kg/day) blocked angiogenesis, low doses of cerivastatin (0.5 mg/kg/day) induced angiogenesis *in vitro* [183]. In contrast, both high- and low-dose statin therapy blocked plaque progression, indicating that pro-angiogenic effects do not lead to pro-atherosclerotic effects, as we anticipated. The inhibition of VV neovascularization may not be the fundamental strategy for plaque stabilizing in atherosclerosis. More attention may be focused on the normalization and maturation of VV to reduce the risk of VV leakage. As a result, studies in pro-angiogenic therapies have been encouraged in recent years.

Nerve growth factor (NGF) is a potent angiogenic factor that is decreased in atherosclerosis-lesioned arteries [184]. NGF application increased the ratio of large matured vessels ($\geq 20$ μm in diameter) from 30% to ~50% compared with the control, whereas VEGF promoted more immature small microvessels (<20 μm) than the control. NGF also enhanced the maturation of VEGF-induced neovessels from 20% to 40% [185]. These studies suggested a therapeutic possibility of pro-angiogenic NGF on atherosclerosis by enhancing VV maturation. bFGF is another predominant angiogenic growth factor that was also required for VV plexus stability in hypercholesterolemic mice [186]. Although it was regarded as a pro-atherogenic factor due to its stimulatory activity on SMC growth [187], recent studies suggested a protective property of bFGF on plaque formation in a hypercholesterolemic rabbit model by reversing endothelial dysfunction and reducing vascular cell adhesion molecule-1 (VCAM-1) expression as well as plaque macrophage content [188]. However, increased endothelial FGF-receptor-2 signaling by EC-selective overexpression of FGF-receptor-2 aggravated atherosclerosis by promoting p21$^{Cip1}$-mediated EC dysfunction [189]. Therefore the use of bFGF for therapeutic angiogenesis such as ischemic injury may have to be considered due to the possible adverse effects in aggravating atherosclerosis. For example, simvastatin with 1.8 mg/kg/day in rabbits increased the expression of HIF-1α and VEGFR-2 in advanced peri-infarcted myocardium, but blocked the protein expression of HIF-1α, VEGF, and VEGFR-2 in early atherosclerotic arteries. Apart from the dosage of this statin, the stage of atherosclerotic disease may influence the outcome of this statin therapy [190]. Therefore, the stage of atherosclerosis and the microenvironment of plaque should be considered to determine whether a pro-angiogenesis or anti-angiogenesis therapy should be employed.

In conclusion, the role of neovescularization in plaque rupture cannot be simply defined. It is instead an intricate process that may exert various physiological effects in different stages of atherosclerosis. Pro-angiogenic factors that improve the maturation and stabilization of VV and reduce the leakage of these neovessels seem essential in further anti-atherosclerosis studies. The stage of plaque development may be a major determinant of whether we should employ anti-angiogenesis or pro-angiogenesis approaches.

## 7. Conclusions

VV neovascularization exists in the process of atherosclerosis, which seems like a compensatory reaction in order to provide adequate oxygen and nourishment for atherosclerotic arteries. However, the imbalance between angiogenic and anti-angiogenic factors along with latter damages leads to the dysfunction of ECs and their surrounding supporting cells, resulting in an immature and fragile VV neovasculature with weak integrity. In addition to the VV leakage itself, further stimulation of inflammation and necrosis of atherosclerotic plaques is another cause of VV-triggered intraplaque hemorrhage. It remains unclear if VV play a causative or reactive role in the atherosclerotic process, yet VV assessment has been suspected as an effective marker in the early detection of vulnerable plaques.

Several imaging modalities have been established to visualize VV neovascularization and to identify plaques with a high risk of rupture. Anatomical imaging techniques such as IVUS and OCT provide high-resolution images of VV and atherosclerotic plaques yet fail to recognize the biological features of the tissue. Recent molecular imaging modalities exhibit both the anatomy and pathological function of VV, which benefit from the development of specific target agents. Molecular imaging methods such as DCE-MRI and CEUS are extensively applied in visualizing VV. The results of these studies are encouraging, but problems remain. The toxicity of enhancement agents, the limited resolution of imaging, and some controversial results from different studies all lead to the need of improving the existing imaging modalities and investigating effective, and practical methods that are safe and precise for bedside use.

Considering the role of VV neovascularization in the process of atherosclerosis, anti-angiogenic therapies seem logical to prevent or attenuate the deterioration of this disease. Several studies have proven that angiogenesis inhibitors preserve the density of intraplaque VV in atherosclerosis. However, a higher risk of arterial cardiovascular events is observed in colorectal cancer patients using angiogenic inhibitors. The use of anti-angiogenic therapy as a proper way to treat atherosclerosis and prevent intraplaque hemorrhage remains complicated. The application of pro-angiogenic NGF and bFGF showed the capacity of enhancing the maturation of VV neovessels. Therefore, pro-angiogenesis factors that improve the maturation of VV and reduce the leakage of these neovessels may be the fundamental solution in reducing intraplaque hemorrhage, and may lead to a new direction for the future study of stabilizing vulnerable plaques.

**Acknowledgments:** The authors thank Chelsea Swallom for her editorial assistance. This study is supported by National Institutes of Health grants HL81090 (GPS).

**Author Contributions:** All three authors contributed to the writing of this manuscript.

**Conflicts of Interest:** The authors declare no conflict of interest.

## References

1. Mozaffarian, D.; Benjamin, E.J.; Go, A.S.; Arnett, D.K.; Blaha, M.J.; Cushman, M.; de Ferranti, S.; Despres, J.P.; Fullerton, H.J.; Howard, V.J.; *et al.* Heart disease and stroke statistics—2015 update: A report from the American Heart Association. *Circulation* **2015**, *131*, e29–e322. [CrossRef] [PubMed]
2. Davies, M.J.; Richardson, P.D.; Woolf, N.; Katz, D.R.; Mann, J. Risk of thrombosis in human atherosclerotic plaques: Role of extracellular lipid, macrophage, and smooth muscle cell content. *Br. Heart J.* **1993**, *69*, 377–381. [CrossRef] [PubMed]
3. De Boer, O.J.; van der Wal, A.C.; Teeling, P.; Becker, A.E. Leucocyte recruitment in rupture prone regions of lipid-rich plaques: A prominent role for neovascularization? *Cardiovasc. Res.* **1999**, *41*, 443–449. [CrossRef] [PubMed]
4. Staub, D.; Patel, M.B.; Tibrewala, A.; Ludden, D.; Johnson, M.; Espinosa, P.; Coll, B.; Jaeger, K.A.; Feinstein, S.B. Vasa vasorum and plaque neovascularization on contrast-enhanced carotid ultrasound imaging correlates with cardiovascular disease and past cardiovascular events. *Stroke J. Cereb. Circ.* **2010**, *41*, 41–47. [CrossRef]

5. Takano, M.; Mizuno, K.; Okamatsu, K.; Yokoyama, S.; Ohba, T.; Sakai, S. Mechanical and structural characteristics of vulnerable plaques: Analysis by coronary angioscopy and intravascular ultrasound. *J. Am. Coll. Cardiol.* **2001**, *38*, 99–104. [CrossRef] [PubMed]

6. Ritman, E.L.; Lerman, A. The dynamic vasa vasorum. *Cardiovasc. Res.* **2007**, *75*, 649–658. [CrossRef] [PubMed]

7. Koester, W. Endareritis and arteritis. *Berl. Klin. Wochenschr.* **1876**, *13*, 454–455.

8. Paterson, J.C. Vascularization and hemorrhage of the intima of arteriosclerotic coronary arteries. *Arch. Pathol.* **1936**, *22*, 313–324.

9. Patterson, J.C. Capillary rupture with intimal hemorrhage as a causative factor in coronary thrombosis. *Arch. Pathol.* **1938**, *25*, 474–487.

10. Barger, A.C.; Beeuwkes, R., 3rd; Lainey, L.L.; Silverman, K.J. Hypothesis: Vasa vasorum and neovascularization of human coronary arteries. A possible role in the pathophysiology of atherosclerosis. *N. Engl. J. Med.* **1984**, *310*, 175–177. [CrossRef] [PubMed]

11. Gossl, M.; Versari, D.; Hildebrandt, H.A.; Bajanowski, T.; Sangiorgi, G.; Erbel, R.; Ritman, E.L.; Lerman, L.O.; Lerman, A. Segmental heterogeneity of vasa vasorum neovascularization in human coronary atherosclerosis. *JACC Cardiovasc. Imaging* **2010**, *3*, 32–40. [CrossRef] [PubMed]

12. Kwon, H.M.; Sangiorgi, G.; Ritman, E.L.; McKenna, C.; Holmes, D.R., Jr.; Schwartz, R.S.; Lerman, A. Enhanced coronary vasa vasorum neovascularization in experimental hypercholesterolemia. *J. Clin. Investig.* **1998**, *101*, 1551–1556. [CrossRef] [PubMed]

13. Dunmore, B.J.; McCarthy, M.J.; Naylor, A.R.; Brindle, N.P. Carotid plaque instability and ischemic symptoms are linked to immaturity of microvessels within plaques. *J. Vasc. Surg.* **2007**, *45*, 155–159. [CrossRef] [PubMed]

14. Sluimer, J.C.; Kolodgie, F.D.; Bijnens, A.P.; Maxfield, K.; Pacheco, E.; Kutys, B.; Duimel, H.; Frederik, P.M.; van Hinsbergh, V.W.; Virmani, R.; *et al.* Thin-walled microvessels in human coronary atherosclerotic plaques show incomplete endothelial junctions relevance of compromised structural integrity for intraplaque microvascular leakage. *J. Am. Coll. Cardiol.* **2009**, *53*, 1517–1527. [CrossRef] [PubMed]

15. Khurana, R.; Zhuang, Z.; Bhardwaj, S.; Murakami, M.; de Muinck, E.; Yla-Herttuala, S.; Ferrara, N.; Martin, J.F.; Zachary, I.; Simons, M. Angiogenesis-dependent and independent phases of intimal hyperplasia. *Circulation* **2004**, *110*, 2436–2443. [CrossRef] [PubMed]

16. Herrmann, J.; Lerman, L.O.; Rodriguez-Porcel, M.; Holmes, D.R., Jr.; Richardson, D.M.; Ritman, E.L.; Lerman, A. Coronary vasa vasorum neovascularization precedes epicardial endothelial dysfunction in experimental hypercholesterolemia. *Cardiovasc. Res.* **2001**, *51*, 762–766. [CrossRef] [PubMed]

17. Maiellaro, K.; Taylor, W.R. The role of the adventitia in vascular inflammation. *Cardiovasc. Res.* **2007**, *75*, 640–648. [CrossRef] [PubMed]

18. Clarke, J.A. An X-ray microscopic study of the postnatal development of the vasa vasorum of normal human coronary arteries. *Acta Anat.* **1966**, *64*, 506–516. [CrossRef] [PubMed]

19. Wolinsky, H.; Glagov, S. Nature of species differences in the medial distribution of aortic vasa vasorum in mammals. *Circ. Res.* **1967**, *20*, 409–421. [CrossRef] [PubMed]

20. Schoenenberger, F.; Mueller, A. On the vascularization of the bovine aortic wall. *Helv. Physiol. Pharmacol. Acta* **1960**, *18*, 136–150. [PubMed]

21. Gossl, M.; Rosol, M.; Malyar, N.M.; Fitzpatrick, L.A.; Beighley, P.E.; Zamir, M.; Ritman, E.L. Functional anatomy and hemodynamic characteristics of vasa vasorum in the walls of porcine coronary arteries. *Anat. Rec. Part A Discov. Mol. Cell. Evol. Biol.* **2003**, *272*, 526–537. [CrossRef]

22. Fleiner, M.; Kummer, M.; Mirlacher, M.; Sauter, G.; Cathomas, G.; Krapf, R.; Biedermann, B.C. Arterial neovascularization and inflammation in vulnerable patients: Early and late signs of symptomatic atherosclerosis. *Circulation* **2004**, *110*, 2843–2850. [CrossRef] [PubMed]

23. Bitar, R.; Moody, A.R.; Leung, G.; Symons, S.; Crisp, S.; Butany, J.; Rowsell, C.; Kiss, A.; Nelson, A.; Maggisano, R. *In vivo* 3D high-spatial-resolution MR imaging of intraplaque hemorrhage. *Radiology* **2008**, *249*, 259–267. [CrossRef] [PubMed]

24. Acoltzin Vidal, C.; Maldonado Villasenor, I.; Rodriguez Cisneros, L.; Muniz Murguia, J.J. Diminished vascular density in the aortic wall. Morphological and functional characteristics of atherosclerosis. *Arch. Cardiol. Mexico* **2004**, *74*, 176–180.

25. Rademakers, T.; Douma, K.; Hackeng, T.M.; Post, M.J.; Sluimer, J.C.; Daemen, M.J.; Biessen, E.A.; Heeneman, S.; van Zandvoort, M.A. Plaque-associated vasa vasorum in aged apolipoprotein E-deficient mice exhibit proatherogenic functional features *in vivo*. *Arterioscler. Thromb. Vasc. Biol.* **2013**, *33*, 249–256. [CrossRef] [PubMed]

26. Gossl, M.; Malyar, N.M.; Rosol, M.; Beighley, P.E.; Ritman, E.L. Impact of coronary vasa vasorum functional structure on coronary vessel wall perfusion distribution. *Am. J. Physiol. Heart Circ. Physiol.* **2003**, *285*, H2019–H2026. [CrossRef] [PubMed]

27. Han, D.G. The innateness of coronary artery: Vasa vasorum. *Med. Hypotheses* **2010**, *74*, 443–444. [CrossRef] [PubMed]

28. Moss, A.J.; Samuelson, P.; Angell, C.; Minken, S.L. Polarographic evaluation of transmural oxygen availabitlity in intact muscular arteries. *J. Atheroscler. Res.* **1968**, *8*, 803–810. [CrossRef] [PubMed]

29. Crawford, D.W.; Back, L.H.; Cole, M.A. *In vivo* oxygen transport in the normal rabbit femoral arterial wall. *J. Clin. Investig.* **1980**, *65*, 1498–1508. [CrossRef] [PubMed]

30. Bratzler, R.L.; Chisolm, G.M.; Colton, C.K.; Smith, K.A.; Lees, R.S. The distribution of labeled low-density lipoproteins across the rabbit thoracic aorta *in vivo*. *Atherosclerosis* **1977**, *28*, 289–307. [CrossRef] [PubMed]

31. Scotland, R.S.; Vallance, P.J.; Ahluwalia, A. Endogenous factors involved in regulation of tone of arterial vasa vasorum: Implications for conduit vessel physiology. *Cardiovasc. Res.* **2000**, *46*, 403–411. [CrossRef] [PubMed]

32. Scotland, R.; Vallance, P.; Ahluwalia, A. Endothelin alters the reactivity of vasa vasorum: Mechanisms and implications for conduit vessel physiology and pathophysiology. *Br. J. Pharmacol.* **1999**, *128*, 1229–1234. [CrossRef] [PubMed]

33. Ohhira, A.; Ohhashi, T. Effects of aortic pressure and vasoactive agents on the vascular resistance of the vasa vasorum in canine isolated thoracic aorta. *J. Physiol.* **1992**, *453*, 233–245. [CrossRef] [PubMed]

34. Vio, A.; Gozzetti, G.; Reggiani, A. Importance of the vasa vasorum in the healing processes of arterial sutures. (Experimental study on the dog). *Boll. Soc. Ital. Biol. Sper.* **1967**, *43*, 88–90. [PubMed]

35. Winternitz, M.C.; Thomas, R.M.; LeCompte, P.M. *The Biology of Arteriosclerosis*; Springfield: Prince George's County, MA, USA, 1938.

36. Ribatti, D.; Levi-Schaffer, F.; Kovanen, P.T. Inflammatory angiogenesis in atherogenesis—A double-edged sword. *Ann. Med.* **2008**, *40*, 606–621. [CrossRef] [PubMed]

37. Williams, J.K.; Armstrong, M.L.; Heistad, D.D. Vasa vasorum in atherosclerotic coronary arteries: Responses to vasoactive stimuli and regression of atherosclerosis. *Circ. Res.* **1988**, *62*, 515–523. [CrossRef] [PubMed]

38. Sano, M.; Sasaki, T.; Hirakawa, S.; Sakabe, J.; Ogawa, M.; Baba, S.; Zaima, N.; Tanaka, H.; Inuzuka, K.; Yamamoto, N.; *et al.* Lymphangiogenesis and angiogenesis in abdominal aortic aneurysm. *PLoS ONE* **2014**, *9*, e89830. [CrossRef] [PubMed]

39. Davie, N.J.; Gerasimovskaya, E.V.; Hofmeister, S.E.; Richman, A.P.; Jones, P.L.; Reeves, J.T.; Stenmark, K.R. Pulmonary artery adventitial fibroblasts cooperate with vasa vasorum endothelial cells to regulate vasa vasorum neovascularization: A process mediated by hypoxia and endothelin-1. *Am. J. Pathol.* **2006**, *168*, 1793–1807. [CrossRef] [PubMed]

40. Bjornheden, T.; Levin, M.; Evaldsson, M.; Wiklund, O. Evidence of hypoxic areas within the arterial wall *in vivo*. *Arterioscler. Thromb. Vasc. Biol.* **1999**, *19*, 870–876. [CrossRef] [PubMed]

41. Nakashima, Y.; Chen, Y.X.; Kinukawa, N.; Sueishi, K. Distributions of diffuse intimal thickening in human arteries: Preferential expression in atherosclerosis-prone arteries from an early age. *Virchows Arch. Int. J. Pathol.* **2002**, *441*, 279–288. [CrossRef]

42. Zemplenyi, T.; Crawford, D.W.; Cole, M.A. Adaptation to arterial wall hypoxia demonstrated *in vivo* with oxygen microcathodes. *Atherosclerosis* **1989**, *76*, 173–179. [CrossRef] [PubMed]

43. Jarvilehto, M.; Tuohimaa, P. Vasa vasorum hypoxia: Initiation of atherosclerosis. *Med. Hypotheses* **2009**, *73*, 40–41. [CrossRef] [PubMed]

44. Den Hartog, J.P. *Strength of Materials*; Dover Publications, Inc.: New York, NY, USA, 1949; p. 323.

45. Bjornheden, T.; Bondjers, G. Oxygen consumption in aortic tissue from rabbits with diet-induced atherosclerosis. *Arteriosclerosis* **1987**, *7*, 238–247. [CrossRef] [PubMed]

46. Sluimer, J.C.; Gasc, J.M.; van Wanroij, J.L.; Kisters, N.; Groeneweg, M.; Sollewijn Gelpke, M.D.; Cleutjens, J.P.; van den Akker, L.H.; Corvol, P.; Wouters, B.G.; *et al.* Hypoxia, hypoxia-inducible transcription factor, and macrophages in human atherosclerotic plaques are correlated with intraplaque angiogenesis. *J. Am. Coll. Cardiol.* **2008**, *51*, 1258–1265. [CrossRef] [PubMed]

47. Kwon, H.M.; Sangiorgi, G.; Ritman, E.L.; Lerman, A.; McKenna, C.; Virmani, R.; Edwards, W.D.; Holmes, D.R.; Schwartz, R.S. Adventitial vasa vasorum in balloon-injured coronary arteries: Visualization and quantitation by a microscopic three-dimensional computed tomography technique. *J. Am. Coll. Cardiol.* **1998**, *32*, 2072–2079. [CrossRef] [PubMed]

48. Gossl, M.; Versari, D.; Mannheim, D.; Ritman, E.L.; Lerman, L.O.; Lerman, A. Increased spatial vasa vasorum density in the proximal LAD in hypercholesterolemia—Implications for vulnerable plaque-development. *Atherosclerosis* **2007**, *192*, 246–252. [CrossRef] [PubMed]

49. Sun, Z. Atherosclerosis and atheroma plaque rupture: Imaging modalities in the visualization of vasa vasorum and atherosclerotic plaques. *Sci. World J.* **2014**, *2014*, 312764.

50. Gossl, M.; Versari, D.; Lerman, L.O.; Chade, A.R.; Beighley, P.E.; Erbel, R.; Ritman, E.L. Low vasa vasorum densities correlate with inflammation and subintimal thickening: Potential role in location–determination of atherogenesis. *Atherosclerosis* **2009**, *206*, 362–368. [CrossRef] [PubMed]

51. Lim, C.S.; Kiriakidis, S.; Sandison, A.; Paleolog, E.M.; Davies, A.H. Hypoxia-inducible factor pathway and diseases of the vascular wall. *J. Vasc. Surg.* **2013**, *58*, 219–230. [CrossRef] [PubMed]

52. Semenza, G.L.; Agani, F.; Booth, G.; Forsythe, J.; Iyer, N.; Jiang, B.H.; Leung, S.; Roe, R.; Wiener, C.; Yu, A. Structural and functional analysis of hypoxia-inducible factor 1. *Kidney Int.* **1997**, *51*, 553–555. [CrossRef] [PubMed]

53. Oikawa, M.; Abe, M.; Kurosawa, H.; Hida, W.; Shirato, K.; Sato, Y. Hypoxia induces transcription factor ETS-1 via the activity of hypoxia-inducible factor-1. *Biochem. Biophys. Res. Commun.* **2001**, *289*, 39–43. [CrossRef] [PubMed]

54. Leung, D.W.; Cachianes, G.; Kuang, W.J.; Goeddel, D.V.; Ferrara, N. Vascular endothelial growth factor is a secreted angiogenic mitogen. *Science* **1989**, *246*, 1306–1309. [CrossRef] [PubMed]

55. Ushio-Fukai, M. VEGF signaling through NADPH oxidase-derived ROS. *Antioxid. Redox Signal.* **2007**, *9*, 731–739. [CrossRef] [PubMed]

56. Shweiki, D.; Itin, A.; Soffer, D.; Keshet, E. Vascular endothelial growth factor induced by hypoxia may mediate hypoxia-initiated angiogenesis. *Nature* **1992**, *359*, 843–845. [CrossRef] [PubMed]

57. Kuwahara, F.; Kai, H.; Tokuda, K.; Shibata, R.; Kusaba, K.; Tahara, N.; Niiyama, H.; Nagata, T.; Imaizumi, T. Hypoxia-inducible factor-1α/vascular endothelial growth factor pathway for adventitial vasa vasorum formation in hypertensive rat aorta. *Hypertension* **2002**, *39*, 46–50. [CrossRef] [PubMed]

58. Joukov, V.; Pajusola, K.; Kaipainen, A.; Chilov, D.; Lahtinen, I.; Kukk, E.; Saksela, O.; Kalkkinen, N.; Alitalo, K. A novel vascular endothelial growth factor, VEGF-C, is a ligand for the Flt4 (VEGFR-3) and KDR (VEGFR-2) receptor tyrosine kinases. *EMBO J.* **1996**, *15*, 290–298. [PubMed]

59. Achen, M.G.; Jeltsch, M.; Kukk, E.; Makinen, T.; Vitali, A.; Wilks, A.F.; Alitalo, K.; Stacker, S.A. Vascular endothelial growth factor D (VEGF-D) is a ligand for the tyrosine kinases VEGF receptor 2 (Flk1) and VEGF receptor 3 (Flt4). *Proc. Natl. Acad. Sci. USA* **1998**, *95*, 548–553. [CrossRef] [PubMed]

60. Kaipainen, A.; Korhonen, J.; Mustonen, T.; van Hinsbergh, V.W.; Fang, G.H.; Dumont, D.; Breitman, M.; Alitalo, K. Expression of the fms-like tyrosine kinase 4 gene becomes restricted to lymphatic endothelium during development. *Proc. Natl. Acad. Sci. USA* **1995**, *92*, 3566–3570. [CrossRef] [PubMed]

61. Kukk, E.; Lymboussaki, A.; Taira, S.; Kaipainen, A.; Jeltsch, M.; Joukov, V.; Alitalo, K. VEGF-C receptor binding and pattern of expression with VEGFR-3 suggests a role in lymphatic vascular development. *Development* **1996**, *122*, 3829–3837. [PubMed]

62. Partanen, T.A.; Arola, J.; Saaristo, A.; Jussila, L.; Ora, A.; Miettinen, M.; Stacker, S.A.; Achen, M.G.; Alitalo, K. VEGF-C and VEGF-D expression in neuroendocrine cells and their receptor, VEGFR-3, in fenestrated blood vessels in human tissues. *FASEB J.* **2000**, *14*, 2087–2096. [CrossRef] [PubMed]

63. Rutanen, J.; Leppanen, P.; Tuomisto, T.T.; Rissanen, T.T.; Hiltunen, M.O.; Vajanto, I.; Niemi, M.; Hakkinen, T.; Karkola, K.; Stacker, S.A.; *et al.* Vascular endothelial growth factor-D expression in human atherosclerotic lesions. *Cardiovasc. Res.* **2003**, *59*, 971–979. [CrossRef] [PubMed]

64. Belgore, F.; Blann, A.; Neil, D.; Ahmed, A.S.; Lip, G.Y. Localisation of members of the vascular endothelial growth factor (VEGF) family and their receptors in human atherosclerotic arteries. *J. Clin. Pathol.* **2004**, *57*, 266–272. [CrossRef] [PubMed]

65. Vuorio, T.; Nurmi, H.; Moulton, K.; Kurkipuro, J.; Robciuc, M.R.; Ohman, M.; Heinonen, S.E.; Samaranayake, H.; Heikura, T.; Alitalo, K.; *et al.* Lymphatic vessel insufficiency in hypercholesterolemic mice alters lipoprotein levels and promotes atherogenesis. *Arterioscler. Thromb. Vasc. Biol.* **2014**, *34*, 1162–1170. [CrossRef] [PubMed]

66. Iwasaka, C.; Tanaka, K.; Abe, M.; Sato, Y. Ets-1 regulates angiogenesis by inducing the expression of urokinase-type plasminogen activator and matrix metalloproteinase-1 and the migration of vascular endothelial cells. *J. Cell. Physiol.* **1996**, *169*, 522–531. [CrossRef] [PubMed]

67. Hashiya, N.; Jo, N.; Aoki, M.; Matsumoto, K.; Nakamura, T.; Sato, Y.; Ogata, N.; Ogihara, T.; Kaneda, Y.; Morishita, R. In vivo evidence of angiogenesis induced by transcription factor Ets-1: Ets-1 is located upstream of angiogenesis cascade. *Circulation* **2004**, *109*, 3035–3041. [CrossRef] [PubMed]

68. Higashida, T.; Kanno, H.; Nakano, M.; Funakoshi, K.; Yamamoto, I. Expression of hypoxia-inducible angiogenic proteins (hypoxia-inducible factor-1α, vascular endothelial growth factor, and E26 transformation-specific-1) and plaque hemorrhage in human carotid atherosclerosis. *J. Neurosurg.* **2008**, *109*, 83–91. [CrossRef] [PubMed]

69. Kitange, G.; Shibata, S.; Tokunaga, Y.; Yagi, N.; Yasunaga, A.; Kishikawa, M.; Naito, S. Ets-1 transcription factor-mediated urokinase-type plasminogen activator expression and invasion in glioma cells stimulated by serum and basic fibroblast growth factors. *Lab. Investig. J. Tech. Methods Pathol.* **1999**, *79*, 407–416.

70. Paumelle, R.; Tulasne, D.; Kherrouche, Z.; Plaza, S.; Leroy, C.; Reveneau, S.; Vandenbunder, B.; Fafeur, V. Hepatocyte growth factor/scatter factor activates the ETS1 transcription factor by a RAS-RAF-MEK-ERK signaling pathway. *Oncogene* **2002**, *21*, 2309–2319. [CrossRef] [PubMed]

71. Langheinrich, A.C.; Kampschulte, M.; Scheiter, F.; Dierkes, C.; Stieger, P.; Bohle, R.M.; Weidner, W. Atherosclerosis, inflammation and lipoprotein glomerulopathy in kidneys of apoE$^{-/-}$/LDL$^{-/-}$ double knockout mice. *BMC Nephrol.* **2010**, *11*. [CrossRef]

72. Kumamoto, M.; Nakashima, Y.; Sueishi, K. Intimal neovascularization in human coronary atherosclerosis: Its origin and pathophysiological significance. *Hum. Pathol.* **1995**, *26*, 450–456. [CrossRef] [PubMed]

73. Yamashita, A.; Shoji, K.; Tsuruda, T.; Furukoji, E.; Takahashi, M.; Nishihira, K.; Tamura, S.; Asada, Y. Medial and adventitial macrophages are associated with expansive atherosclerotic remodeling in rabbit femoral artery. *Histol. Histopathol.* **2008**, *23*, 127–136. [PubMed]

74. Sluimer, J.C.; Daemen, M.J. Novel concepts in atherogenesis: Angiogenesis and hypoxia in atherosclerosis. *J. Pathol.* **2009**, *218*, 7–29. [CrossRef] [PubMed]

75. Brown, A.J.; Dean, R.T.; Jessup, W. Free and esterified oxysterol: Formation during copper-oxidation of low density lipoprotein and uptake by macrophages. *J. Lipid Res.* **1996**, *37*, 320–335. [PubMed]

76. Ho-Tin-Noe, B.; le Dall, J.; Gomez, D.; Louedec, L.; Vranckx, R.; El-Bouchtaoui, M.; Legres, L.; Meilhac, O.; Michel, J.B. Early atheroma-derived agonists of peroxisome proliferator-activated receptor-gamma trigger intramedial angiogenesis in a smooth muscle cell-dependent manner. *Circ. Res.* **2011**, *109*, 1003–1014. [CrossRef] [PubMed]

77. Sahebkar, A.; Watts, G.F. New LDL-cholesterol lowering therapies: Pharmacology, clinical trials, and relevance to acute coronary syndromes. *Clin. Ther.* **2013**, *35*, 1082–1098. [CrossRef] [PubMed]

78. Lusis, A.J. Atherosclerosis. *Nature* **2000**, *407*, 233–241. [CrossRef] [PubMed]

79. Kolodgie, F.D.; Gold, H.K.; Burke, A.P.; Fowler, D.R.; Kruth, H.S.; Weber, D.K.; Farb, A.; Guerrero, L.J.; Hayase, M.; Kutys, R.; *et al.* Intraplaque hemorrhage and progression of coronary atheroma. *N. Engl. J. Med.* **2003**, *349*, 2316–2325. [CrossRef] [PubMed]

80. Tanaka, K.; Nagata, D.; Hirata, Y.; Tabata, Y.; Nagai, R.; Sata, M. Augmented angiogenesis in adventitia promotes growth of atherosclerotic plaque in apolipoprotein E-deficient mice. *Atherosclerosis* **2011**, *215*, 366–373. [CrossRef] [PubMed]

81. Moulton, K.S. Angiogenesis in atherosclerosis: Gathering evidence beyond speculation. *Curr. Opin. Lipidol.* **2006**, *17*, 548–555. [CrossRef] [PubMed]

82. Ricote, M.; Li, A.C.; Willson, T.M.; Kelly, C.J.; Glass, C.K. The peroxisome proliferator-activated receptor-gamma is a negative regulator of macrophage activation. *Nature* **1998**, *391*, 79–82. [CrossRef] [PubMed]

83. Fang, L.; Liu, C.; Miller, Y.I. Zebrafish models of dyslipidemia: Relevance to atherosclerosis and angiogenesis. *Transl. Res. J. Lab. Clin. Med.* **2014**, *163*, 99–108. [CrossRef]

84. Salomon, R.G.; Hong, L.; Hollyfield, J.G. Discovery of carboxyethylpyrroles (CEPs): Critical insights into AMD, autism, cancer, and wound healing from basic research on the chemistry of oxidized phospholipids. *Chem. Res. Toxicol.* **2011**, *24*, 1803–1816. [CrossRef] [PubMed]

85. West, X.Z.; Malinin, N.L.; Merkulova, A.A.; Tischenko, M.; Kerr, B.A.; Borden, E.C.; Podrez, E.A.; Salomon, R.G.; Byzova, T.V. Oxidative stress induces angiogenesis by activating TLR2 with novel endogenous ligands. *Nature* **2010**, *467*, 972–976. [CrossRef] [PubMed]

86. Chatterjee, T.K.; Stoll, L.L.; Denning, G.M.; Harrelson, A.; Blomkalns, A.L.; Idelman, G.; Rothenberg, F.G.; Neltner, B.; Romig-Martin, S.A.; Dickson, E.W.; *et al.* Proinflammatory phenotype of perivascular adipocytes: Influence of high-fat feeding. *Circ. Res.* **2009**, *104*, 541–549. [CrossRef] [PubMed]

87. Barandier, C.; Montani, J.P.; Yang, Z. Mature adipocytes and perivascular adipose tissue stimulate vascular smooth muscle cell proliferation: Effects of aging and obesity. *Am. J. Physiol. Heart Circ. Physiol.* **2005**, *289*, H1807–H1813. [CrossRef] [PubMed]

88. Wang, P.; Xu, T.Y.; Guan, Y.F.; Su, D.F.; Fan, G.R.; Miao, C.Y. Perivascular adipose tissue-derived visfatin is a vascular smooth muscle cell growth factor: Role of nicotinamide mononucleotide. *Cardiovasc. Res.* **2009**, *81*, 370–380. [CrossRef] [PubMed]

89. Manka, D.; Chatterjee, T.K.; Stoll, L.L.; Basford, J.E.; Konaniah, E.S.; Srinivasan, R.; Bogdanov, V.Y.; Tang, Y.; Blomkalns, A.L.; Hui, D.Y.; Weintraub, N.L. Transplanted perivascular adipose tissue accelerates injury-induced neointimal hyperplasia: Role of monocyte chemoattractant protein-1. *Arterioscler. Thromb. Vasc. Biol.* **2014**, *34*, 1723–1730. [CrossRef] [PubMed]

90. Rajsheker, S.; Manka, D.; Blomkalns, A.L.; Chatterjee, T.K.; Stoll, L.L.; Weintraub, N.L. Crosstalk between perivascular adipose tissue and blood vessels. *Curr. Opin. Pharmacol.* **2010**, *10*, 191–196. [CrossRef] [PubMed]

91. Garcia-Touchard, A.; Henry, T.D.; Sangiorgi, G.; Spagnoli, L.G.; Mauriello, A.; Conover, C.; Schwartz, R.S. Extracellular proteases in atherosclerosis and restenosis. *Arterioscler. Thromb. Vasc. Biol.* **2005**, *25*, 1119–1127. [CrossRef] [PubMed]

92. Liu, X.Q.; Mao, Y.; Wang, B.; Lu, X.T.; Bai, W.W.; Sun, Y.Y.; Liu, Y.; Liu, H.M.; Zhang, L.; Zhao, Y.X.; *et al.* Specific matrix metalloproteinases play different roles in intraplaque angiogenesis and plaque instability in rabbits. *PLoS ONE* **2014**, *9*, e107851. [CrossRef] [PubMed]

93. Svensson, P.A.; Olson, F.J.; Hagg, D.A.; Ryndel, M.; Wiklund, O.; Karlstrom, L.; Hulthe, J.; Carlsson, L.M.; Fagerberg, B. Urokinase-type plasminogen activator receptor is associated with macrophages and plaque rupture in symptomatic carotid atherosclerosis. *Int. J. Mol. Med.* **2008**, *22*, 459–464. [PubMed]

94. Mattock, K.L.; Gough, P.J.; Humphries, J.; Burnand, K.; Patel, L.; Suckling, K.E.; Cuello, F.; Watts, C.; Gautel, M.; Avkiran, M.; *et al.* Legumain and cathepsin-L expression in human unstable carotid plaque. *Atherosclerosis* **2010**, *208*, 83–89. [CrossRef] [PubMed]

95. Li, W.; Kornmark, L.; Jonasson, L.; Forssell, C.; Yuan, X.M. Cathepsin L is significantly associated with apoptosis and plaque destabilization in human atherosclerosis. *Atherosclerosis* **2009**, *202*, 92–102. [CrossRef] [PubMed]

96. Le Dall, J.; Ho-Tin-Noe, B.; Louedec, L.; Meilhac, O.; Roncal, C.; Carmeliet, P.; Germain, S.; Michel, J.B.; Houard, X. Immaturity of microvessels in haemorrhagic plaques is associated with proteolytic degradation of angiogenic factors. *Cardiovasc. Res.* **2010**, *85*, 184–193.

97. Leclercq, A.; Houard, X.; Philippe, M.; Ollivier, V.; Sebbag, U.; Meilhac, O.; Michel, J.B. Involvement of intraplaque hemorrhage in atherothrombosis evolution via neutrophil protease enrichment. *J. Leukoc. Biol.* **2007**, *82*, 1420–1429. [CrossRef] [PubMed]

98. Maisonpierre, P.C.; Suri, C.; Jones, P.F.; Bartunkova, S.; Wiegand, S.J.; Radziejewski, C.; Compton, D.; McClain, J.; Aldrich, T.H.; Papadopoulos, N.; *et al.* Angiopoietin-2, a natural antagonist for Tie2 that disrupts *in vivo* angiogenesis. *Science* **1997**, *277*, 55–60. [CrossRef] [PubMed]

99. Hashimoto, T.; Lam, T.; Boudreau, N.J.; Bollen, A.W.; Lawton, M.T.; Young, W.L. Abnormal balance in the angiopoietin-tie2 system in human brain arteriovenous malformations. *Circ. Res.* **2001**, *89*, 111–113. [CrossRef] [PubMed]

100. Anagnostopoulos, A.; Eleftherakis-Papaiakovou, V.; Kastritis, E.; Tsionos, K.; Bamias, A.; Meletis, J.; Dimopoulos, M.A.; Terpos, E. Serum concentrations of angiogenic cytokines in Waldenstrom macroglobulinaemia: The ration of angiopoietin-1 to angiopoietin-2 and angiogenin correlate with disease severity. *Br. J. Haematol.* **2007**, *137*, 560–568. [CrossRef] [PubMed]

101. Post, S.; Peeters, W.; Busser, E.; Lamers, D.; Sluijter, J.P.; Goumans, M.J.; de Weger, R.A.; Moll, F.L.; Doevendans, P.A.; Pasterkamp, G.; *et al.* Balance between angiopoietin-1 and angiopoietin-2 is in favor of angiopoietin-2 in atherosclerotic plaques with high microvessel density. *J. Vasc. Res.* **2008**, *45*, 244–250. [CrossRef] [PubMed]

102. Chistiakov, D.A.; Orekhov, A.N.; Bobryshev, Y.V. Contribution of neovascularization and intraplaque haemorrhage to atherosclerotic plaque progression and instability. *Acta Physiol.* **2015**, *213*, 539–553. [CrossRef]

103. Nagy, E.; Eaton, J.W.; Jeney, V.; Soares, M.P.; Varga, Z.; Galajda, Z.; Szentmiklosi, J.; Mehes, G.; Csonka, T.; Smith, A.; *et al.* Red cells, hemoglobin, heme, iron, and atherogenesis. *Arterioscler. Thromb. Vasc. Biol.* **2010**, *30*, 1347–1353. [CrossRef] [PubMed]

104. Balla, J.; Jacob, H.S.; Balla, G.; Nath, K.; Eaton, J.W.; Vercellotti, G.M. Endothelial-cell heme uptake from heme proteins: Induction of sensitization and desensitization to oxidant damage. *Proc. Natl. Acad. Sci. USA* **1993**, *90*, 9285–9289. [CrossRef] [PubMed]

105. Potor, L.; Banyai, E.; Becs, G.; Soares, M.P.; Balla, G.; Balla, J.; Jeney, V. Atherogenesis may involve the prooxidant and proinflammatory effects of ferryl hemoglobin. *Oxid. Med. Cell. Longev.* **2013**, *2013*. [CrossRef]

106. Juckett, M.B.; Balla, J.; Balla, G.; Jessurun, J.; Jacob, H.S.; Vercellotti, G.M. Ferritin protects endothelial cells from oxidized low density lipoprotein *in vitro*. *Am. J. Pathol.* **1995**, *147*, 782–789. [PubMed]

107. Lee, F.Y.; Lee, T.S.; Pan, C.C.; Huang, A.L.; Chau, L.Y. Colocalization of iron and ceroid in human atherosclerotic lesions. *Atherosclerosis* **1998**, *138*, 281–288. [CrossRef] [PubMed]

108. Baldwin, A.L. Modified hemoglobins produce venular interendothelial gaps and albumin leakage in the rat mesentery. *Am. J. Physiol.* **1999**, *277*, H650–H659. [PubMed]

109. Abela, G.S.; Aziz, K.; Vedre, A.; Pathak, D.R.; Talbott, J.D.; Dejong, J. Effect of cholesterol crystals on plaques and intima in arteries of patients with acute coronary and cerebrovascular syndromes. *Am. J. Cardiol.* **2009**, *103*, 959–968. [CrossRef] [PubMed]

110. Kaartinen, M.; Penttila, A.; Kovanen, P.T. Mast cells accompany microvessels in human coronary atheromas: Implications for intimal neovascularization and hemorrhage. *Atherosclerosis* **1996**, *123*, 123–131. [CrossRef] [PubMed]

111. Ambrose, J.A.; Tannenbaum, M.A.; Alexopoulos, D.; Hjemdahl-Monsen, C.E.; Leavy, J.; Weiss, M.; Borrico, S.; Gorlin, R.; Fuster, V. Angiographic progression of coronary artery disease and the development of myocardial infarction. *J. Am. Coll. Cardiol.* **1988**, *12*, 56–62. [CrossRef] [PubMed]

112. Hyafil, F.; Cornily, J.C.; Feig, J.E.; Gordon, R.; Vucic, E.; Amirbekian, V.; Fisher, E.A.; Fuster, V.; Feldman, L.J.; Fayad, Z.A. Noninvasive detection of macrophages using a nanoparticulate contrast agent for computed tomography. *Nat. Med.* **2007**, *13*, 636–641. [CrossRef] [PubMed]

113. Romero, J.M.; Pizzolato, R.; Atkinson, W.; Meader, A.; Jaimes, C.; Lamuraglia, G.; Jaff, M.R.; Buonanno, F.; Delgado Almandoz, J.; Gonzalez, R.G. Vasa vasorum enhancement on computerized tomographic angiography correlates with symptomatic patients with 50% to 70% carotid artery stenosis. *Stroke J. Cereb. Circ.* **2013**, *44*, 3344–3349. [CrossRef]

114. Sadeghi, M.M.; Glover, D.K.; Lanza, G.M.; Fayad, Z.A.; Johnson, L.L. Imaging atherosclerosis and vulnerable plaque. *J. Nucl. Med.* **2010**, *51*, 51S–65S. [CrossRef] [PubMed]

115. Li, W.; van der Steen, A.F.; Lancee, C.T.; Cespedes, I.; Bom, N. Blood flow imaging and volume flow quantitation with intravascular ultrasound. *Ultrasound Med. Biol.* **1998**, *24*, 203–214. [CrossRef] [PubMed]

116. Moritz, R.; Eaker, D.R.; Anderson, J.L.; Kline, T.L.; Jorgensen, S.M.; Lerman, A.; Ritman, E.L. IVUS detection of vasa vasorum blood flow distribution in coronary artery vessel wall. *JACC Cardiovasc. Imaging* **2012**, *5*, 935–940. [CrossRef] [PubMed]

117. Papaioannou, T.G.; Vavuranakis, M.; Androulakis, A.; Lazaros, G.; Kakadiaris, I.; Vlaseros, I.; Naghavi, M.; Kallikazaros, I.; Stefanadis, C. *In-vivo* imaging of carotid plaque neoangiogenesis with contrast-enhanced harmonic ultrasound. *Int. J. Cardiol.* **2009**, *134*, e110–e112. [CrossRef] [PubMed]

118. O'Malley, S.M.; Vavuranakis, M.; Naghavi, M.; Kakadiaris, I.A. Intravascular ultrasound-based imaging of vasa vasorum for the detection of vulnerable atherosclerotic plaque. *Med. Image Comput. Comput. Assist. Interv. MICCAI* **2005**, *8 Pt 1*, 343–351.

119. Vavuranakis, M.; Kakadiaris, I.A.; O'Malley, S.M.; Papaioannou, T.G.; Sanidas, E.A.; Naghavi, M.; Carlier, S.; Tousoulis, D.; Stefanadis, C. A new method for assessment of plaque vulnerability based on vasa vasorum imaging, by using contrast-enhanced intravascular ultrasound and differential image analysis. *Int. J. Cardiol.* **2008**, *130*, 23–29. [CrossRef] [PubMed]

120. Goertz, D.E.; Frijlink, M.E.; Tempel, D.; van Damme, L.C.; Krams, R.; Schaar, J.A.; Ten Cate, F.J.; Serruys, P.W.; de Jong, N.; van der Steen, A.F. Contrast harmonic intravascular ultrasound: A feasibility study for vasa vasorum imaging. *Investig. Radiol.* **2006**, *41*, 631–638. [CrossRef]

121. Goertz, D.E.; Frijlink, M.E.; Tempel, D.; Bhagwandas, V.; Gisolf, A.; Krams, R.; de Jong, N.; van der Steen, A.F. Subharmonic contrast intravascular ultrasound for vasa vasorum imaging. *Ultrasound Med. Biol.* **2007**, *33*, 1859–1872. [CrossRef] [PubMed]

122. Goertz, D.E.; Frijlink, M.E.; de Jong, N.; van der Steen, A.F. Nonlinear intravascular ultrasound contrast imaging. *Ultrasound Med. Biol.* **2006**, *32*, 491–502. [CrossRef] [PubMed]

123. Magnoni, M.; Coli, S.; Marrocco-Trischitta, M.M.; Melisurgo, G.; de Dominicis, D.; Cianflone, D.; Chiesa, R.; Feinstein, S.B.; Maseri, A. Contrast-enhanced ultrasound imaging of periadventitial vasa vasorum in human carotid arteries. *Eur. J. Echocardiogr.* **2009**, *10*, 260–264. [CrossRef] [PubMed]

124. Shah, F.; Balan, P.; Weinberg, M.; Reddy, V.; Neems, R.; Feinstein, M.; Dainauskas, J.; Meyer, P.; Goldin, M.; Feinstein, S.B. Contrast-enhanced ultrasound imaging of atherosclerotic carotid plaque neovascularization: A new surrogate marker of atherosclerosis? *Vasc. Med.* **2007**, *12*, 291–297. [CrossRef] [PubMed]

125. Schinkel, A.F.; Krueger, C.G.; Tellez, A.; Granada, J.F.; Reed, J.D.; Hall, A.; Zang, W.; Owens, C.; Kaluza, G.L.; Staub, D.; *et al.* Contrast-enhanced ultrasound for imaging vasa vasorum: Comparison with histopathology in a swine model of atherosclerosis. *Eur. J. Echocardiogr.* **2010**, *11*, 659–664. [CrossRef] [PubMed]

126. Coli, S.; Magnoni, M.; Sangiorgi, G.; Marrocco-Trischitta, M.M.; Melisurgo, G.; Mauriello, A.; Spagnoli, L.; Chiesa, R.; Cianflone, D.; Maseri, A. Contrast-enhanced ultrasound imaging of intraplaque neovascularization in carotid arteries: Correlation with histology and plaque echogenicity. *J. Am. Coll. Cardiol.* **2008**, *52*, 223–230. [CrossRef] [PubMed]

127. Lee, S.C.; Carr, C.L.; Davidson, B.P.; Ellegala, D.; Xie, A.; Ammi, A.; Belcik, T.; Lindner, J.R. Temporal characterization of the functional density of the vasa vasorum by contrast-enhanced ultrasonography maximum intensity projection imaging. *JACC Cardiovasc. Imaging* **2010**, *3*, 1265–1272. [CrossRef] [PubMed]

128. Moguillansky, D.; Leng, X.; Carson, A.; Lavery, L.; Schwartz, A.; Chen, X.; Villanueva, F.S. Quantification of plaque neovascularization using contrast ultrasound: A histologic validation. *Eur. Heart J.* **2011**, *32*, 646–653. [CrossRef] [PubMed]

129. Tian, J.; Hu, S.; Sun, Y.; Yu, H.; Han, X.; Cheng, W.; Ban, X.; Zhang, S.; Yu, B.; Jang, I.K. Vasa vasorum and plaque progression, and responses to atorvastatin in a rabbit model of atherosclerosis: Contrast-enhanced ultrasound imaging and intravascular ultrasound study. *Heart* **2013**, *99*, 48–54. [CrossRef] [PubMed]

130. Deyama, J.; Nakamura, T.; Takishima, I.; Fujioka, D.; Kawabata, K.; Obata, J.E.; Watanabe, K.; Watanabe, Y.; Saito, Y.; Mishina, H.; *et al.* Contrast-enhanced ultrasound imaging of carotid plaque neovascularization is useful for identifying high-risk patients with coronary artery disease. *Circ. J.* **2013**, *77*, 1499–1507. [CrossRef] [PubMed]

131. Vavuranakis, M.; Sigala, F.; Vrachatis, D.A.; Papaioannou, T.G.; Filis, K.; Kavantzas, N.; Kalogeras, K.I.; Massoura, C.; Toufektzian, L.; Kariori, M.G.; *et al.* Quantitative analysis of carotid plaque vasa vasorum by CEUS and correlation with histology after endarterectomy. *VASA Z. Gefasskrankh.* **2013**, *42*, 184–195. [CrossRef]

132. Kubo, T.; Tanaka, A.; Ino, Y.; Kitabata, H.; Shiono, Y.; Akasaka, T. Assessment of coronary atherosclerosis using optical coherence tomography. *J. Atheroscler. Thromb.* **2014**, *21*, 895–903. [CrossRef] [PubMed]

133. Kume, T.; Akasaka, T.; Kawamoto, T.; Watanabe, N.; Toyota, E.; Neishi, Y.; Sukmawan, R.; Sadahira, Y.; Yoshida, K. Assessment of coronary intima—Media thickness by optical coherence tomography: Comparison with intravascular ultrasound. *Circ. J.* **2005**, *69*, 903–907. [CrossRef] [PubMed]

134. Kitabata, H.; Tanaka, A.; Kubo, T.; Takarada, S.; Kashiwagi, M.; Tsujioka, H.; Ikejima, H.; Kuroi, A.; Kataiwa, H.; Ishibashi, K.; *et al.* Relation of microchannel structure identified by optical coherence tomography to plaque vulnerability in patients with coronary artery disease. *Am. J. Cardiol.* **2010**, *105*, 1673–1678. [CrossRef] [PubMed]
135. Cheng, K.H.; Sun, C.; Vuong, B.; Lee, K.K.; Mariampillai, A.; Marotta, T.R.; Spears, J.; Montanera, W.J.; Herman, P.R.; Kiehl, T.R.; *et al.* Endovascular optical coherence tomography intensity kurtosis: Visualization of vasa vasorum in porcine carotid artery. *Biomed. Opt. Express* **2012**, *3*, 388–399. [CrossRef] [PubMed]
136. Tearney, G.J.; Regar, E.; Akasaka, T.; Adriaenssens, T.; Barlis, P.; Bezerra, H.G.; Bouma, B.; Bruining, N.; Cho, J.M.; Chowdhary, S.; *et al.* Consensus standards for acquisition, measurement, and reporting of intravascular optical coherence tomography studies: A report from the International Working Group for Intravascular Optical Coherence Tomography Standardization and Validation. *J. Am. Coll. Cardiol.* **2012**, *59*, 1058–1072. [CrossRef] [PubMed]
137. Nishimiya, K.; Matsumoto, Y.; Takahashi, J.; Uzuka, H.; Odaka, Y.; Nihei, T.; Hao, K.; Tsuburaya, R.; Ito, K.; Shimokawa, H. *In vivo* visualization of adventitial vasa vasorum of the human coronary artery on optical frequency domain imaging. Validation study. *Circ. J.* **2014**, *78*, 2516–2518. [CrossRef] [PubMed]
138. Aoki, T.; Rodriguez-Porcel, M.; Matsuo, Y.; Cassar, A.; Kwon, T.G.; Franchi, F.; Gulati, R.; Kushwaha, S.S.; Lennon, R.J.; Lerman, L.O.; *et al.* Evaluation of coronary adventitial vasa vasorum using 3D optical coherence tomography—Animal and human studies. *Atherosclerosis* **2015**, *239*, 203–208. [CrossRef] [PubMed]
139. Dobrucki, L.W.; Sinusas, A.J. PET and SPECT in cardiovascular molecular imaging. *Nat. Rev. Cardiol.* **2010**, *7*, 38–47. [CrossRef] [PubMed]
140. Tawakol, A.; Migrino, R.Q.; Bashian, G.G.; Bedri, S.; Vermylen, D.; Cury, R.C.; Yates, D.; LaMuraglia, G.M.; Furie, K.; Houser, S.; *et al.* In vivo $^{18}$F-fluorodeoxyglucose positron emission tomography imaging provides a noninvasive measure of carotid plaque inflammation in patients. *J. Am. Coll. Cardiol.* **2006**, *48*, 1818–1824. [CrossRef] [PubMed]
141. Folco, E.J.; Sheikine, Y.; Rocha, V.Z.; Christen, T.; Shvartz, E.; Sukhova, G.K.; di Carli, M.F.; Libby, P. Hypoxia but not inflammation augments glucose uptake in human macrophages: Implications for imaging atherosclerosis with $^{18}$Fluorine-labeled 2-deoxy-D-glucose positron emission tomography. *J. Am. Coll. Cardiol.* **2011**, *58*, 603–614. [CrossRef] [PubMed]
142. Rodriguez-Porcel, M.; Cai, W.; Gheysens, O.; Willmann, J.K.; Chen, K.; Wang, H.; Chen, I.Y.; He, L.; Wu, J.C.; Li, Z.B.; *et al.* Imaging of VEGF receptor in a rat myocardial infarction model using PET. *J. Nucl. Med.* **2008**, *49*, 667–673. [CrossRef] [PubMed]
143. Meoli, D.F.; Sadeghi, M.M.; Krassilnikova, S.; Bourke, B.N.; Giordano, F.J.; Dione, D.P.; Su, H.; Edwards, D.S.; Liu, S.; Harris, T.D.; *et al.* Noninvasive imaging of myocardial angiogenesis following experimental myocardial infarction. *J. Clin. Investig.* **2004**, *113*, 1684–1691. [CrossRef] [PubMed]
144. Janssen, M.L.; Oyen, W.J.; Dijkgraaf, I.; Massuger, L.F.; Frielink, C.; Edwards, D.S.; Rajopadhye, M.; Boonstra, H.; Corstens, F.H.; Boerman, O.C. Tumor targeting with radiolabeled $\alpha_v \beta_3$ integrin binding peptides in a nude mouse model. *Cancer Res.* **2002**, *62*, 6146–6151. [PubMed]
145. Kopka, K.; Breyholz, H.J.; Wagner, S.; Law, M.P.; Riemann, B.; Schroer, S.; Trub, M.; Guilbert, B.; Levkau, B.; Schober, O.; *et al.* Synthesis and preliminary biological evaluation of new radioiodinated MMP inhibitors for imaging MMP activity *in vivo*. *Nucl. Med. Biol.* **2004**, *31*, 257–267. [CrossRef] [PubMed]
146. Furumoto, S.; Takashima, K.; Kubota, K.; Ido, T.; Iwata, R.; Fukuda, H. Tumor detection using $^{18}$F-labeled matrix metalloproteinase-2 inhibitor. *Nucl.Med. Biol.* **2003**, *30*, 119–125. [CrossRef] [PubMed]
147. Schafers, M.; Riemann, B.; Kopka, K.; Breyholz, H.J.; Wagner, S.; Schafers, K.P.; Law, M.P.; Schober, O.; Levkau, B. Scintigraphic imaging of matrix metalloproteinase activity in the arterial wall *in vivo*. *Circulation* **2004**, *109*, 2554–2559. [CrossRef] [PubMed]
148. Matter, C.M.; Schuler, P.K.; Alessi, P.; Meier, P.; Ricci, R.; Zhang, D.; Halin, C.; Castellani, P.; Zardi, L.; Hofer, C.K.; *et al.* Molecular imaging of atherosclerotic plaques using a human antibody against the extra-domain B of fibronectin. *Circ. Res.* **2004**, *95*, 1225–1233. [CrossRef] [PubMed]
149. Kuge, Y.; Takai, N.; Ogawa, Y.; Temma, T.; Zhao, Y.; Nishigori, K.; Ishino, S.; Kamihashi, J.; Kiyono, Y.; Shiomi, M.; *et al.* Imaging with radiolabelled anti-membrane type 1 matrix metalloproteinase (MT1-MMP) antibody: Potentials for characterizing atherosclerotic plaques. *Eur. J. Nucl. Med. Mol. Imaging* **2010**, *37*, 2093–2104. [CrossRef] [PubMed]

150. Aoki, S.; Aoki, K.; Ohsawa, S.; Nakajima, H.; Kumagai, H.; Araki, T. Dynamic MR imaging of the carotid wall. *J. Magn. Reson. Imaging JMRI* **1999**, *9*, 420–427. [CrossRef]
151. Winter, P.M.; Morawski, A.M.; Caruthers, S.D.; Fuhrhop, R.W.; Zhang, H.; Williams, T.A.; Allen, J.S.; Lacy, E.K.; Robertson, J.D.; Lanza, G.M.; *et al.* Molecular imaging of angiogenesis in early-stage atherosclerosis with $\alpha_v \beta_3$-integrin-targeted nanoparticles. *Circulation* **2003**, *108*, 2270–2274. [CrossRef] [PubMed]
152. Cai, K.; Caruthers, S.D.; Huang, W.; Williams, T.A.; Zhang, H.; Wickline, S.A.; Lanza, G.M.; Winter, P.M. MR molecular imaging of aortic angiogenesis. *JACC Cardiovasc. Imaging* **2010**, *3*, 824–832. [CrossRef] [PubMed]
153. Kerwin, W.; Hooker, A.; Spilker, M.; Vicini, P.; Ferguson, M.; Hatsukami, T.; Yuan, C. Quantitative magnetic resonance imaging analysis of neovasculature volume in carotid atherosclerotic plaque. *Circulation* **2003**, *107*, 851–856. [CrossRef] [PubMed]
154. Kerwin, W.S.; Oikawa, M.; Yuan, C.; Jarvik, G.P.; Hatsukami, T.S. MR imaging of adventitial vasa vasorum in carotid atherosclerosis. *Magn. Reson. Med.* **2008**, *59*, 507–514. [CrossRef] [PubMed]
155. Sun, J.; Song, Y.; Chen, H.; Kerwin, W.S.; Hippe, D.S.; Dong, L.; Chen, M.; Zhou, C.; Hatsukami, T.S.; Yuan, C. Adventitial perfusion and intraplaque hemorrhage: A dynamic contrast-enhanced MRI study in the carotid artery. *Stroke J. Cereb. Circ.* **2013**, *44*, 1031–1036. [CrossRef]
156. Grobner, T. Gadolinium—A specific trigger for the development of nephrogenic fibrosing dermopathy and nephrogenic systemic fibrosis? *Nephrol. Dial. Transpl.* **2006**, *21*, 1104–1108. [CrossRef]
157. Feinstein, S.B. Contrast ultrasound imaging of the carotid artery vasa vasorum and atherosclerotic plaque neovascularization. *J. Am. Coll. Cardiol.* **2006**, *48*, 236–243. [CrossRef] [PubMed]
158. Decano, J.L.; Moran, A.M.; Ruiz-Opazo, N.; Herrera, V.L. Molecular imaging of vasa vasorum neovascularization via DEspR-targeted contrast-enhanced ultrasound micro-imaging in transgenic atherosclerosis rat model. *Mol. Imaging Biol. MIB* **2011**, *13*, 1096–1106. [CrossRef]
159. Baron Toaldo, M.; Salvatore, V.; Marinelli, S.; Palama, C.; Milazzo, M.; Croci, L.; Venerandi, L.; Cipone, M.; Bolondi, L.; Piscaglia, F. Use of VEGFR-2 targeted ultrasound contrast agent for the early evaluation of response to sorafenib in a mouse model of hepatocellular carcinoma. *Mol. Imaging Biol. MIB* **2015**, *17*, 29–37. [CrossRef]
160. Araldi, E.; Chamorro-Jorganes, A.; van Solingen, C.; Fernandez-Hernando, C.; Suarez, Y. Therapeutic potential of modulating microRNAs in atherosclerotic vascular disease. *Curr. Vasc. Pharmacol.* **2013**, in press.
161. O'Reilly, M.S.; Holmgren, L.; Shing, Y.; Chen, C.; Rosenthal, R.A.; Moses, M.; Lane, W.S.; Cao, Y.; Sage, E.H.; Folkman, J. Angiostatin: A novel angiogenesis inhibitor that mediates the suppression of metastases by a Lewis lung carcinoma. *Cell* **1994**, *79*, 315–328. [CrossRef] [PubMed]
162. O'Reilly, M.S.; Boehm, T.; Shing, Y.; Fukai, N.; Vasios, G.; Lane, W.S.; Flynn, E.; Birkhead, J.R.; Olsen, B.R.; Folkman, J. Endostatin: An endogenous inhibitor of angiogenesis and tumor growth. *Cell* **1997**, *88*, 277–285. [CrossRef] [PubMed]
163. Moulton, K.S.; Heller, E.; Konerding, M.A.; Flynn, E.; Palinski, W.; Folkman, J. Angiogenesis inhibitors endostatin or TNP-470 reduce intimal neovascularization and plaque growth in apolipoprotein E-deficient mice. *Circulation* **1999**, *99*, 1726–1732. [CrossRef] [PubMed]
164. Moulton, K.S.; Vakili, K.; Zurakowski, D.; Soliman, M.; Butterfield, C.; Sylvin, E.; Lo, K.M.; Gillies, S.; Javaherian, K.; Folkman, J. Inhibition of plaque neovascularization reduces macrophage accumulation and progression of advanced atherosclerosis. *Proc. Natl. Acad. Sci. USA* **2003**, *100*, 4736–4741. [CrossRef] [PubMed]
165. Eriksson, K.; Magnusson, P.; Dixelius, J.; Claesson-Welsh, L.; Cross, M.J. Angiostatin and endostatin inhibit endothelial cell migration in response to FGF and VEGF without interfering with specific intracellular signal transduction pathways. *FEBS Lett.* **2003**, *536*, 19–24. [CrossRef] [PubMed]
166. Stefanadis, C.; Toutouzas, K.; Stefanadi, E.; Kolodgie, F.; Virmani, R.; Kipshidze, N. First experimental application of bevacizumab-eluting PC coated stent for inhibition of vasa vasorum of atherosclerotic plaque: Angiographic results in a rabbit atheromatic model. *Hell. J. Cardiol. HJC* **2006**, *47*, 7–10.
167. Luttun, A.; Tjwa, M.; Moons, L.; Wu, Y.; Angelillo-Scherrer, A.; Liao, F.; Nagy, J.A.; Hooper, A.; Priller, J.; de Klerck, B.; *et al.* Revascularization of ischemic tissues by PlGF treatment, and inhibition of tumor angiogenesis, arthritis and atherosclerosis by anti-Flt1. *Nat. Med.* **2002**, *8*, 831–840. [PubMed]

168. Khurana, R.; Moons, L.; Shafi, S.; Luttun, A.; Collen, D.; Martin, J.F.; Carmeliet, P.; Zachary, I.C. Placental growth factor promotes atherosclerotic intimal thickening and macrophage accumulation. *Circulation* **2005**, *111*, 2828–2836. [CrossRef] [PubMed]

169. Fischer, C.; Jonckx, B.; Mazzone, M.; Zacchigna, S.; Loges, S.; Pattarini, L.; Chorianopoulos, E.; Liesenborghs, L.; Koch, M.; de Mol, M.; *et al.* Anti-PlGF inhibits growth of VEGF(R)-inhibitor-resistant tumors without affecting healthy vessels. *Cell* **2007**, *131*, 463–475. [CrossRef] [PubMed]

170. Johnson, M.D.; Kim, H.R.; Chesler, L.; Tsao-Wu, G.; Bouck, N.; Polverini, P.J. Inhibition of angiogenesis by tissue inhibitor of metalloproteinase. *J. Cell. Physiol.* **1994**, *160*, 194–202. [CrossRef] [PubMed]

171. Gossl, M.; Herrmann, J.; Tang, H.; Versari, D.; Galili, O.; Mannheim, D.; Rajkumar, S.V.; Lerman, L.O.; Lerman, A. Prevention of vasa vasorum neovascularization attenuates early neointima formation in experimental hypercholesterolemia. *Basic Res. Cardiol.* **2009**, *104*, 695–706. [CrossRef] [PubMed]

172. Kampschulte, M.; Gunkel, I.; Stieger, P.; Sedding, D.G.; Brinkmann, A.; Ritman, E.L.; Krombach, G.A.; Langheinrich, A.C. Thalidomide influences atherogenesis in aortas of ApoE$^{-/-}$/LDLR$^{-/-}$ double knockout mice: A nano-CT study. *Int. J. Cardiovasc. Imaging* **2014**, *30*, 795–802. [CrossRef] [PubMed]

173. Dashwood, M.R.; Barker, S.G.; Muddle, J.R.; Yacoub, M.H.; Martin, J.F. [125I]-endothelin-1 binding to vasa vasorum and regions of neovascularization in human and porcine blood vessels: A possible role for endothelin in intimal hyperplasia and atherosclerosis. *J. Cardiovasc. Pharmacol.* **1993**, *22*, S343–S347. [CrossRef] [PubMed]

174. Herrmann, J.; Best, P.J.; Ritman, E.L.; Holmes, D.R.; Lerman, L.O.; Lerman, A. Chronic endothelin receptor antagonism prevents coronary vasa vasorum neovascularization in experimental hypercholesterolemia. *J. Am. Coll. Cardiol.* **2002**, *39*, 1555–1561. [CrossRef] [PubMed]

175. Winter, P.M.; Neubauer, A.M.; Caruthers, S.D.; Harris, T.D.; Robertson, J.D.; Williams, T.A.; Schmieder, A.H.; Hu, G.; Allen, J.S.; Lacy, E.K.; *et al.* Endothelial $\alpha_v\beta_3$ integrin-targeted fumagillin nanoparticles inhibit angiogenesis in atherosclerosis. *Arterioscler. Thromb. Vasc. Biol.* **2006**, *26*, 2103–2109. [CrossRef] [PubMed]

176. Drinane, M.; Mollmark, J.; Zagorchev, L.; Moodie, K.; Sun, B.; Hall, A.; Shipman, S.; Morganelli, P.; Simons, M.; Mulligan-Kehoe, M.J. The antiangiogenic activity of rPAI-1$_{23}$ inhibits vasa vasorum and growth of atherosclerotic plaque. *Circ. Res.* **2009**, *104*, 337–345. [CrossRef] [PubMed]

177. Mollmark, J.; Ravi, S.; Sun, B.; Shipman, S.; Buitendijk, M.; Simons, M.; Mulligan-Kehoe, M.J. Antiangiogenic activity of rPAI-1$_{23}$ promotes vasa vasorum regression in hypercholesterolemic mice through a plasmin-dependent mechanism. *Circ. Res.* **2011**, *108*, 1419–1428. [CrossRef] [PubMed]

178. Langheinrich, A.C.; Sedding, D.G.; Kampschulte, M.; Moritz, R.; Wilhelm, J.; Haberbosch, W.G.; Ritman, E.L.; Bohle, R.M. 3-Deazaadenosine inhibits vasa vasorum neovascularization in aortas of ApoE$^{-/-}$/LDLR$^{-/-}$ double knockout mice. *Atherosclerosis* **2009**, *202*, 103–110. [CrossRef] [PubMed]

179. Tsai, H.T.; Marshall, J.L.; Weiss, S.R.; Huang, C.Y.; Warren, J.L.; Freedman, A.N.; Fu, A.Z.; Sansbury, L.B.; Potosky, A.L. Bevacizumab use and risk of cardiovascular adverse events among elderly patients with colorectal cancer receiving chemotherapy: A population-based study. *Ann. Oncol.* **2013**, *24*, 1574–1579. [CrossRef] [PubMed]

180. Colon, G.P.; Deveikis, J.P.; Dickinson, L.D. Revascularization of occluded internal carotid arteries by hypertrophied vasa vasorum: Report of four cases. *Neurosurgery* **1999**, *45*, 634–637. [CrossRef] [PubMed]

181. Shanahan, C.M.; Weissberg, P.L. Smooth muscle cell phenotypes in atherosclerotic lesions. *Curr. Opin. Lipidol.* **1999**, *10*, 507–513. [CrossRef] [PubMed]

182. Wilson, S.H.; Herrmann, J.; Lerman, L.O.; Holmes, D.R., Jr.; Napoli, C.; Ritman, E.L.; Lerman, A. Simvastatin preserves the structure of coronary adventitial vasa vasorum in experimental hypercholesterolemia independent of lipid lowering. *Circulation* **2002**, *105*, 415–418. [CrossRef] [PubMed]

183. Urbich, C.; Dernbach, E.; Zeiher, A.M.; Dimmeler, S. Double-edged role of statins in angiogenesis signaling. *Circ. Res.* **2002**, *90*, 737–744. [CrossRef] [PubMed]

184. Chaldakov, G.N.; Stankulov, I.S.; Fiore, M.; Ghenev, P.I.; Aloe, L. Nerve growth factor levels and mast cell distribution in human coronary atherosclerosis. *Atherosclerosis* **2001**, *159*, 57–66. [CrossRef] [PubMed]

185. Asanome, A.; Kawabe, J.; Matsuki, M.; Kabara, M.; Hira, Y.; Bochimoto, H.; Yamauchi, A.; Aonuma, T.; Takehara, N.; Watanabe, T.; *et al.* Nerve growth factor stimulates regeneration of perivascular nerve, and induces the maturation of microvessels around the injured artery. *Biochem. Biophys. Res. Commun.* **2014**, *443*, 150–155. [CrossRef] [PubMed]

186. Mollmark, J.I.; Park, A.J.; Kim, J.; Wang, T.Z.; Katzenell, S.; Shipman, S.L.; Zagorchev, L.G.; Simons, M.; Mulligan-Kehoe, M.J. Fibroblast growth factor-2 is required for vasa vasorum plexus stability in hypercholesterolemic mice. *Arterioscler. Thromb. Vasc. Biol.* **2012**, *32*, 2644–2651. [CrossRef] [PubMed]

187. Lindner, V.; Lappi, D.A.; Baird, A.; Majack, R.A.; Reidy, M.A. Role of basic fibroblast growth factor in vascular lesion formation. *Circ. Res.* **1991**, *68*, 106–113. [CrossRef] [PubMed]

188. Six, I.; Mouquet, F.; Corseaux, D.; Bordet, R.; Letourneau, T.; Vallet, B.; Dosquet, C.C.; Dupuis, B.; Jude, B.; Bertrand, M.E.; *et al.* Protective effects of basic fibroblast growth factor in early atherosclerosis. *Growth Factors* **2004**, *22*, 157–167. [CrossRef] [PubMed]

189. Che, J.; Okigaki, M.; Takahashi, T.; Katsume, A.; Adachi, Y.; Yamaguchi, S.; Matsunaga, S.; Takeda, M.; Matsui, A.; Kishita, E.; *et al.* Endothelial FGF receptor signaling accelerates atherosclerosis. *Am. J. Physiol. Heart Circ. Physiol.* **2011**, *300*, H154–H161. [CrossRef] [PubMed]

190. Shen, W.; Shi, H.M.; Fan, W.H.; Luo, X.P.; Jin, B.; Li, Y. The effects of simvastatin on angiogenesis: Studied by an original model of atherosclerosis and acute myocardial infarction in rabbit. *Mol. Biol. Rep.* **2011**, *38*, 3821–3828. [CrossRef] [PubMed]

International Journal of
*Molecular Sciences*

MDPI

*Article*

# The Arginine/ADMA Ratio Is Related to the Prevention of Atherosclerotic Plaques in Hypercholesterolemic Rabbits When Giving a Combined Therapy with Atorvastatine and Arginine

Saskia J. H. Brinkmann [1], Elisabeth A. Wörner [1], Nikki Buijs [1,2], Milan Richir [1,2], Luc Cynober [3,4], Paul A. M. van Leeuwen [1,*] and Rémy Couderc [4,5]

[1]   Department of Surgery, VU University Medical Center, 1081 HV Amsterdam, The Netherlands;
      sj.brinkmann@vumc.nl (S.J.H.B.); lisaworner@gmail.com (E.A.W.); n.buijs@vumc.nl (N.B.);
      m.richir@vumc.nl (M.R.)
[2]   Department of Surgery, Medical Center Alkmaar, 1815 JD Alkmaar, The Netherlands
[3]   Laboratory of Nutrition Biology EA 4466, Faculty of Pharmacy, Paris Descartes University, 75006 Paris,
      France; luc.cynober@parisdescartes.fr
[4]   Clinical Chemistry Department, Hopital Cochin et Hôtel-Dieu, AP-HP, 75014/75004 Paris, France;
      remy.couderc@trs.aphp.fr
[5]   Clinical chemistry laboratory, Armand Trousseau hospital, AP-HP, 75012 Paris, France
*    Correspondence: pam.vleeuwen@vumc.nl; Tel.: +31-20-444-3601; Fax: +31-20-444-3620

Academic Editor: Michael Henein
Received: 8 April 2015; Accepted: 26 May 2015; Published: 29 May 2015

**Abstract:** Supplementation with arginine in combination with atorvastatin is more efficient in reducing the size of an atherosclerotic plaque than treatment with a statin or arginine alone in homozygous Watanabe heritable hyperlipidemic (WHHL) rabbits. We evaluated the mechanism behind this feature by exploring the role of the arginine/asymmetric dimethylarginine (ADMA) ratio, which is the substrate and inhibitor of nitric oxide synthase (NOS) and thereby nitric oxide (NO), respectively. Methods: Rabbits were fed either an arginine diet (group A, $n = 9$), standard rabbit chow plus atorvastatin (group S, $n = 8$), standard rabbit chow plus an arginine diet with atorvastatin (group SA, $n = 8$) or standard rabbit chow (group C, $n = 9$) as control. Blood was sampled and the aorta was harvested for topographic and histological analysis. Plasma levels of arginine, ADMA, cholesterol and nitric oxide were determined and the arginine/ADMA ratio was calculated. Results: The decrease in ADMA levels over time was significantly correlated to fewer aortic lesions in the distal aorta and total aorta. The arginine/ADMA ratio was correlated to cholesterol levels and decrease in cholesterol levels over time in the SA group. A lower arginine/ADMA ratio was significantly correlated to lower NO levels in the S and C group. Discussion: A balance between arginine and ADMA is an important indicator in the prevention of the development of atherosclerotic plaques.

**Keywords:** atherosclerosis; arginine; ADMA; ratio; statins; nitric oxide; cholesterol

## 1. Introduction

Atherogenic risk factors, such as smoking, diabetes mellitus, hypertension and dyslipidemia, harm the endothelium and make it dysfunctional [1]. Endothelial nitric oxide (NO), released by the intact and healthy endothelium plays a very important role in the maintenance of vascular tone and structure [1]. NO has a number of intracellular effects that lead to vasorelaxation, endothelial regeneration, inhibition of leukocyte chemotaxis, and platelet adhesion. Therefore, decreased NO levels lead to endothelial dysfunction and this is an initial event in the development of atherosclerosis [2,3].

The formation of atherosclerosis can be overcome by increasing the synthesis of NO and providing its precursor arginine as a substrate to the endothelial cell [3]. Arginine is a semi-conditionally amino acid, which in its turn is the substrate for nitric oxide synthase (NOS) (see Figure 1). NOS converts arginine into NO and citrulline [4]. It has been shown that arginine supplementation increases NO availability, improves vascular responsiveness and reduces atherosclerosis in animals and patients [3,5]. Furthermore, endogenously produced inhibitors of the enzyme NOS, in particular asymmetric dimethylarginine (ADMA), also have an important role in NO metabolism [6,7]. ADMA and other methylarginines are continuously formed from intracellular proteolysis of methylated arginine residues in the nucleus of the cell, by enzymes called protein arginine methyltransferases (PRMTs) [8]. Whereas the structure of ADMA is similar to arginine, it competes with arginine for NOS binding, hereby blocking the formation of NO from arginine by NOS directly. NOS is mainly localized in the cell, thus the intracellular ADMA and arginine levels regulate NOS activity [9]. In addition, extracellular ADMA is an antagonist to extracellular arginine on cell membrane transporter level, whereas they are both transported into the cell via the cell membrane by the cationic amino acid transporter (y+ system), a high-affinity, $Na^+$-independent transporter of the basic amino acids and therefore also compete with each other on this level [10]. Since ADMA competes with arginine for NOS and for cell transport via CAT-2, the bioavailability of NO depends on the balance between the two, the so-called arginine/ADMA ratio [11]. The arginine/ADMA ratio is an important indicator of NO bioavailability and therefore of the risk of formation of atherosclerotic plaques [11].

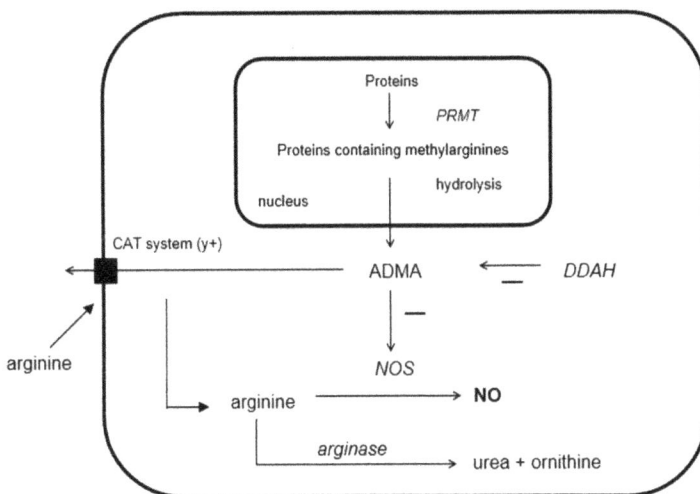

**Figure 1.** Schematic overview of the interactions between arginine, asymmetric dimethylarginine (ADMA), dimethylarginine dimethylaminohydrolase (DDAH), and nitric oxide synthase (NOS). PRMTs, protein arginine methyltransferases.

Statins are widely prescribed all over the world to treat hypercholesterolemia in current clinical practice. Statins are inhibitors of 3-hydroxy-3-methylglutaryl-coenzyme A reductase (HMG-CoA reductase), the rate limiting enzyme in the biosynthesis of cholesterol. Prospective clinical trials have demonstrated that reductions in low-density lipoprotein (LDL) cholesterol with statins decrease morbidity and mortality rates in particular in coronary artery disease and stroke [12,13]. A therapy that up-regulates NOS and subsequently NO through dietary manipulation of arginine, combined with a statin, would be the exquisite method to prevent and treat the formation of atherosclerotic plaques. Rasmusen *et al.* were the first to demonstrate that diet supplementation with arginine associated with atorvastatin was more efficient in reducing lesion size than treatment with arginine or statin alone in hypercholesterolemic rabbits [14]. The mechanism behind this feature remains unclear. The arginine/ADMA ratio is gaining more interest in the field of research as a potential marker of those of cardiovascular diseases [15–17]. Therefore, we hypothesized, as an ancillary study of Rasmusen *et al.*, that this ratio and changes in NO availability, could be the underlying factors of the positive effects of arginine combined with a statin on the development of atherosclerotic plaques in hypercholesterolemic rabbits. We investigated whether there was a (cor)relation between the ratio, levels of arginine and ADMA and the occurrence of atherosclerotic plaques as demonstrated in the study of Rasmusen *et al.* [14].

## 2. Results and Discussion

### 2.1. Results

#### 2.1.1. Effect of Treatment on L-Arginine Levels

At baseline (T0), mean plasma levels of arginine did not significantly differ between the groups. After eight weeks of treatment, arginine plasma levels increased significantly compared to T0 in the groups supplied with arginine (group A and SA, $p < 0.001$) (see Table 1).

#### 2.1.2. Effect of Treatment on ADMA and NO Levels

At T0 and T8, ADMA and NO levels did not significantly differ between groups. At the end of treatment (T8) ADMA levels decreased in all groups, but not significantly. The decrease in ADMA levels over time (T0–T8), when analyzing all groups together, showed to be significantly correlated to less aortic lesions in the distal aorta ($r = 0.677$, $p = 0.01$) and total aorta ($r = 0.599$, $p = 0.03$). Thus, the bigger the decrease in ADMA levels over time, the smaller the amount of arteriosclerotic lesions in the distal aorta (see Figure 2).

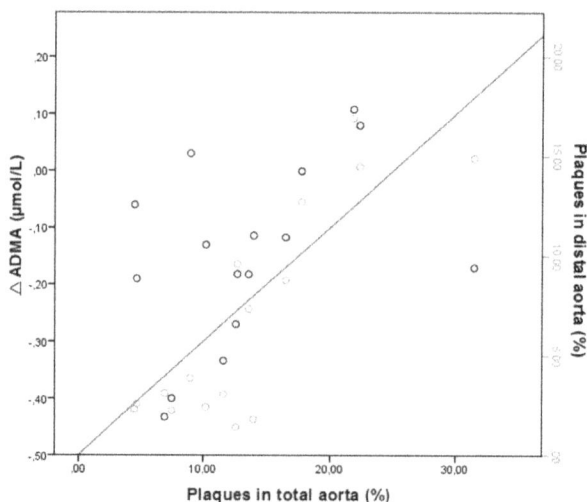

**Figure 2.** Pearson correlation between ΔADMA (T0–T8, $n = 4$ per group) and aortic lesions in the distal aorta ($r = 0.677$, $p = 0.01$) and total aorta ($r = 0.599$, $p = 0.03$).

#### 2.1.3. Effect of Treatment on Arginine/ADMA Ratio and Relation with Other Parameters

At T0, no significant difference between groups was found in arginine/ADMA ratio levels. The ratio was significantly increased at T8 in group A and SA ($p < 0.05$). A Pearson's correlation test revealed the correlation between the arginine/ADMA ratio and cholesterol levels at T8, most pronounced in the SA group ($r = -0.462$). The arginine/ADMA ratio and cholesterol levels at T8 correlated positively ($r = 0.279$) in group A. In addition, the decrease in cholesterol over time was strongly correlated to the arginine/ADMA ratio in the S and SA group (S: $r = 0.461$, SA: 0.699) (see Figure 3). A lower arginine/ADMA ratio was significantly correlated to lower NO levels in the S and C group (S: $r = 0.709$, $p = 0.049$, C: $r = 0.697$, $p = 0.056$) (see Figure 4).

**Table 1.** Effect of different treatments on arginine levels, ADMA levels, arginine/ADMA ratio, and atherosclerotic lesions in the aorta.

| Group | A (n = 9) | | S (n = 8) | | SA (n = 8) | | C (n = 9) | |
|---|---|---|---|---|---|---|---|---|
| | Mean | SEM | Mean | SEM | Mean | SEM | Mean | SEM |
| Plasma levels L-arginine (μmol/L) | | | | | | | | |
| T0 | 463 | 39 | 367 | 33 | 371 | 13 | 460 | 51 |
| T8 | 388 [a] | 47 | 243 [b] | 28 | 476 [a] | 58 | 265 [b] | 14 |
| Plasma levels ADMA (μmol/L) | | | | | | | | |
| T0 [∞] | 1.14 | 0.11 | 1.02 | 0.03 | 1.15 | 0.05 | 1.11 | 0.05 |
| T8 | 0.88 | 0.05 | 0.97 | 0.05 | 1.03 | 0.05 | 0.99 | 0.03 |
| Arginine/ADMA ratio | | | | | | | | |
| T0 [∞] | 347 | 19 | 304 | 42 | 317 | 23 | 374 | 43 |
| T8 | 442 [a] | 46 | 252 [b] | 26 | 476 [a] | 54 | 261 [b] | 16 |
| NO levels | | | | | | | | |
| T0 | 110 | 20 | 88 | 9 | 103 | 10 | 108 | 16 |
| T8 | 156 | 35 | 82 | 8 | 104 | 10 | 105 | 10 |
| Total lesions in aorta (%) | | | | | | | | |
| – | 15.1 | 3.2 | 12.2 | 1.4 | 9.0 | 1.1 | 13.8 | 1.6 |
| Distal lesions in aorta (%) | | | | | | | | |
| – | 9.8 | 2.6 | 5.4 | 1.3 | 3.0 | 0.7 | 6.8 | 1.3 |

Mean values with their standard errors. A, L-arginine group; S, statin group; SA, statin-L-arginine group; C, control group. T0, at the beginning of treatment; T8, after 8 weeks of treatment; NO, nitric oxide. [a,b] mean values within a row with different superscript letters are significantly different ($p < 0.05$) (independent samples test). Data about plasma arginine and NO are adapted from [14]. [∞] these plasma levels are based on $n = 4$ per group.

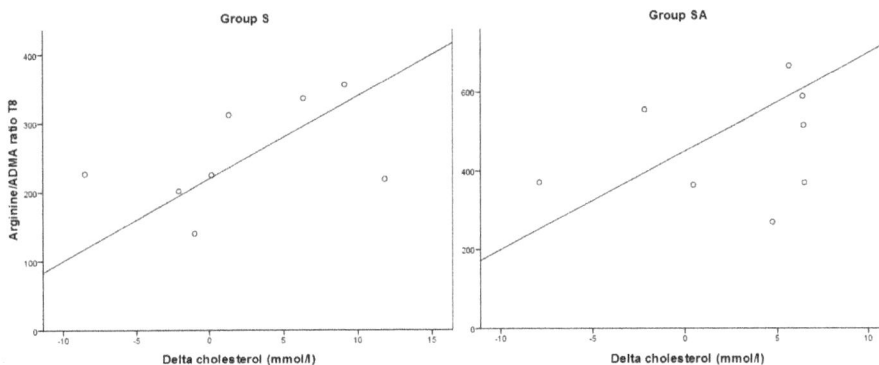

**Figure 3.** Pearson correlation between arginine/ADMA ratio and the difference in cholesterol levels over time in the statine ($n = 8$) and statine-arginine group ($n = 8$) (S: $r = 0.461$, SA: $r = 0.699$).

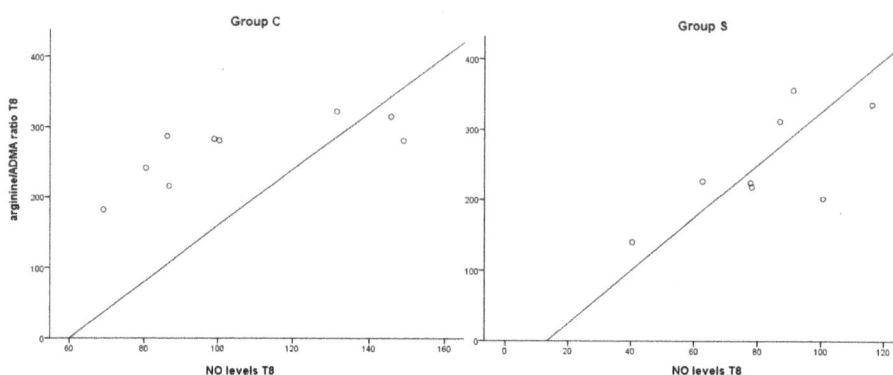

**Figure 4.** Pearson correlation between arginine/ADMA ratio T8 and NO levels at T8 in the statine ($n = 8$) and control group ($n = 9$) (S: $r = 0.709$, $p = 0.049$, C: $r = 0.694$, $p = 0.056$).

## 2.2. Discussion

The purpose of the present study was to determine the contribution of the arginine/ADMA ratio in the explanation of the positive effect from the combined therapy of arginine and a statin in the prevention of atherosclerosis as we reported previously [14]. In the present part of the study, we showed that arginine/ADMA ratio has a correlation to cholesterol, development of plaques and levels of NO in this model and could be a sensitive marker in the prevention of atherosclerosis by arginine and statin.

The arginine/ADMA ratio is gaining more interest in the field of research as a potential marker of cardiovascular diseases [15–17]. It is well-known that arginine is an important mediator in vascular flow and the integrity of the vascular wall by being the substrate of NO. Since ADMA inhibits NO production by competing with arginine for NOS binding, the net amount of NO production is indicated by the ratio between substrate and inhibitor, the arginine/ADMA ratio. Endothelium-derived NO plays a central role in normal vascular homeostasis. Impaired NO bioavailability causes endothelial dysfunction and not only contributes to the initiation and progression of atherosclerosis but is also associated with long-term risk of cardiovascular events [18]. NO is essential for flow-mediated dilatation. Richir *et al.* demonstrated that low arginine plasma levels in combination with high ADMA plasma levels deteriorates systemic hemodynamics and reduces blood flow through the kidney and spleen and liver [19].

Supplementation of arginine in a diet or intravenously, may contribute to higher levels of NO and maybe a nullification of the detrimental effects of ADMA. It was shown that dietary L-arginine reduces the progression of atherosclerosis and improves endothelium-dependent vasodilatation in cholesterol-fed rabbits [5,20–22]. Oral supplementation of L-arginine in hypercholesterolemic patients increases endothelial dependent vasodilatation in forearm conduit arteries [20,23]. Although L-arginine deficiency has never been documented in hypercholesterolemia, it seems that NO bioavailability is partly restored after L-arginine administration in hypercholesterolemic subjects [11,24]. Supplementation of dietary L-arginine can increase NO-mediated bloodflow [25,26]. In our study, arginine supplementation caused a two-fold increase in plasma arginine, which is consistent with previous studies [22,23].

However, just providing a substrate for NOS by supplementation of arginine cannot explain the positive effects on vasculature. All cells produce ADMA, the inhibitor of NO. It was shown that ADMA directly affects the integrity and function of vasculature itself by damaging the endothelial gap junction function, induction of smooth muscle cell migration, foam cell formation and apoptosis of smooth muscle cells and endothelial cells [27]. This was confirmed in our study setting in which we demonstrated that a decrease in ADMA levels over time showed to be significantly correlated to less aortic lesions in the distal aorta and total aorta. Thus, decreased ADMA levels may prevent the development of arteriosclerotic lesions in the aorta. It was shown that levels of ADMA are high in hypercholesterolemia, whereas levels of arginine have been found to be in the normal range [28]. This causes a shift in the arginine/ADMA ratio and results in diminished NOS activity and subsequently less NO bioavailability. This implicates the importance of the ratio above the sole levels of arginine in plasma. In the current study, lower levels of NO in the groups not supplied with arginine and/or statin, were correlated to the arginine/ADMA ratio. This confirms that a low arginine/ADMA low ratio deteriorates NO metabolism and therefore vascular flow in hypercholesterolemic rabbits.

The endothelial cell transporter that facilitates uptake of both arginine and ADMA is cationic amino transporter 1 (CAT1). ADMA inhibits not only NOS, but also this CAT transporter, that mediates cellular uptake of arginine. However, in normal physiological concentrations, ADMA cannot impair the CAT1-mediated transport of arginine [9,29]. Conversely, high (but still physiological) concentrations of arginine can inhibit CAT1-mediated cellular uptake of ADMA [20].

Rasmusen *et al.* confirmed the positive effects of statins on atherosclerotic plaques, because lower cholesterol levels after 8 weeks of treatment correlated significantly to less aortic lesions in the total and distal aorta [14]. In the present ancillary study, we revealed a positive correlation between the arginine/ADMA ratio and decrease of cholesterol levels over time and at the end of the treatment, primarily in the SA group. So when the ratio increases, cholesterol levels decrease, or inversely. In addition, a decrease in ADMA levels over time was significantly correlated to less aortic lesions in the distal aorta and total aorta. Statins are inhibitors of 3-hydroxy-3-methylglutaryl-coenzyme A reductase (HMG-CoA reductase) and reduce the amount of circulating LDL-cholesterol. Currently, seven types of statins are available: atorvastatin, fluvastatin, lovastatin, pitavastatin, pravastatin, rosuvastatin and simvastatin. Besides lowering levels of circulating LDL-cholesterol, statins exhibit different effects on NO-mediators. One study found that treatment with statins elevates levels of DDAH, which metabolizes ADMA [30]. Moreover, statins have shown to be able to decrease plasma levels of ADMA in different types of study settings [31–36]. Atorvastatin specifically is demonstrated to increase levels of DDAH in rats and humans and thereby to decrease plasma ADMA levels [37–39]. However, the effect of statins on ADMA concentrations is ambiguous, whereas some studies found that statins had no effect on ADMA levels in hypercholesterolemic subjects [40–44]. However, other intriguing experimental data have shown that statins exhibits pleiotropic effects beyond their lipid-lowering actions, including up-regulation of eNOS and thereby enhancement of endothelial NO production, inhibition of smooth muscle proliferation and anti-inflammatory and anti-oxidative actions [45,46]. Therefore, the exact combination of statins and arginine supplementation could have caused the correlation between the arginine/ADMA ratio and the decrease in cholesterol over time. Moreover, we did not find a correlation of arginine and ADMA alone with levels of cholesterol, so the combination of the two seems to determine the level of cholesterol in the arteries. This is supported by recently published data in which the arginine/ADMA ratio seemed to be a sensitive marker in the in the progression of atherosclerosis in means of intima-media thickness, rather than arginine or ADMA alone [16].

The limitation of the present study needs to be addressed. The number of Watanabe rabbits was relatively small in order to be a representative group. In order to transfer the effects measured to a clinical benefit, it needs to be researched in a clinical trial. We also suggest for upcoming studies to measure NOS and ADMA levels in response to the combination of statin and arginine in the vascular wall.

## 3. Experimental Section

This study is based on the results and dataset from the study of Rasmusen *et al.* [14].

### 3.1. Treatment of Animals

Thirty-four six-week-old homozygous Watanabe heritable hyperlipidaemic rabbits were assigned to one of four treatment groups: during 8 weeks, rabbits in the L-arginine group (group A; *n* = 9) were fed by a 1.5% L-arginine in 1 g/kg of bodyweight/d chow diet. This diet contained 16% crude protein, 3.2% fat, 49.3% carbohydrates, 13.6% fibre and 8.03 MJ gross energy/kg. Details of feeding regime were as described by Rasmusen *et al.* [14]. Arginine content in the diet was 0.98%. The statin group (group S; $n = 8$) consisted of rabbits receiving standard rabbit chow diet plus 2.5 mg/kg/d of atorvastatin in their drinking water [47]. In the L-arginine plus statin group (group SA; $n = 8$), rabbits received the L-arginine enriched rabbit chow together with atorvastatin in their drinking water. The control group (group C; $n = 9$) was fed with standard rabbit chow. Blood samples were collected at the vein of the ear of the rabbit into sodium heparinate (Sanofi, Winthrop Industry, Gentilly, France) and immediately centrifuged at +4 °C for 15 min at $4500 \times g$. Blood samples were taken at the beginning of the treatment (T0) and at the end of the treatment period after eight weeks (T8). Due to ethical reasons and to prevent a hypovolemic situation, we were only allowed to take a limited amount of volume at T0. Because the inter- and intra essay variations are the smallest for ADMA, the number of plasma ADMA levels at T0 are determined in 4 rabbits per group.

Animal care and experimentation and the Council of Europe Guidelines agreed with the rules of the laboratory of Nutrition Biology of the Descartes University, Paris, France. Animal care and experimentation complied with the rules of our institution and with Council of Europe Guidelines. Chantal Martin (No. 75-10) and Christophe Moinard (No. 75-522) are licensed to conduct experimental studies on living animals, and we had received approval for the use of animal facilities (agreement No. A75-06-02, February 2006, Veterinary Service Directorate, Prefecture de Police de Paris, France).

### 3.2. Anatomy and Histological Analysis

For topographic and histological analysis the full length of the cervico-thoracic region of the aorta of the rabbits were obtained by Rasmusen *et al.* [14]. In brief, the areas of lesions were determined by using NIH Scion Image Software (Scion, Frederick, MD, USA). The areas were measured three times. The aorta was then prepared to 2 mm-thick fragments transversely cut into four segments (aortic arch, thoracic aorta, abdominal aorta and bifurcation), which were classified into distal lesions and lesions of the total aorta.

### 3.3. Biochemical Analysis

The concentrations of ADMA were determined simultaneously by high-performance liquid chromatography (HPLC). Plasma L-arginine levels were assessed on a JEOL automated amino acid analyzer (Tokyo, Japan) using ion-exchange chromatography [48]. The L-arginine over ADMA ratio was calculated. At T0, the arginine/ADMA ratio was only calculated from the rabbits of which ADMA was determined in plasma ($n = 4$ per group). To indicate the total NO production the products of NO, nitrite and nitrate, were measured in plasma as described in the Griess method [49]. Plasma total cholesterol was determined by routine enzymatic method on a Hitachi 911 analyzer (Roche, Meylan, France) [50].

### 3.4. Statistical Analysis

Results were tested upon distribution using Kolmogorov-Smirnov Test and QQ-plots. When normally distributed, results are presented in mean ± standard error of the mean (SEM) and in median and interquartile range when data were not normally distributed. The Student's *t*-test or Mann Whitney U (according to data distribution) test was used to determine significant differences in arginine and ADMA concentrations between groups. The one-sample *t*-test was used to test whether the values were significantly different from the beginning of the treatment. Correlations between the continuous variables were assessed by using bivariate analysis to estimate the Pearson's coefficient. SPSS for Windows software was used to perform statistical analysis. $p < 0.05$ was considered to indicate a significant difference.

*Int. J. Mol. Sci.* **2015**, *16*, 12230–12242

## 4. Conclusions

In conclusion, supplementation of both arginine and a statin in hypercholesterolemic rabbits induces a high arginine/ADMA ratio and this is correlated to lower cholesterol levels after 8 weeks of treatment. The decrease in ADMA levels over time correlated to less aortic lesions in the distal aorta and total aorta. When hypercholesterolemic rabbits were not fed with arginine, a lower arginine/ADMA ratio was significantly correlated to lower NO levels. These results support the hypothesis that a balance between arginine and ADMA, being the substrate and inhibitor of NOS respectively, might contribute to better understanding of the development of atherosclerotic plaques. Further research on the combination of statins and L-arginine, preferably in clinical trials, is necessary to confirm our findings and translate them to a clinical setting.

**Author Contributions:** Author contributions: Saskia J. H. Brinkmann and Elisabeth A. Wörner: data acquisition and analysis; Saskia J. H. Brinkmann, Elisabeth A. Wörner, Nikki Buijs and Milan Richir: drafting of the manuscript; Luc Cynober, Paul A. M. van Leeuwen, and Rémy Couderc: study concept and design; Saskia J. H. Brinkmann, Elisabeth A. Wörner and Nikki Buijs: statistical analysis. All authors contributed to manuscript revision.

**Conflicts of Interest:** The authors declare no conflict of interest.

## References

1. Tousoulis, D.; Kampoli, A.M.; Tentolouris, C.; Papageorgiou, N.; Stefanadis, C. The role of nitric oxide on endothelial function. *Curr. Vasc. Pharmacol.* **2012**, *10*, 4–18. [CrossRef] [PubMed]
2. Dias, R.G.; Negrao, C.E.; Krieger, M.H. Nitric oxide and the cardiovascular system: Cell activation, vascular reactivity and genetic variant. *Arq. Bras. Cardiol.* **2011**, *96*, 68–75. [PubMed]
3. Napoli, C.; de Nigris, F.; Williams-Ignarro, S.; Pignalosa, O.; Sica, V.; Ignarro, L.J. Nitric oxide and atherosclerosis: An update. *Nitric Oxide* **2006**, *15*, 265–279. [CrossRef] [PubMed]
4. Morris, S.M., Jr. Arginine metabolism: Boundaries of our knowledge. *J. Nutr.* **2007**, *137*, 1602S–1609S. [PubMed]
5. Boger, R.H.; Bode-Boger, S.M.; Brandes, R.P.; Phivthong-Ngam, L.; Böhme, M.; Nafe, R.; Mügge, A.; Frölich, J.C. Dietary L-arginine reduces the progression of atherosclerosis in cholesterol-fed rabbits: Comparison with lovastatin. *Circulation* **1997**, *96*, 1282–1290. [CrossRef] [PubMed]
6. Vallance, P.; Leone, A.; Calver, A.; Collier, J.; Moncada, S. Endogenous dimethylarginine as an inhibitor of nitric oxide synthesis. *J. Cardiovasc. Pharmacol.* **1992**, *20* (Suppl. 12), S60–S62. [CrossRef] [PubMed]
7. Vallance, P.; Leiper, J. Cardiovascular biology of the asymmetric dimethylarginine: Dimethylarginine dimethylaminohydrolase pathway. *Arterioscler. Thromb. Vasc. Biol.* **2004**, *24*, 1023–1030. [CrossRef] [PubMed]
8. Teerlink, T. ADMA metabolism and clearance. *Vasc. Med.* **2005**, *10* (Suppl. 1), S73–S81. [CrossRef] [PubMed]
9. Blackwell, S. The biochemistry, measurement and current clinical significance of asymmetric dimethylarginine. *Ann. Clin. Biochem.* **2010**, *47*, 17–28. [CrossRef] [PubMed]
10. Brinkmann, S.J.; de Boer, M.C.; Buijs, N.; van Leeuwen, P.A. Asymmetric dimethylarginine and critical illness. *Curr. Opin. Clin. Nutr. Metab. Care* **2014**, *17*, 90–97. [CrossRef] [PubMed]
11. Bode-Boger, S.M.; Boger, R.H.; Kienke, S.; Junker, W.; Frolich, J.C. Elevated L-arginine/dimethylarginine ratio contributes to enhanced systemic NO production by dietary L-arginine in hypercholesterolemic rabbits. *Biochem. Biophys. Res. Commun.* **1996**, *219*, 598–603. [CrossRef] [PubMed]
12. Mihaylova, B.; Emberson, J.; Blackwell, L.; Keech, A.; Simes, J.; Barnes, E.H.; Voysey, M.; Gray, A.; Collins, R.; Baigent, C.; *et al.* The effects of lowering LDL cholesterol with statin therapy in people at low risk of vascular disease: Meta-analysis of individual data from 27 randomised trials. *Lancet* **2012**, *380*, 581–590. [PubMed]
13. Baigent, C.; Blackwell, L.; Emberson, J.; Holland, L.E.; Reith, C.; Bhala, N.; Peto, R.; Barnes, E.H.; Keech, A.; Simes, J.; *et al.* Efficacy and safety of more intensive lowering of LDL cholesterol: A meta-analysis of data from 170,000 participants in 26 randomised trials. *Lancet* **2010**, *376*, 1670–1681. [PubMed]
14. Rasmusen, C.; Moinard, C.; Martin, C.; Tricottet, V.; Cynober, L.; Couderc, R. L-Arginine plus atorvastatin for prevention of atheroma formation in genetically hypercholesterolaemic rabbits. *Br. J. Nutr.* **2007**, *97*, 1083–1089. [CrossRef] [PubMed]
15. Bode-Boger, S.M.; Scalera, F.; Ignarro, L.J. The L-arginine paradox: Importance of the L-arginine/asymmetrical dimethylarginine ratio. *Pharmacol. Ther.* **2007**, *114*, 295–306. [CrossRef] [PubMed]

16. Notsu, Y.; Yano, S.; Shibata, H.; Nagai, A.; Nabika, T. Plasma arginine/ADMA ratio as a sensitive risk marker for atherosclerosis: Shimane CoHRE study. *Atherosclerosis* **2014**, *239*, 61–66. [CrossRef] [PubMed]

17. Visser, M.; Vermeulen, M.A.; Richir, M.C.; Teerlink, T.; Houdijk, A.P.; Kostense, P.J.; Wisselink, W.; de Mol, B.A.; van Leeuwen, P.A.; Oudemans-van Straaten, H.M. Imbalance of arginine and asymmetric dimethylarginine is associated with markers of circulatory failure, organ failure and mortality in shock patients. *Br. J. Nutr.* **2012**, *107*, 1458–1465. [CrossRef] [PubMed]

18. Lu, T.M.; Ding, Y.A.; Charng, M.J.; Lin, S.J. Asymmetrical dimethylarginine: A novel risk factor for coronary artery disease. *Clin. Cardiol.* **2003**, *26*, 458–464. [CrossRef] [PubMed]

19. Richir, M.C.; van Lambalgen, A.A.; Teerlink, T.; Wisselink, W.; Bloemena, E.; Prins, H.A.; de Vries, T.P.; van Leeuwen, P.A. Low arginine/asymmetric dimethylarginine ratio deteriorates systemic hemodynamics and organ blood flow in a rat model. *Crit. Care Med.* **2009**, *37*, 2010–2017. [CrossRef] [PubMed]

20. Clarkson, P.; Adams, M.R.; Powe, A.J.; Donald, A.E.; McCredie, R.; Robinson, J.; McCarthy, S.N.; Keech, A.; Celermajer, D.S.; Deanfield, J.E. Oral L-arginine improves endothelium-dependent dilation in hypercholesterolemic young adults. *J. Clin. Investig.* **1996**, *97*, 1989–1994. [CrossRef] [PubMed]

21. Boger, R.H.; Bode-Boger, S.M.; Mugge, A.; Kienke, S.; Brandes, R.; Dwenger, A.; Frölich, J.C. Supplementation of hypercholesterolaemic rabbits with L-arginine reduces the vascular release of superoxide anions and restores NO production. *Atherosclerosis* **1995**, *117*, 273–284. [CrossRef]

22. Cooke, J.P.; Singer, A.H.; Tsao, P.; Zera, P.; Rowan, R.A.; Billingham, M.E. Antiatherogenic effects of L-arginine in the hypercholesterolemic rabbit. *J. Clin. Investig.* **1992**, *90*, 1168–1172. [CrossRef] [PubMed]

23. Cooke, J.P.; Dzau, J.; Creager, A. Endothelial dysfunction in hypercholesterolemia is corrected by L-arginine. *Basic Res. Cardiol.* **1991**, *86* (Suppl. 2), 173–181. [PubMed]

24. Boger, R.H.; Bode-Boger, S.M.; Szuba, A.; Tsao, P.S; Chan, J.R.; Tangphao, O.; Blaschke, T.F.; Cooke, J.P. Asymmetric dimethylarginine (ADMA): A novel risk factor for endothelial dysfunction: Its role in hypercholesterolemia. *Circulation* **1998**, *98*, 1842–1847. [CrossRef] [PubMed]

25. Bode-Boger, S.M.; Muke, J.; Surdacki, A.; Brabant, G.; Boger, R.H.; Frolich, J.C. Oral L-arginine improves endothelial function in healthy individuals older than 70 years. *Vasc. Med.* **2003**, *8*, 77–81. [CrossRef] [PubMed]

26. Piatti, P.M.; Monti, L.D.; Valsecchi, G.; Magni, F.; Setola, E.; Marchesi, F.; Galli-Kienle, M.; Pozza, G.; Alberti, K.G. Long-term oral L-arginine administration improves peripheral and hepatic insulin sensitivity in type 2 diabetic patients. *Diabetes Care* **2001**, *24*, 875–880. [CrossRef] [PubMed]

27. Chen, S.; Li, N.; Deb-Chatterji, M.; Dong, Q.; Kielstein, J.T.; Weissenborn, K.; Worthmann, H. Asymmetric dimethyarginine as marker and mediator in ischemic stroke. *Int. J. Mol. Sci.* **2012**, *13*, 15983–16004. [CrossRef] [PubMed]

28. Boger, R.H.; Bode-Boger, S.M.; Sydow, K.; Heistad, D.D.; Lentz, S.R. Plasma concentration of asymmetric dimethylarginine, an endogenous inhibitor of nitric oxide synthase, is elevated in monkeys with hyperhomocyst(e)inemia or hypercholesterolemia. *Arterioscler. Thromb. Vasc. Biol.* **2000**, *20*, 1557–1564. [CrossRef] [PubMed]

29. Betz, B.; Moller-Ehrlich, K.; Kress, T.; Kniepert, J.; Schwedhelm, E.; Böger, R.H.; Wanner, C.; Sauvant, C.; Schneider, R. Increased symmetrical dimethylarginine in ischemic acute kidney injury as a causative factor of renal L-arginine deficiency. *Transl. Res.* **2013**, *162*, 67–76. [CrossRef] [PubMed]

30. Ivashchenko, C.Y.; Bradley, B.T.; Ao, Z.; Leiper, J.; Vallance, P.; Johns, D.G. Regulation of the ADMA-DDAH system in endothelial cells: A novel mechanism for the sterol response element binding proteins, SREBP1c and -2. *Am. J. Physiol. Heart Circ. Physiol.* **2010**, *298*, H251–H258. [CrossRef] [PubMed]

31. Li, J.; Xia, W.; Feng, W.; Qu, X. Effects of rosuvastatin on serum asymmetric dimethylarginine levels and atrial structural remodeling in atrial fibrillation dogs. *Pacing Clin. Electrophysiol.* **2012**, *35*, 456–464. [CrossRef] [PubMed]

32. Vladimirova-Kitova, L.G.; Deneva-Koycheva, T.I. The effect of simvastatin on asymmetric dimethylarginine and flow-mediated vasodilation after optimizing the LDL level: A randomized, placebo-controlled study. *Vascul. Pharmacol.* **2012**, *56*, 122–130. [CrossRef] [PubMed]

33. Vladimirova-Kitova, L.G.; Deneva, T.I. Simvastatin and asymmetric dimethylarginine-homocysteine metabolic pathways in patients with newly detected severe hypercholesterolemia. *Clin. Lab.* **2010**, *56*, 291–302. [PubMed]

34. Xia, W.; Yin, Z.; Li, J.; Song, Y.; Qu, X. Effects of rosuvastatin on asymmetric dimethylarginine levels and early atrial fibrillation recurrence after electrical cardioversion. *Pacing Clin. Electrophysiol.* **2009**, *32*, 1562–1566. [CrossRef] [PubMed]

35. Oguz, A.; Uzunlulu, M. Short term fluvastatin treatment lowers serum asymmetric dimethylarginine levels in patients with metabolic syndrome. *Int. Heart J.* **2008**, *49*, 303–311. [CrossRef] [PubMed]

36. Sicard, P.; Delemasure, S.; Korandji, C.; Segueira-Le Grand, A.; Lauzier, B.; Guilland, J.C.; Duvillard, L.; Zeller, M.; Cottin, Y.; Vergely, C.; *et al.* Anti-hypertensive effects of Rosuvastatin are associated with decreased inflammation and oxidative stress markers in hypertensive rats. *Free Radic. Res.* **2008**, *42*, 226–236. [CrossRef] [PubMed]

37. Tanaka, N.; Katayama, Y.; Katsumata, T.; Otori, T.; Nishiyama, Y. Effects of long-term administration of HMG-CoA reductase inhibitor, atorvastatin, on stroke events and local cerebral blood flow in stroke-prone spontaneously hypertensive rats. *Brain Res.* **2007**, *1169*, 125–132. [CrossRef] [PubMed]

38. Chen, P.; Xia, K.; Zhao, Z.; Deng, X.; Yang, T. Atorvastatin modulates the DDAH1/ADMA system in high-fat diet-induced insulin-resistant rats with endothelial dysfunction. *Vasc. Med.* **2012**, *17*, 416–423. [CrossRef] [PubMed]

39. Nishiyama, Y.; Ueda, M.; Otsuka, T.; Katsura, K.; Abe, A.; Nagayama, H.; Katayama, Y. Statin treatment decreased serum asymmetric dimethylarginine (ADMA) levels in ischemic stroke patients. *J. Atheroscler. Thromb.* **2011**, *18*, 131–137. [CrossRef] [PubMed]

40. Lu, T.M.; Ding, Y.A.; Leu, H.B.; Yin, W.H.; Sheu, W.H.; Chu, K.M. Effect of rosuvastatin on plasma levels of asymmetric dimethylarginine in patients with hypercholesterolemia. *Am. J. Cardiol.* **2004**, *94*, 157–161. [CrossRef] [PubMed]

41. Eid, H.M.; Eritsland, J.; Larsen, J.; Arnesen, H.; Seljeflot, I. Increased levels of asymmetric dimethylarginine in populations at risk for atherosclerotic disease. Effects of pravastatin. *Atherosclerosis* **2003**, *166*, 279–284. [CrossRef]

42. Young, J.M.; Strey, C.H.; George, P.M.; Florkowski, C.M.; Sies, C.W.; Frampton, C.M.; Scott, R.S. Effect of atorvastatin on plasma levels of asymmetric dimethylarginine in patients with non-ischaemic heart failure. *Eur. J. Heart Fail.* **2008**, *10*, 463–466. [CrossRef] [PubMed]

43. Valkonen, V.P.; Laakso, J.; Paiva, H.; Lehtimäki, T.; Lakka, T.A.; Isomustajärvi, M.; Ruokonen, I.; Salonen, J.T.; Laaksonen, R. Asymmetrical dimethylarginine (ADMA) and risk of acute coronary events. Does statin treatment influence plasma ADMA levels? *Atheroscler. Suppl.* **2003**, *4*, 19–22. [CrossRef]

44. Paiva, H.; Laakso, J.; Lehtimaki, T.; Isomustajarvi, M.; Ruokonen, I.; Laaksonen, R. Effect of high-dose statin treatment on plasma concentrations of endogenous nitric oxide synthase inhibitors. *J. Cardiovasc. Pharmacol.* **2003**, *41*, 219–222. [CrossRef] [PubMed]

45. Laufs, U. Beyond lipid-lowering: Effects of statins on endothelial nitric oxide. *Eur. J. Clin. Pharmacol.* **2003**, *58*, 719–731. [PubMed]

46. John, S.; Schneider, M.P.; Delles, C.; Jacobi, J.; Schmieder, R.E. Lipid-independent effects of statins on endothelial function and bioavailability of nitric oxide in hypercholesterolemic patients. *Am. Heart J.* **2005**, *149*, 473. [CrossRef] [PubMed]

47. Maeso, R.; Aragoncillo, P.; Navarro-Cid, J.; Ruilope, L.M.; Diaz, C.; Hernández, G.; Lahera, V.; Cachofeiro, V. Effect of atorvastatin on endothelium-dependent constrictor factors in dyslipidemic rabbits. *Gen. Pharmacol.* **2000**, *34*, 263–272. [CrossRef]

48. Neveux, N.; David, P.; Cynober, L. Measurement of amino acid concentrations in biological fluids using ion exchange chromatography. In *Metabolic and Therapeutic Aspects of Amino Acids in Clinical Nutrition*, 2rd ed.; Cynober, L., Ed.; CRC Press: BocaRaton, FL, USA, 2004; pp. 17–28.

49. Ricart-Jane, D.; Llobera, M.; Lopez-Tejero, M.D. Anticoagulants and other preanalytical factors interfere in plasma nitrate/nitrite quantification by the Griess method. *Nitric Oxide* **2002**, *6*, 178–185. [CrossRef] [PubMed]

50. Allain, C.C.; Poon, L.S.; Chan, C.S.; Richmond, W.; Fu, P.C. Enzymatic determination of total serum cholesterol. *Clin. Chem.* **1974**, *20*, 470–475. [PubMed]

International Journal of
*Molecular Sciences*

MDPI

*Article*

# Meta-Analysis of miR-146a Polymorphisms Association with Coronary Artery Diseases and Ischemic Stroke

Mei-Hua Bao [1,*], Yan Xiao [2], Qing-Song Zhang [1], Huai-Qing Luo [1], Ji Luo [1], Juan Zhao [1], Guang-Yi Li [1], Jie Zeng [1] and Jian-Ming Li [1,*]

[1] Department of Anatomy, Histology and Embryology, Changsha Medical University, Changsha 410219, China; zhangqingsong@whut.edu.cn (Q.-S.Z.); luohuaiqing@163.com (H.-Q.L.); luoji927@163.com (J.L.); zhaojuannanfang2@163.com (J.Z.); liguangyi1977@163.com (G.-Y.L.); zengjie84117@163.com (J.Z.)
[2] Qingdao Science & Standard Chemicals Analysing and Testing Co., Ltd., Qingdao 266000, China; y_xiao@sscta.cn
* Correspondence: mhbao78@163.com (M.-H.B.); ljming0901@sina.com (J.-M.L.); Tel.: +86-731-8488-4488 (M.-H.B.); Fax: +86-731-8849-8866 (M.-H.B.)

Academic Editor: Michael Henein
Received: 21 April 2015; Accepted: 9 June 2015; Published: 24 June 2015

**Abstract:** Coronary artery disease (CAD) and ischemic stroke (IS) are manifestations of atherosclerosis, with a high death rate. miR-146a is a microRNA that participates in the progress of CAD and IS. A single nucleotide polymorphism (SNP) in the precursor of miR-146a, rs2910164, was found to be associated with the risks of CAD and IS. However, the results were inconsistent and inconclusive. A meta-analysis was performed to assess the relationship of rs2910164 and CAD as well as IS susceptibility. The database Pubmed, Embase, Cochrane Central Register of Controlled Trials (CENTRAL), Chinese National Knowledge Infrastructure (CNKI), and Chinese Biomedical Literature Database (CBM) were searched for related studies. Crude odds ratios with 95% confidence intervals were used to investigate the strength of the association by random- or fixed-effect model. A total of eight studies, with 3138 cases and 3097 controls were identified for the meta-analysis. The results shows that rs2910164 is associated with the risk of CAD significantly in allelic model (OR = 0.86), homozygous model (OR = 0.70), heterozygous model (OR = 0.80) and dominant model (OR = 0.76). The subjects carrying the GG genotype, GG + GC genotype or G allele are at lower risks of CAD. For the susceptibility of IS, there are no significant associations between rs2910164 and total studies. However, in subgroup analysis by sample size and ethnicity, the GG, GG + GC and G allele of rs2910164 are found to be associated with higher risks of IS in large sample size group and in Koreans, under homozygous and dominant models. In conclusion, the current meta-analysis suggests lower risks of CAD for GG, GG + GC genotype and G allele of rs2910164, while rs2910164 is not associated with the risk of IS. Thus rs2910164 might be recommended as a predictor for susceptibility of CAD, but not IS.

**Keywords:** miR-146a polymorphism; rs2910164; coronary artery disease; ischemic stroke; meta-analysis

---

## 1. Introduction

Cardiovascular diseases are the major cause of death and disability worldwide. According to the World Health Organization report in 2011, more than 17 million people died of cardiovascular diseases. Coronary artery disease (CAD) is characterized by occlusive epicardial coronary artery stenosis. Ischemic stroke (IS) is a major kind of stroke, which causes a high death rate and adult

*Int. J. Mol. Sci.* **2015**, *16*, 14305–14317

disability in the world. According to the report of Liu, the annual stroke mortality rate was 120–180 per 100,000 in China [1]. In the United States, stroke ranks as the third leading cause of death [2]. IS has exceeded heart diseases to become the most frequent cause of death. IS and CAD are principal clinical manifestations of atherosclerosis, and have caused a huge burden for society. Previous studies demonstrated that single nucleotide polymorphisms (SNPs) were associated with the risks of CAD and IS [3,4].

MicroRNAs are small, non-coding RNAs which regulate the gene expression in post-transcriptional levels. The regulation effects of microRNAs are obtained by binding to the 3′-UTR of target mRNAs, and lead to degradation or translation repression of target genes. Previous reports predicted that 1/3 of human genes were regulated by microRNAs [5]. Thus, the microRNAs interfere with many physiologic and pathological processes. The changes in the sequences of microRNAs, such as single nucleotide polymorphisms (SNPs), may result in diseases.

miR-146a is a microRNA located at the human chromosome 5q33. Previous studies have demonstrated that miR-146a participates in inflammatory processes, thus interferes with the pathology of cardiovascular diseases [6]. A SNP has been found to exist at the precursor miRNA-146a, which mutates G:U to C:U and results in a low production of mature miR-146a [7]. Studies have reported the relationship between miR-146a polymorphism and susceptibility of coronary artery disease and ischemic stroke. However, the results are somehow inconsistent. For example, some studies demonstrated that miR-146a polymorphism was associated with the risk of CAD and IS [8–12], while others thought it was not [13,14].

Thus it is necessary to make a precise and comprehensive estimation of the association between miR-146a and the risks of CAD and IS. In the present study, eight studies, including 3138 cases and 3097 controls, were included in the meta-analysis.

## 2. Results

### 2.1. Characteristics of Eligible Studies

A total of 19 studies were obtained from the literature search after duplicates were removed. Among them, three studies were excluded for irrelevance, two for being reviews, three for being master degree theses and three were only abstracts. Finally, eight studies meeting the criteria were preserved, which included 3138 cases and 3097 controls. The Preferred reporting items for systematic reviews and meta-analyses (PRISMA) flow chart is shown in Figure 1 and the information for the selected studies was presented in Table 1.

### 2.2. Results of Meta-Analysis

The results of meta-analysis for the association between miR-146a (rs2910164) and CAD, IS risks were shown in Table 2 and Figures 2 and 3.

For CAD, significantly decreased risks were found to be associated with rs2910164 under all genetic models, allelic model (OR = 0.86, 95% CI = 0.77–0.96, $p$ = 0.01), homozygous model (OR = 0.70, 95% CI = 0.55–0.88, $p$ = 0.003), heterozygous model (OR = 0.80, 95% CI = 0.65–0.98, $p$ = 0.03) and dominant model (OR = 0.76, 95% CI = 0.63–0.93, $p$ = 0.007). When we conducted a subgroup analysis by sample size and ethnicity, the same significant associations were observed in the large sample size group and the Chinese group. However, in the small sample size group, a significant decrease in CAD risks was only found in the homozygous model.

On the other hand, no significant association was found between rs2910164 and IS susceptibility in the analysis as a whole. However, subgroup analysis by sample size indicated a significant association between rs2910164 and IS susceptibility in large sample size groups and in Koreans, under homozygous and dominant models (Table 2). No other evident associations between rs2910164 and risk of IS were observed among subgroup analysis by Trial of Org 10,172 in Acute Stroke Treatment (TOAST), gender, smoking, hypertension, diabetes mellitus and hyperlipidemia in the dominant model (Table 3).

**Table 1.** Characteristics of eligible studies included in the meta-analysis.

| Author | Year | Country | Ethnicity | Disease | Genotyping Methods | Sex Ratio (Male:Female) (Case/Control) | Age (Case/Control) | Quality Score | Sample Size (Case/Control) | GG (Case/Control) | GC (Case/Control) | CC (Case/Control) | HWE of Control |
|---|---|---|---|---|---|---|---|---|---|---|---|---|---|
| Chen *et al.* | 2013 | China | Asian | CAD | Taqman | 1:1.79/1:1.184 | 64.57 ± 10.33/64.08 ± 11.21 | 14 | 658/658 | 181/194 | 305/330 | 172/134 | 0.769 |
| Hamann *et al.* | 2014 | German | Caucasian | CAD | HRM | Unavailable | Unavailable | 12 | 206/200 | 120/117 | 74/73 | 12/10 | 0.748 |
| Ramkaran *et al.* | 2014 | South Africa | Indian | CAD | PCR-RFLP | 1:0/1:0 | 37.6 ± 0.40/37.5 ± 0.44 | 12 | 106/100 | 50/45 | 43/46 | 13/9 | 0.569 |
| Xiong *et al.* | 2014 | Chian | Asian | CAD | PCR-RFLP | 1:0.661/1:0.601 | 65.13 ± 11.86/62.59 ± 12.89 | 11 | 295/283 | 41/61 | 141/125 | 113/97 | 0.086 |
| Huang *et al.* | 2015 | China | Asian | IS | Taqman | 1:0.616/1:0.616 | 61.0 (54, 68)/63.0 (54, 70) | 14 | 531/531 | 81/55 | 261/257 | 189/219 | 0.106 |
| Jeon *et al.* | 2013 | Korea | Korean | IS | Taqman | 1:0.441/1:0.496 | 64.16 ± 11.90/63.14 ± 10.19 | 15 | 678/553 | 128/76 | 327/266 | 223/211 | 0.589 |
| Liu *et al.* | 2013 | China | Asian | IS | PCR-RFLP | 1:0.608/1:0.581 | 67.52 ± 10.29/66.34 ± 11.07 | 12 | 296/391 | 52/77 | 159/198 | 85/116 | 0.65 |
| Zhu *et al.* | 2014 | China | Asian | IS | PCR-LDR | 1:0.688/1:0.685 | 61.62 ± 0.986/62.05 ± 0.982 | 12 | 368/381 | 50/64 | 173/185 | 145/132 | 0.952 |

HWE: Hardy-Weinberg equilibrium; CAD: Coronary Artery Disease; HRM: High-resolution melting; PCR-RFLP: polymerase chain reaction-restriction fragment length polymorphism; IS: ischemic stroke.

**Table 2.** Pooled ORs and 95% CIs of the association between miR-146a (rs2910164) and coronary artery disease (CAD) and ischemic stroke (IS).

| Genetic Model | Overall or Subgroup | CAD | | | | IS | | | |
|---|---|---|---|---|---|---|---|---|---|
| | | *n* | OR (95% CI) | *p*-Value | $I^2$ (%) | *n* | OR (95% CI) | *p*-Value | $I^2$ (%) |
| G *vs.* C | Overall | 5 | 0.86 (0.77, 0.96) | 0.01 | 0% | 4 | 1.07 (0.89, 1.29) | 0.45 | 74% |
| | Large size | 2 | 0.86 (0.73, 1.00) | 0.05 | NA | 2 | 1.25 (1.11, 1.41) | 0.0003 | 0% |
| | Small size | 3 | 0.87 (0.73, 1.03) | 0.10 | 0% | 2 | 0.91 (0.78, 1.05) | 0.20 | 0% |
| | Chinese | 2 | 0.83 (0.73, 0.95) | 0.006 | 0% | 3 | 1.02 (0.80, 1.29) | 0.89 | 76% |
| | Other (Caucasian, Indian, Korean) | 2 | 0.97 (0.75, 0.83) | 0.83 | 0% | 1 | 1.24 (1.06, 1.46) | 0.009 | NA |
| GG *vs.* CC | Overall | 4 | 0.70 (0.55, 0.88) | 0.003 | 0% | 4 | 1.17 (0.78, 1.77) | 0.44 | 76% |
| | Large size | 1 | 0.73 (0.54, 0.98) | 0.04 | NA | 2 | 1.64 (1.27, 2.12) | 0.0002 | 0% |
| | Small size | 3 | 0.65 (0.44, 0.96) | 0.03 | 0% | 2 | 0.81 (0.59, 1.10) | 0.18 | 0% |
| | Chinese | 2 | 0.68 (0.53, 0.88) | 0.003 | 0% | 3 | 1.05 (0.62, 1.77) | 0.87 | 78% |
| | Other (Caucasian, Indian, Korean) | 2 | 0.81 (0.43, 1.54) | 0.53 | 0% | 1 | 1.59 (1.13, 2.24) | 0.007 | NA |
| GC *vs.* CC | Overall | 4 | 0.80 (0.65, 0.98) | 0.03 | 0% | 4 | 1.08 (0.94, 1.25) | 0.27 | 0% |
| | Large size | 1 | 0.72 (0.55, 1.39) | 0.02 | NA | 2 | 1.17 (0.98, 1.40) | 0.09 | 0% |
| | Small size | 3 | 0.91 (0.66, 1.25) | 0.56 | 0% | 2 | 0.95 (0.75, 1.22) | 0.71 | 10% |
| | Chinese | 2 | 0.80 (0.64, 1.00) | 0.05 | 39% | 3 | 1.04 (0.86, 1.27) | 0.67 | 20% |
| | Other (Caucasian, Indian, Korean) | 2 | 0.74 (0.39, 1.43) | 0.37 | 0% | 1 | 1.16 (0.91, 1.49) | 0.23 | NA |
| GG/GC *vs.* CC | Overall | 4 | 0.76 (0.63, 0.93) | 0.007 | 0% | 4 | 1.10 (0.90, 1.34) | 0.36 | 54% |
| | Large size | 1 | 0.72 (0.56, 0.93) | 0.01 | NA | 2 | 1.26 (1.07, 1.50) | 0.007 | 0% |
| | Small size | 3 | 0.83 (0.61, 1.11) | 0.21 | 0% | 2 | 0.91 (0.73, 1.14) | 0.41 | 17% |
| | Chinese | 2 | 0.76 (0.62, 0.94) | 0.01 | 0% | 3 | 1.04 (0.79, 1.35) | 0.79 | 60% |
| | Other (Caucasian, Indian, Korean) | 2 | 0.78 (0.42, 1.45) | 0.43 | 0% | 1 | 1.26 (1.00, 1.59) | 0.05 | NA |

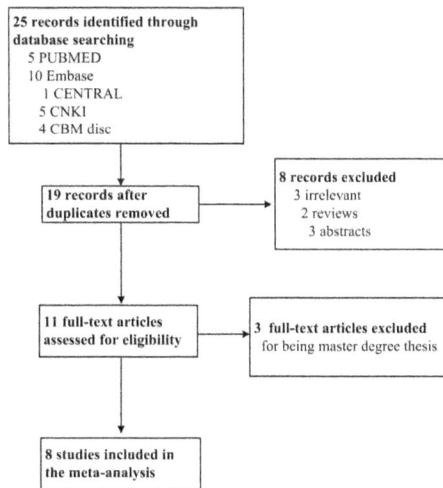

**Figure 1.** PRISMA flow chart of studies inclusion and exclusion.

**Table 3.** Stratified effects of miR-146a (rs2910164) on ischemic stroke risk under dominant genetic model (GG/GC *vs.* CC).

| Selected Variables | | Number of Studies | OR (95% CI) | *p*-Value | $I^2$ (%) |
|---|---|---|---|---|---|
| TOAST | LAA | 3 | 0.80 (0.51, 1.27) | 0.34 | 71% |
| | SVD | 3 | 1.29 (0.90, 1.84) | 0.17 | 58% |
| Gender | Male | 2 | 1.06 (0.82, 1.38) | 0.65 | 0% |
| | Female | 2 | 1.16 (0.45, 2.97) | 0.76 | 86% |
| Smoker | Yes | 2 | 1.36 (0.91, 2.03) | 0.14 | 0% |
| | No | 2 | 1.09 (0.86, 1.38 ) | 0.86 | 69% |
| Hypertension | Yes | 2 | 0.87 (0.42, 1.82) | 0.72 | 78% |
| | No | 2 | 1.30 (0.97, 1.75) | 0.08 | 0% |
| Diabetes mellitus | Yes | 2 | 1.76 (0.96, 3.21) | 0.07 | 0% |
| | No | 2 | 1.25 (0.95, 1.63) | 0.11 | 0% |
| Hyperlipidemia | Yes | 2 | 1.16 (0.78, 1.71) | 0.47 | 17% |
| | No | 2 | 1.23 (0.80, 1.88) | 0.34 | 42% |

TOAST: Trial of Org 10172 in Acute Stroke Treatment; LAA, Large artery atherosclerosis; SVD, Small vessel Disease.

**(A) G *vs.* C**

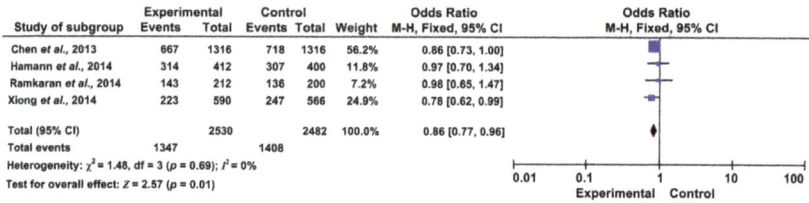

**(B) GG *vs.* CC**

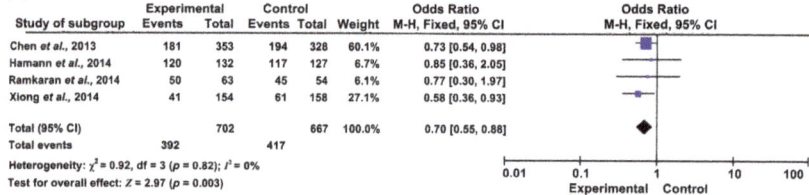

**(C) GC *vs.* CC**

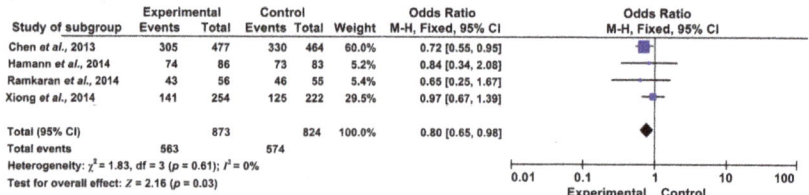

**(D) GG/GC *vs.* CC**

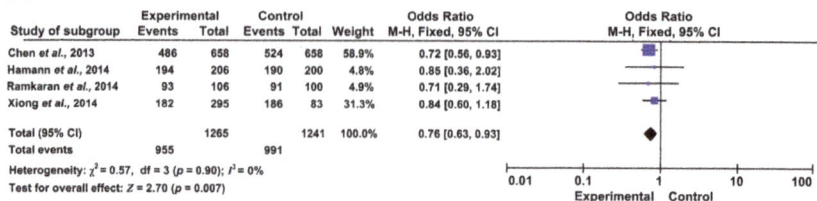

**Figure 2.** Forest plots of odds ratios for the association between microRNA-146a rs2910164 and the risk of CAD. (**A**) G *vs.* C; (**B**) GG *vs.* CC; (**C**) GC *vs.* CC; (**D**) GG/GC *vs.* CC.

**(A) G *vs.* C**

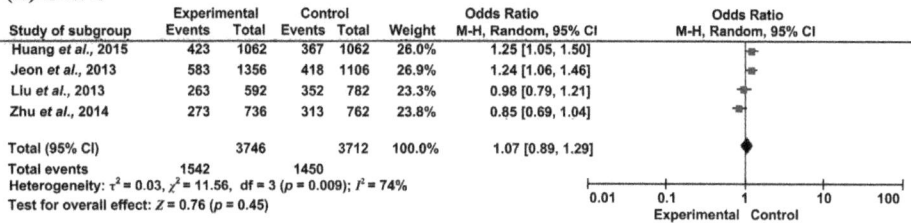

**(B) GG *vs.* CC**

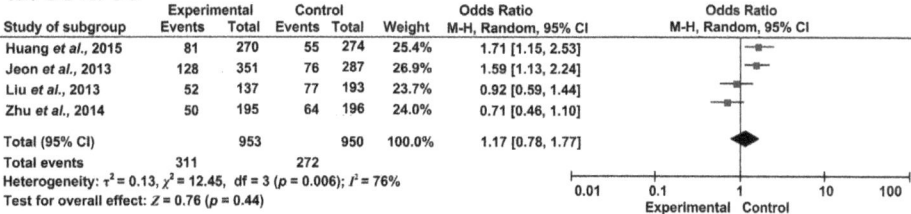

**(C) GC *vs.* CC**

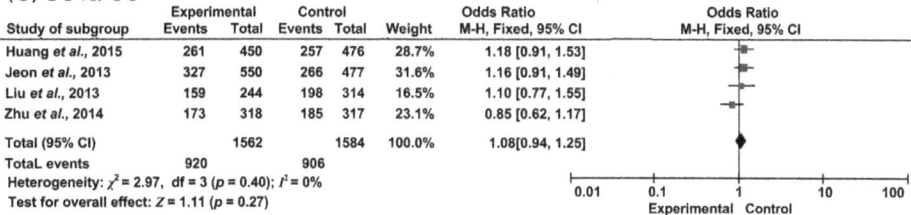

**(D) GG/GC *vs.* CC**

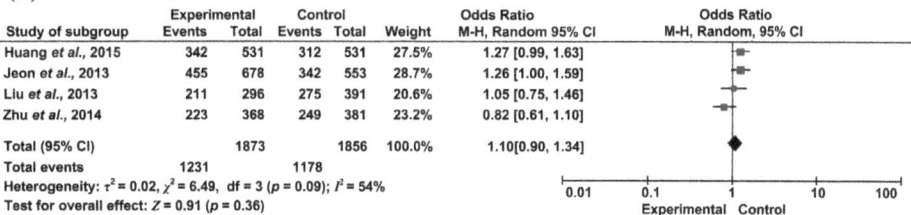

**Figure 3.** Forest plots of odds ratios for the association between microRNA-146a rs2910164 and risk of IS. (**A**) G *vs.* C; (**B**) GG *vs.* CC; (**C**) GC *vs.* CC; (**D**) GG/GC *vs.* CC.

## *2.3. Sources of Heterogeneity*

For IS, the subgroup analysis indicated the sample size was the source of the heterogeneity among studies ($p = 0.001$ for allelic model; $p = 0.0006$ for homozygous model; and $p = 0.02$ for dominant model) (data not shown). For CAD, there are no significant heterogeneity existing among all studies.

## *2.4. Sensitivity Analysis*

The influence of each study on the pooled ORs and 95% CIs were evaluated by excluding one single study at a time. The corresponding pooled OR was not significantly altered in all genetic models (Figure 4).

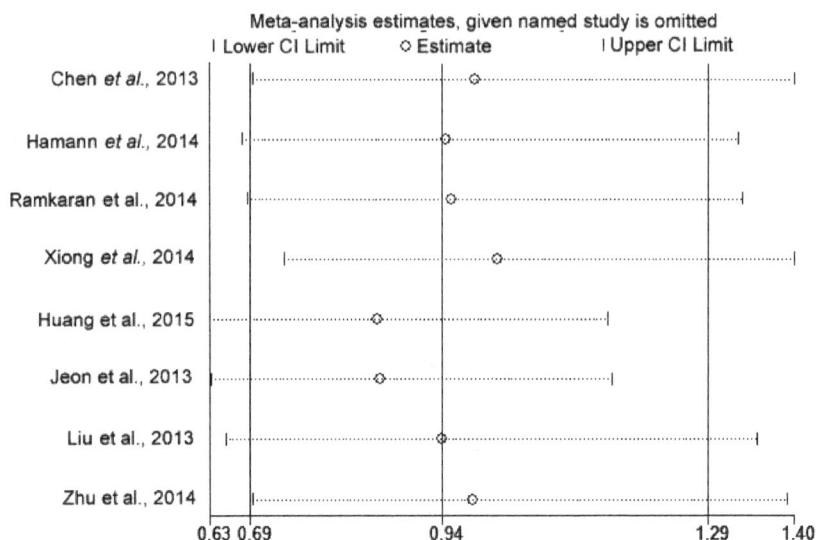

**Figure 4.** The influence of each study by removal of individual studies in China for G *vs.* C model.

*2.5. Publication Bias*

We performed the Begg's funnel plot and Egger's test to evaluate the publication bias. The *p* value for Egger's linear regression tests is shown in Table 4. As the results indicated, no obvious publication bias was observed. And these results were also demonstrated by the shape of funnel plot (not shown).

**Table 4.** Egger's linear regression test for funnel plot asymmetries.

| Group | *p*-Value | | | |
|---|---|---|---|---|
| | G *vs.* C | GG *vs.* CC | GC *vs.* CC | GG/GC *vs.* CC |
| CAD | 0.240 | 0.137 | 0.965 | 0.967 |
| IS | 0.110 | 0.191 | 0.420 | 0.276 |

**3. Discussion**

The main findings of our meta-analysis show that rs2910164 is associated with the risk of CAD significantly in allelic model (OR = 0.86), homozygous model (OR = 0.70), heterozygous model (OR = 0.80) and dominant model (OR = 0.76). The subjects carrying GG genotype, GG + GC genotype or G allele are at lower risks of CAD. For the susceptibility of IS, there are no significant associations between rs2910164 and total studies. However, in subgroup analysis by sample size and ethnicity, the GG, GG + GC and G allele of rs2910164 are found to be associated with higher risks of IS in large size group and in Koreans, under homozygous and dominant model.

Although a very recent meta-analysis was performed to reveal the relationship of rs2910164 and cardio-cerebrovascular diseases, He *et al.* [15], they included fewer studies on CAD and IS than those in our meta-analysis. In our meta-analysis, we included two more studies, with 737 cases and 731 controls, for CAD and IS analysis. We further conducted a more comprehensive subgroup analysis for IS by TOAST, gender, smoking, hypertension, diabetes mellitus and hyperlipidemia. Furthermore, we found some surprisingly different results between our meta-analysis and He *et al.*'s in CAD. Although we found a similar lower CAD risks for G allele *vs.* C allele, our meta-analysis also demonstrated a lower risks of CAD for GG/GC and GG genotype *vs.* CC carriers (Figure 2 and Table 2). However, He *et al.*'s meta-analysis stated a higher risk for GG carriers *vs.* CC carriers.

In the sensitivity analysis, no significant changes were found after omitting each study one at a time, indicating the relative stability and credibility of the results of our meta-analysis.

We conducted a subgroup analysis by sample size and ethnicity, and observed that rs2910164 was associated with the risks of CAD and IS in the large sample group. However, only one study was included in the large size subgroup in CAD, and two studies were included in IS. We also found a higher risk of IS in Korean for G allele, GG/GC and GG genotypes of miR-146a, and once again, only one study is included in this subgroup. Thus, further studies are needed to confirm these results.

Since no associations were found between rs2910164 and IS, we performed subgroup analysis by TOAST, gender, smoking, hypertension, diabetes mellitus and hyperlipidemia (Table 3). No significant relations were found in all the analyses. But some studies lack sufficient data for the subgroup analysis. For example, the study of Huang *et al.* [11] lacks TOAST data, while Jeon *et al.* and Zhu *et al.* lack gender, smoking, hypertension, hyperlipidemia and diabetes mellitus data [12,16]. Interestingly, in the study of Zhu *et al.* [16], no significant correlations were found between rs2910164 and overall subjects, though when they divided the patients to LAA and SVD according to TOAST typing, significantly higher frequencies in LAA-caused IS were found in the CC genotype and C allele. However, the results of Zhu *et al.* [16], focusing on LAA-caused IS was contradictory to the study of Jeon *et al.* [12]. This may be the result of different target genes, geographical ethnic groups, sample size, *etc.* Further studies are still needed to resolve this discrepancy.

CAD and IS are both manifestations of atherosclerosis. Most studies included in the present meta-analysis found dyslipidemia in the case group. Previous studies have indicated atherosclerosis as an inflammatory process. Thus, anti-inflammatory treatments decrease the risk of atherosclerosis. miR-146a regulates the NF-κB-induced inflammatory process by targeting interleukin-1 receptor-associated kinase 1 (IRAK-1) and TNF receptor-associated factor 6 (TRAF-6). IRAK-1 and TRAF-6 are upstream regulators of NF-κB activation, and major signal transducers of the Toll-like receptor (TLR) system [9]. The SNP in the pre-miR-146a changed the G to C allele, thus influence the mature miR-146a production, and subsequently influence the inflammatory process of atherosclerosis. Interestingly, the GG, GG + GC genotype and G allele are associated with lower susceptibility of CAD in overall subjects and large sample size group. On the contrary, GG, GG + GC genotype and G allele are related to higher risk of IS in large sample size group. This could be explained by the target gene selection of miR-146a and the pathological differences of these two diseases [10,17–20]. Furthermore, the endothelial dysfunction induced by the inflammatory process is considered to be the first step of atherosclerosis. Further investigations are needed to reveal whether or not rs2910164 affects endothelial mature miR-146a production and the effects of this variation might have on endothelial functions.

The results of the present meta-analysis should be interpreted carefully because of the following limitations. Firstly, the number of patients was relatively small, and may influence the outcomes. After a very comprehensive literature search from several different databases, only a total of eight studies were included in the present meta-analysis. Among them, four are related to CAD (1265 cases and 1241 controls) and four are related to IS (1873 cases and 1856 controls). The overall OR only indicated the associations between rs2910164 and CAD. However, no associations were found between rs2910164 and IS risk. The second limitation is that the clinicopathological characteristics or disease subtypes are limited in most of these studies. Although we conducted subgroup analysis by clinicopathological characteristics in IS, only half of the included studies provide the necessary data. And for CAD, no such characteristics were available. Thirdly, CAD and IS are both multi-factorial diseases influenced by both genetic and environmental factors. The gene-gene and gene-environment interactions may play important roles in the function of rs2910164, but most studies lack information about environmental exposure and multiple SNPs in miRNA-encoding genes. The fourth limitation lies in the ethnicity of the subjects. Most of the patients were Asians in the present study and this limited the general application of the results to other populations.

In conclusion, the current meta-analysis suggests a decreased risk of CAD for GG, GG + GC genotype and G allele of rs2910164, while rs2910164 is not associated with the risk of IS. Thus rs2910164 might be recommended as a predictor for susceptibility of CAD, but not IS. However, the results of this meta-analysis should be interpreted with caution because of the heterogeneity among study designs. Further study is needed to evaluate the association of rs2910164 and these two diseases, especially in a large sample size, in Caucasians, and with clinicopathological characteristics.

## 4. Methods

### 4.1. Publication Search Strategy and Inclusion Criteria

Published studies were systematically searched by Mei-hua Bao and Yan Xiao; the electronic databases Pubmed, Embase, Cochrane Central Register of Controlled Trials (CENTRAL), Chinese National Knowledge Infrastructure (CNKI) and Chinese Biomedical Literature Database (CBM) were searched for the following terms: "Coronary artery disease (CAD)" or "ischemic stroke (IS)" and ("miR-146a" or "miRNA-146a" or "microRNA-146a") and ("polymorphism" or "mutation" or "variant" or "SNP" or "single nucleotide polymorphism"), without restriction on language. The deadline for publication was 15 March, 2015. All the results from the databases were screened. First, we screened the title. If the titles fulfilled our criteria, we then screened the abstract. We retrieved the full text if the abstract was interesting. All eligible studies were retrieved manually for other potentially relevant studies from their references. We contacted the authors for related data when they were unavailable in the original publications.

Inclusion Criteria: (a) Case-control design; (b) The association of miR-146a (rs2910164) and CAD or IS risks should be evaluated; (c) The genotype in the control group should be agreed with the Hardy-Weinberg equilibrium (HWE); (d) The data in the publication are sufficient for present estimation. Studies were excluded if any of the following applies: (1) Repeat publications, abstracts, letters or reviews; (2) Studies not meeting all of the inclusion criteria.

### 4.2. Data Extraction

We extracted the information from each eligible publication manually by two investigators independently (Qing-song Zhang and Ji Luo). For each study, the extracted information included: First authors's name, year of publication, country, ethnicity, genotype method, sex ratio, age, source of controls, diseases, genotype numbers of cases and controls. If we encountered discrepancies during data extraction, it was resolved by a consensus achieved by the third author (Jian-ming Li).

### 4.3. Quality Assessment

The quality of the included studies was evaluated by the following aspects: source of cases, source of controls, specimens used for determining genotypes, total sample size and evidence of HWE. The quality scores ranged from 0–15, higher scores indicating better quality [21]. The quality evaluation was performed by two authors independently (Juan Zhao and Guang-yi Li). Meetings were held to resolve any discrepancies in the assessment process.

### 4.4. Statistical Methods

$\chi^2$-test was used to evaluate the HWE of the control group polymorphism. If $p < 0.05$, it was considered to be deviated from HWE. To evaluate the association between miR-146a (rs2910164) and disease (CAD and IS) risk, the crude odds ratio (OR) with 95% confidence interval (CI) was used. The pooled ORs were calculated using genetic model of allelic model (G *vs.* C), homozygous (GG *vs.* CC), heterozygous (GC *vs.* CC) and dominant (GG/GC *vs.* CC) model, and the statistical significance was determined by the Z-test, and $p < 0.05$ was considered to be statistically significant. Subgroup analysis was conducted according to the sample size. Total samples less than 1000 was treated as small, and otherwise as large.

The statistical heterogeneity between studies was evaluated by $I$-square statistical test, which was not dependent on the number of studies in the meta-analysis [22]. If there was an obvious heterogeneity among the studies ($I^2 > 50\%$), the random-effects model (the DerSimonian and Laird method) was used for the meta-analysis [23]. Otherwise, the fixed-effect model using the Mantel-Haenszel method was used [24]. Sensitivity analysis was performed to assess the effects of individual study on pooled results and the stability of results. The publication bias was detected with Begg's funnel plot, and Egger's linear regression method, and $p < 0.05$ was considered to be statistically significant [25]. All statistical analysis was performed using the STATA 12.0 software (StataCorp, College Station, TX, USA) and Revman 5.3.

**Acknowledgments:** This work was supported by the Key Discipline Construction Program of Hunan Province, National Natural Science Foundation of China (No. 81300231), and Hunan Provincial Natural Science Foundation of China (No. 13JJ4112).

**Author Contributions:** Conceived of and designed the experiments: Mei-Hua Bao and Jian-Ming Li; Performed the experiments: Mei-Hua Bao, Yan Xiao, Juan Zhao and Guang-Yi Li; Analysis of the data: Qing-Song Zhang, Huai-Qing Luo and Ji Luo; Contributed reagents/materials/analysis tools: Jie Zeng; Wrote the pater: Mei-Hua Bao and Jian-Ming Li.

**Conflicts of Interest:** The authors declare no conflict of interest.

# References

1. Liu, L.; Wang, D.; Wong, K.S.; Wang, Y. Stroke and stroke care in China: Huge burden, significant workload, and a national priority. *Stroke* **2011**, *42*, 3651–3654. [CrossRef] [PubMed]
2. Goldstein, L.B.; Adams, R.; Becker, K.; Furberg, C.D.; Gorelick, P.B.; Hademenos, G.; Hill, M.; Howard, G.; Howard, V.J.; Jacobs, B.; *et al.* Primary prevention of ischemic stroke: A statement for healthcare professionals from the Stroke Council of the American Heart Association. *Stroke* **2001**, *32*, 280–299. [CrossRef] [PubMed]
3. Marshall, H.W.; Morrison, L.C.; Wu, L.L. Apolipoprotein polymorphisms pail to define risk of coronary-artery disease-results of a prospective, angiographically controlled-study. *Circulation* **1994**, *89*, 567–577. [CrossRef] [PubMed]
4. Fedele, F.; Mancone, M.; Chilian, W.M.; Severino, P.; Canali, E.; Logan, S.; de Marchis, M.L.; Volterrani, M.; Palmirotta, R.; Guadagni, F. Role of genetic polymorphisms of ion channels in the pathophysiology of coronary microvascular dysfunction and ischemic heart disease. *Basic Res. Cardiol.* **2013**, *108*, 387. [CrossRef] [PubMed]
5. Lewis, B.P.; Burge, C.B.; Bartel, D.P. Conserved seed pairing, often flanked by adenosines, indicates that thousands of human genes are microRNA targets. *Cell* **2005**, *120*, 15–20. [CrossRef] [PubMed]
6. Roldan, V.; Arroyo, A.B.; Salloum-Asfar, S.; Manzano-Fernandez, S.; Garcia-Barbera, N.; Marin, F.; Vicente, V.; Gonzalez-Conejero, R.; Martinez, C. Prognostic role of miR146a polymorphisms for cardiovascular events in atrial fibrillation. *Thromb. Haemost.* **2014**, *112*, 781–788. [CrossRef] [PubMed]
7. Jazdzewski, K.; Murray, E.L.; Franssila, K.; Jarzab, B.; Schoenberg, D.R.; de la Chapelle, A. Common SNP in pre-miR-146a decreases mature miR expression and predisposes to papillary thyroid carcinoma. *Proc. Natl. Acad. Sci. USA* **2008**, *105*, 7269–7274. [CrossRef] [PubMed]
8. Chen, L.; Wu, Y.T. Association of genetic polymorphisms in microRNAs precursor with the risk and prognosis of coronary heart diseases. *J. Xi'an Jiaotong Univ.* **2013**, *34*, 495–499.
9. Ramkaran, P.; Khan, S.; Phulukdaree, A.; Moodley, D.; Chuturgoon, A.A. miR-146a polymorphism influences levels of miR-146a, IRAK-1, and TRAF-6 in young patients with coronary artery disease. *Cell Biochem. Biophys.* **2014**, *68*, 259–266. [CrossRef] [PubMed]
10. Xiong, X.D.; Cho, M.; Cai, X.P.; Cheng, J.; Jing, X.; Cen, J.M.; Liu, X.; Yang, X.L.; Suh, Y. A common variant in pre-miR-146 is associated with coronary artery disease risk and its mature miRNA expression. *Mutat. Res.* **2014**, *761*, 15–20. [CrossRef] [PubMed]
11. Huang, S.; Zhou, S.; Zhang, Y.; Lv, Z.; Li, S.; Xie, C.; Ke, Y.; Deng, P.; Geng, Y.; Zhang, Q.; *et al.* Association of the genetic polymorphisms in pre-microRNAs with risk of ischemic stroke in a Chinese population. *PLoS ONE* **2015**, *10*, e0117007. [CrossRef] [PubMed]

12. Jeon, Y.J.J.; Kim, O.J.; Kim, S.Y.; Oh, S.H.; Oh, D.; Kim, O.J.; Shin, B.S.; Kim, N.K. Association of the miR-146a, miR-149, miR-196a2, and miR-499 polymorphisms with ischemic stroke and silent brain infarction risk Genetic polymorphisms in pre-microRNAs and risk of ischemic stroke in a Chinese population. *Arterioscler. Thromb. Vasc. Biol.* **2013**, *33*, 420–430. [CrossRef] [PubMed]

13. Liu, Y.; Ma, Y.; Zhang, B.; Wang, S.X.; Wang, X.M.; Yu, J.M. Genetic polymorphisms in pre-microRNAs and risk of ischemic stroke in a Chinese population. *J. Mol. Neurosci.* **2014**, *52*, 473–480. [CrossRef] [PubMed]

14. Hamann, L.; Glaeser, C.; Schulz, S.; Gross, M.; Franke, A.; Nothlings, U.; Schumann, R.R. A micro RNA-146a polymorphism is associated with coronary restenosis. *Int. J. Immunogenet.* **2014**, *41*, 393–396. [CrossRef] [PubMed]

15. He, Y.; Yang, J.; Kong, D.; Lin, J.; Xu, C.; Ren, H.; Ouyang, P.; Ding, Y.; Wang, K. Association of miR-146a rs2910164 polymorphism with cardio-cerebrovascular diseases: A systematic review and meta-analysis. *Gene* **2011**, *565*, 171–179. [CrossRef] [PubMed]

16. Zhu, R.; Liu, X.; He, Z.; Li, Q. miR-146a and miR-196a2 polymorphisms in patients with ischemic stroke in the northern Chinese Han population. *Neurochem. Res.* **2014**, *39*, 1709–1716. [CrossRef] [PubMed]

17. Guo, M.; Mao, X.; Ji, Q.; Lang, M.; Li, S.; Peng, Y.; Zhou, W.; Xiong, B.; Zeng, Q. miR-146a in PBMCs modulates Th1 function in patients with acute coronary syndrome. *Immunol. Cell Biol.* **2010**, *88*, 555–564. [CrossRef] [PubMed]

18. Cheng, H.S.; Sivachandran, N.; Lau, A.; Boudreau, E.; Zhao, J.L.; Baltimore, D.; Delgado-Olguin, P.; Cybulsky, M.I.; Fish, J.E. MicroRNA-146 represses endothelial activation by inhibiting pro-inflammatory pathways. *EMBO Mol. Med.* **2013**, *5*, 949–966. [CrossRef] [PubMed]

19. El, G.M.; Church, A.; Liu, T.; McCall, C.E. MicroRNA-146a regulates both transcription silencing and translation disruption of TNF-α during TLR4-induced gene reprogramming. *J. Leukoc. Biol.* **2011**, *90*, 509–519.

20. Cui, G.; Wang, H.; Li, R.; Zhang, L.; Li, Z.; Wang, Y.; Hui, R.; Ding, H.; Wang, D.W. Polymorphism of tumor necrosis factor α (TNF-α) gene promoter, circulating TNF-α level, and cardiovascular risk factor for ischemic stroke. *J. Neuroinflamm.* **2012**, *9*, 235. [CrossRef] [PubMed]

21. Jiang, D.K.; Wang, W.Z.; Ren, W.H.; Yao, L.; Peng, B.; Yu, L. *TP53* Arg72Pro polymorphism and skin cancer risk: A meta-analysis. *J. Investig. Dermatol.* **2011**, *131*, 220–228. [CrossRef] [PubMed]

22. Higgins, J.P.; Thompson, S.G.; Deeks, J.J.; Altman, D.G. Measuring inconsistency in meta-analyses. *BMJ* **2003**, *327*, 557–560. [CrossRef] [PubMed]

23. DerSimonian, R.; Laird, N. Meta-analysis in clinical trials. *Control Clin. Trials* **1986**, *7*, 177–188. [CrossRef]

24. Mantel, N.; Haenszel, W. Statistical aspects of the analysis of data from retrospective studies of disease. *J. Natl. Cancer Inst.* **1959**, *22*, 719–748. [PubMed]

25. Peters, J.L.; Sutton, A.J.; Jones, D.R.; Abrams, K.R.; Rushton, L. Comparison of two methods to detect publication bias in meta-analysis. *JAMA* **2006**, *295*, 676–680. [CrossRef] [PubMed]

International Journal of
*Molecular Sciences*

MDPI

*Review*

# Advances in the Study of the Antiatherogenic Function and Novel Therapies for HDL

Peiqiu Cao [1], Haitao Pan [1], Tiancun Xiao [2,3], Ting Zhou [3], Jiao Guo [1,*] and Zhengquan Su [1,*]

[1] Key Research Center of Liver Regulation for Hyperlipemia SATCM/Class III, Laboratory of Metabolism SATCM, Guangdong TCM Key Laboratory for Metabolic Diseases, Guangdong Pharmaceutical University, Guangzhou 510006, China; cpq_520@126.com (P.C.); pangel7835001@163.com (H.P.)

[2] Inorganic Chemistry Laboratory, University of Oxford, South Parks Road, Oxford OX1 3QR, UK; xiao.tiancun@chem.ox.ac.uk

[3] Guangzhou Boxabio Ltd., D-106 Guangzhou International Business Incubator, Guangzhou 510530, China; ting4677@126.com

* Correspondence: suzhq@scnu.edu.cn (Z.S.); wshxalb@163.com (J.G.); Tel./Fax: +86-20-3935-2067 (Z.S. & J.G.)

Academic Editor: Michael Henein
Received: 23 April 2015; Accepted: 16 July 2015; Published: 28 July 2015

**Abstract:** The hypothesis that raising high-density lipoprotein cholesterol (HDL-C) levels could improve the risk for cardiovascular disease (CVD) is facing challenges. There is multitudinous clear clinical evidence that the latest failures of HDL-C-raising drugs show no clear association with risks for CVD. At the genetic level, recent research indicates that steady-state HDL-C concentrations may provide limited information regarding the potential antiatherogenic functions of HDL. It is evident that the newer strategies may replace therapeutic approaches to simply raise plasma HDL-C levels. There is an urgent need to identify an efficient biomarker that accurately predicts the increased risk of atherosclerosis (AS) in patients and that may be used for exploring newer therapeutic targets. Studies from recent decades show that the composition, structure and function of circulating HDL are closely associated with high cardiovascular risk. A vast amount of data demonstrates that the most important mechanism through which HDL antagonizes AS involves the reverse cholesterol transport (RCT) process. Clinical trials of drugs that specifically target HDL have so far proven disappointing, so it is necessary to carry out review on the HDL therapeutics.

**Keywords:** HDL; biomarker; HDL function; reverse cholesterol transport; HDL therapies

## 1. Introduction

Hyperlipidemia, a risk factor for atherosclerosis (AS), is a serious consequence for people who have experienced coronary heart disease, stroke, and artery stenosis disease. AS, which is the leading cause of cardiovascular disease (CVD), is responsible for 50% of all mortality in many developed countries [1]. A persistent increase in circulating low-density lipoprotein cholesterol (LDL-C) levels in the body is one of the most important causes for the initiation and progression of AS [2]. It has been shown in epidemiological studies and clinical trials that LDL-C levels are directly related to the rate at which CVD events occur [2,3]. There are abundant antilipemic agents (Table 1) on the market, but the meta-analysis of intervention trials has shown that a per mmol/L decrease in LDL-C is associated with an approximate 22% reduction of CVD events and a 10% reduction of all-cause mortality [3]. There is a wealth of evidence showing that statins that play a beneficial role in lowering LDL-C levels and are efficient in preventing first cardiovascular events. However, a large residual disease burden remains, even in patients treated with a high dose of statins and other CVD risk-modifying interventions [1]. Furthermore, treatment with statins may lead to a significant increase of muscle toxicity and liver transaminase, and may not be suitable for all CVDs. Investigators are eagerly searching for novel therapeutic targets. Because high-density lipoprotein cholesterol (HDL-C) levels, a predictor of major cardiovascular events in patients, are inversely associated with risk for CVD [2,4], strategies for increasing HDL levels have been explored as a new approach for combatting CVD, which may overcome the significant residual cardiovascular risk remaining after treatment with statins [4]. However, a recent genetic analysis failed to show a causal association between genetically raised plasma HDL cholesterol levels and risk for myocardial infarction [5]. In addition, nicotinic acid and fibrates currently are used to increase HDL-C levels, both of which have some weaknesses (e.g., uricosuria, increased glucose tolerance and flushing for nicotinic acid and problematic pharmacokinetic interactions for fibrates) that impose restrictions on their use [6]. Moreover, raising HDL-cholesterol by the cholesteryl-ester transfer protein (CETP) inhibitor did not play its expected role in protection from CVD [7,8]. The goals of previous HDL therapy are currently being reassessed due to numerous difficulties validating the hypothesis. Therefore, HDL structure, composition and function are the focus of ongoing research efforts, as they might provide more valuable information than steady-state HDL-cholesterol levels [9]. From the study of the structure and composition of HDL, novel strategies for the treatment of hyperlipemia-induced AS are being developed. Apolipoprotein A-I (apoA-I) is the major structural protein component of HDL particles [10], and it has been shown to play a pivotal role in reverse cholesterol transport (RCT) [11]. ApoA-I has also been shown to exert direct anti-inflammatory effects [12]. Stimulating increased synthesis of endogenous apoA-I may be a promising approach. Regarding the function of HDL, the atheroprotective activities of HDL particles are attributed to their central role in anti-inflammatory, antithrombotic, and antioxidant processes and their ability to improve endothelial function [13,14]. In addition, RCT is currently understood as the physiological process by which cholesterol in peripheral tissues is transported by HDL to the liver for excretion in the bile and feces [15]. This process is complex and beneficial, and it has been known as a widely accepted mechanism for the protective effect of HDL.

This paper is aimed at looking for a better HDL biomarker to predict AS precisely and explaining how HDL functions in detail, then providing an overview of novel therapies that target HDL to enhance HDL's ability to reduce residual cardiovascular risk in the population. Finally, we carry out an assessment of a natural drug that has been linked to HDL, which represents a unique field in the treatment of AS that warrants exploration.

**Table 1.** Drugs of anti-hyperlipidemia in the current market. In recent years many cholesterol-lowering drugs are commonly used on the market, which mainly includes stains, fibrates, nicotinic acids, cholesterol absorption inhibitor and polyene unsaturated fatty acids.

| Classification | Drug | Mechanism | TC | TG | VLDL | LDL | HDL | Advantage | Disadvantage |
|---|---|---|---|---|---|---|---|---|---|
| Stains | Atorvastatin Lovastatin | As inhibition of HMG CoA reductase, reduce cholesterol synthesis | | ↓ | ↓ | ↓↓ | ↑ | The advantage of these drugs is a low incidence of adverse reaction, and can be suitable for a variety of hypercholesterolemia, except hypertriglyceridemia, is a lipid-lowering drug with rapid development in recent years | Gastrointestinal symptoms and rash, and the residual risk |
| Fibrates | Gemfibrozil Fenofibrate | The drug can increase Lp(a) Lipase activity to remove VLDL, TG; Thus reducing VLDL and TG, TC and LDL can also be reduced | | ↓ ↓ | ↓ | ↓ | ↑ | These drugs do not cause the increase of diabetic insulin resistance or affect the control of blood sugar, therefore this kind of drugs is the first choice for treating the diabetic patients with hyperlipidemia | Gastrointestinal reactions, allergic reaction, due the drugs increase the concentration of cholesterol in bile, it may cause gallstones, occasional eyesight obstacle and hematological abnormalities |
| Nicotinic Acids | Niacin; Inositol Aluminum Aluminum Nicotinate | The drug can prevent fat decomposition, prevent free fatty acid formation, inhibit synthesis of TG and secretion of VLDL in liver | | ↓ | ↓ | ↓ | ↑ | Cheap, and it is the only lipid-lowering drug can also reduce risk and mortality of cardiovascular disease | It is not suitable for diabetes patients, overdose adverse reactions (toxic to the liver, high blood sugar) has a high incidence common adverse reactions are skin flushing, itching, rash |
| Cholesterol Absorption Inhibitor | Ezetimibe | The drug can combine with bile acid to block the bile acid absorption; Prompt the translation of the cholesterol into bile in the gallbladder, then bile binding to drug is eliminated from the body | ↓ | ↓ | | ↓ | ↑ | This kind of medicine is recognized as TC lowering drugs, when treats together with statins, the risk of accidental heart disease related to decrease the occurrence of 50% or more | The common adverse reactions are mild nausea and abdominal distension, constipation, therefore, it is not suitable for intestinal diseases and intractable constipation patients |
| Polyene Unsaturated Fatty Acids | Duoxikang Ecosapeatanolic acid | This drug can combine with total cholesterol to be ester; Then promotes the degradation of bile acid excreted along with the bile, decreases plasma total cholesterol concentration | ↓ | ↓ ↓ | | ↓ | | These drugs in combination with statins can reduce the level of TG, and play an effective role in the prevention and treatment of coronary heart disease | This kind of medicaments is easy to be oxidized to atherogenic substance, has inhibitory effect on platelet aggregation, so it needs to be used with caution |

133

## 2. HDL and AS

AS can be considered to be a form of chronic inflammation, beginning with increased endothelial permeability of monocytes under the influence of adhesion molecules [16]. It promotes endocytosis of oxidized LDL into the arterial wall and intima, which can be devoured by intimal macrophages to become foam cells. This is a key process in the development of atherosclerotic plaque. Along with an increase in macrophages and foam cells, the plaque becomes unstable. This progression can ultimately lead to the development of complex lesions, or plaques, that protrude into the arterial lumen. Plaque rupture and thrombosis result in the acute clinical complications of myocardial infarction and stroke [17,18]. Cardiovascular diseases are the leading cause of death and illness in developed countries, with AS being the most important contributor [13]. In recent years, there has been growing interest in finding a cardiovascular biomarker that provides prognostic and predictive information to act as a tool to influence treatment strategies [19]. The use of HDL-related indexes as biomarkers is undergoing predictive tests, and they represent the greatest promise of this technology and the shortest and most effective path to furthering our understanding.

### 2.1. HDL as the Biomarker for AS

#### 2.1.1. HDL-C

Because trends relating high plasma levels of HDL-C and decreased incidence of CVD endpoints were observed in prospective epidemiological studies conducted in several countries many years ago, HDL-C levels have served as a significant predictor for relieving AS in the clinic. There have been several attempts to increase HDL-C levels using pharmacological intervention [3,17]. HDL-C levels have been reported to increase upon chronic administration of fibrates as agonists of the peroxisome proliferator activated receptor $\alpha$ (PPAR$\alpha$) in animals and in humans. Niacin is the first antidyslipidemic agent identified that has been available for patients to raise HDL-C. However, recent data from a Global Health outcomes (AIM-HIGH) trial with niacin did not show any significant improvement in the cardiovascular risk over statins, which are commonly used drugs for lipid management [20,21]. Recently a few attempts have been made to inhibit CETP to raise the levels of HDL-C. Torcetrapib, a cholesterol ester transferase inhibitor, reliably increased HDL (without countervailing effects on LDL) but ultimately increased cardiovascular mortality. Dalcetrapib, a new CETP inhibitor, also has unintended effects, and the outcomes are the same as those for torcetrapib [7,8]. As a result, CETP inhibitors continue to face safety hurdles, and no significant clinical benefits have resulted from these pharmacological regimens. A recent report suggested that the sole increase of HDL-C in humans would not necessarily improve the rate of cardiovascular events (all-cause mortality, coronary heart disease mortality, non-fatal myocardial infarction, and stroke) [22–25]. In justification for the use of statins in prevention: an Intervention Trial Evaluating Rosuvastatin (JUPITER), on-treatment HDL-C was not predictive of residual risk among statin-treated individuals, whereas HDL-C was predictive among those taking placebo [26]. Similarly, on-treatment apoA-I and triglycerides were not predictive of residual risk [27]. For these reasons, the use of the HDL-C biomarker as a surrogate end point remains a difficult and distant goal.

#### 2.1.2. HDL Particle Size

Dyslipidemia may influence enzymes and transfer proteins needed for lipoprotein particle remodeling. HDL particle size may differ in the number of molecules of apo and free cholesterol, esterified cholesterol, and phospholipids content on the lipoprotein surface [28,29] due to the changes in activity of Lecithin-cholesterol acyltransferase (LCAT), CETP, phospholipid transfer protein (PLTP) plasma transfer proteins, and enzymes (e.g., lipoprotein lipase (LPL), hepatic lipase (HL)) in the process. Calculated indices and the evaluation of lipoprotein particle size have been widely used to predict cardiovascular risk. An evaluation of the association between HDL particle size and risk of incident coronary artery disease (CHD) in apparently healthy volunteers indicated that decreased

HDL particle size is associated with an adverse cardiometabolic risk profile [30]. Small HDL particle size was also associated with an increased CHD risk, which resulted from the different lipid transfer ability [31]. Another study found that the high TG-low HDL cholesterol dyslipidemia, which is found in viscerally obese subjects and characterized by hyperinsulinemia, was strongly correlated with reduced HDL particle size [32]. However, the latest finding is not consistent with previous conclusions. The HDL particle size (nm) values were not different between the dyslipidemia, normolipidemic and dyslipdemic groups without treatment with lipid-lowering drugs [33]. In addition, several studies using GGE-measured HDL size have reported that patients with CHD tend to have smaller HDL particles and that large HDL particles may protect against the development of AS. There are also other studies that show that only small HDL particles are atheroprotective, this is supported by the analysis of HDL subfraction data and carotid artery disease (CAAD) [34]. It is found that among HDL-C, $HDL_2$, and $HDL_3$, $HDL_3$ best predicts CAAD risk, and the remaining phenotypes do not add significant predictive power [34]. Although public opinions about HDL size are divergent, the propagation rate and maximal diene formation during total HDL oxidation correlated significantly with HDL mean particle size [30,32]. It is clear that proper HDL particle size may have advantages in the reduction of CVD events. In the future, we believe that HDL particle size will play an essential role in predicting the risk of AS.

### 2.1.3. HDL Particles Concentration (HDL-P)

Given the extreme heterogeneity of HDL, measuring the content of HDL particles will, at best, only partially reflect the potential role of HDL in cardiovascular risk assessment and therapeutic drug development. In this regard, the HDL-P may be a better marker of residual vascular risk after potent statin therapy in the JUPITER trial than chemically measured HDL-C or apoA-I [35]. HDL-P may be a promising metric of HDL that is more independent of other metabolic and lipoprotein risk factors than HDL cholesterol or HDL size [36]. In the European Prospective Investigation into Cancer and Nutrition (EPIC)-Norfolk study, HDL-P was inversely associated with CVD, consistent with the above-mentioned observations [37]. However, in the Women's Health Study, HDL-P was not associated with incident CVD events among healthy low-risk women in contrast with inverse associations seen for HDL size and HDL-C [38]. Moreover, an emerging outcome from a determinant of residual risk among statin-treated individuals suggests that overall HDL-P is not as important as the sub-type of HDL-P, which contains the lysosphingolipid sphingosine-1-phosphate (S1P) [36]. This conclusion is supported by the evidence that it is the S1P content of HDL, as opposed to the HDL particle itself, which is responsible for the beneficial antiatherothrombotic, anti-inflammatory, antioxidant, antiglycation, and profibrinolytic activities of these lipoproteins [36]. At present, two techniques, nuclear magnetic resonance (NMR) and ion mobility analysis (IMA) have been described for quantifying HDL-P in human plasma. However, the final HDL-P yield determined using these two methods differ by up to >5-fold. With the progress of science and technology, the accurate and reasonable use of $^1H$ NMR [39] and application of termed calibrated ion mobility analysis (calibrated IMA) [40] will greatly determine the feasibility of HDL-P as biomarker.

### 2.1.4. Other Biomarkers

The HDL-associated Apolipoprotein M (apoM) plays a role in the anti-atherogenic function in a variety of atherosclerotic models, making it interesting to investigate whether apoM is a predictor of AS [41–44]. With the discovery of apoM as an important carrier of S1P in HDL particles, it forms a new basis for investigations of apoM biology. An improved understanding of the role of the apoM/S1P axis in relation to AS may unravel new avenues for treatment or use of biomarkers for disease or risk evaluation [45].

The expression of serum amyloid A (SAA) protein was greater in the patients with AS, and it can be explained by high serum levels of SAA-predicted AS [45–49]. By means of two-dimensional electrophoresis (2-DE) coupled with mass spectrometry (MS) analysis on plasma-purified VLDL, LDL and HDL fractions from patients undergoing carotid endarterectomy, increased levels of acute-phase SAA (AP SAA) in all lipoprotein fractions helped to identify AP SAA as a potential marker of advanced carotid AS [50]. Studies in both mice and humans suggest that proinflammatory HDL may be a novel biomarker for increased risk of AS in patients with systemic lupus erythematosus (SLE) and rheumatoid arthritis (RA) [51]. Moreover, proinflammatory HDL-associated hemoglobin (Hb) was also found to be differentially associated with HDL from coronary heart disease patients compared with healthy controls [52,53]. Hb contributes to the proinflammatory nature of HDL in mouse and human models of AS and may serve as a novel biomarker for AS [52–54].

In addition, HDL has a wide range of functions, some of which are independent of its cholesterol content. Tests for HDL function as a biomarker may be useful for diagnosing the risk for patients with CHD. As a result, we elaborate the functions of HDL in the next section with the purpose of identifying the best biomarker of AS.

*2.2. HDL in Anti-Atherosclerosis*

HDLs have several well-documented functions with the potential to protect against AS. These functions include an ability to promote the efflux of cholesterol from macrophages in the artery wall, inhibit the oxidative modification of LDLs, inhibit vascular inflammation, inhibit thrombosis, promote endothelial repairing, promote angiogenesis, enhance endothelial function, improve diabetic control, and inhibit hematopoietic stem cell proliferation. Some of these functions have been mechanistically linked to the well-known ability of HDLs to induce the activation of cellular cholesterol efflux pathways, whereas many other functions of HDLs are independent of the effects of HDLs on cellular cholesterol homeostasis. Below, we will discuss the most relevant functions of HDLs for AS more specifically (Figure 1).

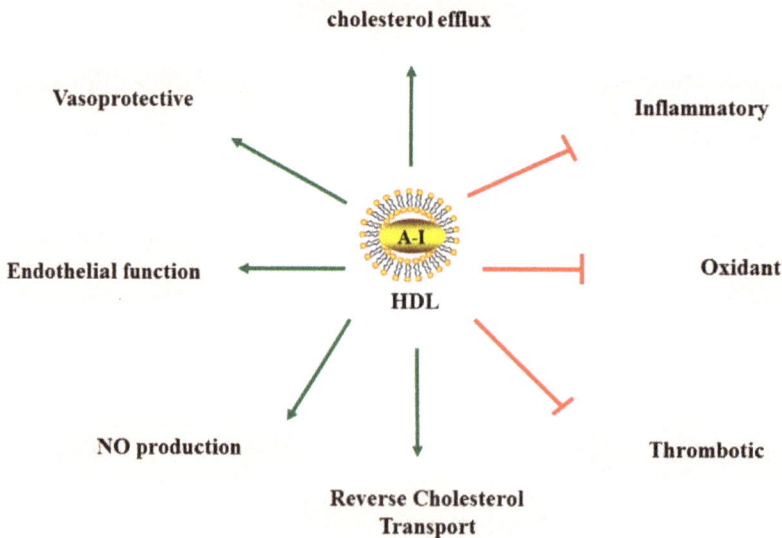

**Figure 1.** HDL antiatherosclerotic functions. ⬅ (promotion), ⊢ (inhibition).

2.2.1. Reverse Cholesterol Transport

In many studies, we found that the ability of HDL to promote cholesterol efflux from macrophages was strongly and inversely associated with both subclinical AS and obstructive coronary artery disease [55,56]. The efflux of cholesterol from a variety of cell types, including macrophages, to HDLs in the extracellular space is mediated by two distinct processes. One is the efflux of cholesterol induced by a specific cellular transporter [57–62], and the other is passive aqueous diffusion of cholesterol from cell membranes to HDLs [63,64]. Then, excess cholesterol from peripheral tissues will be transported back to the liver for excretion in the bile and ultimately the feces by HDL via a process called RCT. We herein would like to describe the importance of RCT in detail, and therefore, the multiplex RCT pathway has been described in the following four parts (Figure 2) [65].

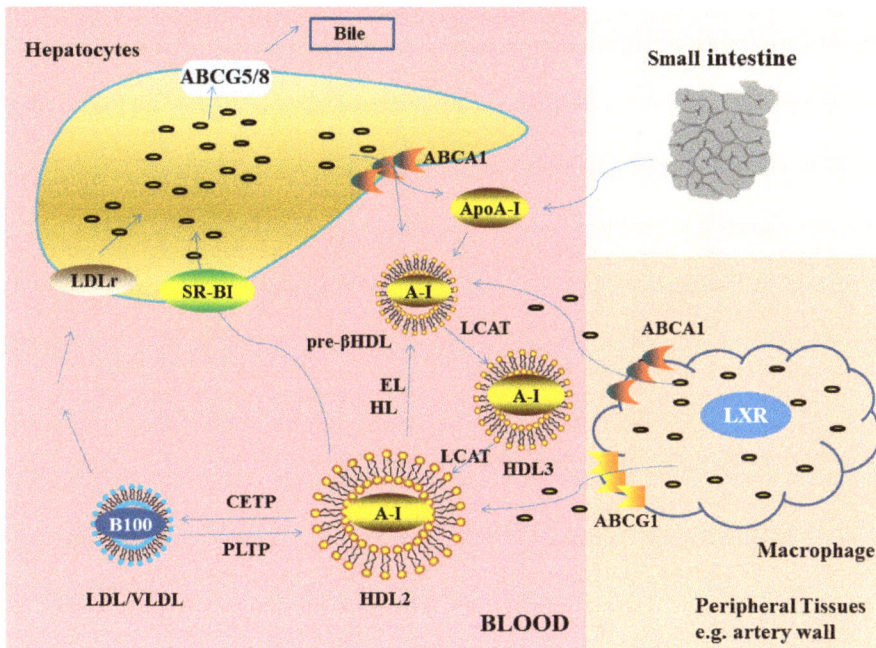

**Figure 2.** HDL in reverse cholesterol transport pathway. Lipid-poor apoA-I also promotes the efflux of free cholesterol from macrophages via ABCA1. LCAT esterifies free cholesterol to cholesteryl esters to form mature HDL, which promotes cholesterol efflux from macrophages via the ABCG1 transporter, as well as from other peripheral tissues by processes not fully defined. In macrophages, both ABCA1 and ABCG1 are regulated by Liver X receptors (LXR). Mature HDL can transfer its cholesterol to the liver directly via scavenger receptor class B type I (SR-BI) or indirectly via CETP-mediated transfer to ApoB-containing lipoproteins, with subsequent uptake by the liver via the LDL-r. Hepatic cholesterol can be excreted directly into the bile as cholesterol or after conversion to bile acids and, unless reabsorbed by the intestine, is ultimately excreted in the feces. HL, EL, and PLTP, play an indispensable role in remodeling HDL, thus, the RCT pathway is dependent on interaction with them [9].

Part One: The Formation of Nascent HDL

Lipid-free or lipid-poor apoA-I secreted in the liver can mediate cellular efflux of both cholesterol and phospholipids from macrophages via the ATP-binding cassette transporter A1 (ABCA1) and gather them on the surface of pre-β HDL, resulting in the rapid lipidation of apoA-I to generate mature α HDL, which is called nascent HDL particles [66,67]. Once loss of two ABCA1 genes occurs in the body, such as in the case of Tangier disease, the result is abnormally low levels of serum HDL-C, an acceleration of the accumulation of cholesterol in peripheral tissues, and the formation of premature AS [68]. The mature HDL particles can then serve as acceptors of cholesterol provided by ATP-binding cassette transporter G1 (ABCG1) [69] or SR-BI [70]. ABCG1 is another member of the ATP-binding cassette family that plays a critical role in the efflux of cellular phospholipid (PL) and free cholesterol (FC) to mature HDL, but not pre-β HDL. By intraperitoneal injection of mice with [$^3$H]-cholesterol-labeled J774 macrophages with either increased or reduced ABCG1 expression and primary macrophages lacking ABCG1 expression, followed by measurement of the macrophage-derived [$^3$H]-cholesterol levels in plasma and feces [71], it was shown that macrophages lacking ABCG1 cause damage and secretion of FC and excessive cholesterol accumulation in macrophages. Therefore, ABCG1 mediates cholesterol efflux [72]. In an investigation of extracellular cholesterol microdomains that form during the enrichment of macrophages with cholesterol, extracellular cholesterol microdomains did not develop when ABCG1-deficient mouse bone marrow-derived macrophages were enriched with cholesterol [69]. Many studies have demonstrated that ABCA1 and ABCG1 play a role in many aspects of cholesterol efflux from macrophages [69,71–73].

Liver X receptors (LXR), the members of the steroid nuclear receptor superfamily, are oxysterol-activated transcription factors that, after heterodimerization with the 9-*cis*-retinoic acid receptor (RXR), bind to specific LXR response elements (LXREs), thus regulating the expression of target genes involved in intra- and extracellular lipid metabolism [74]. In part by modulating cholesterol efflux from macrophages to apoA-I and HDL, LXRs induce the direct target genes ABCA1 and ABCG1/ABCG4 to promote reverse cholesterol transport. The oxidation of steroids from FC can activate LXR and regulate gene expression of ABCA1 and ABCG1 to strengthen the peripheral tissue cholesterol secretion. Meanwhile, the LXRs are also readily oxidized by PPARα. PPARαcontrols lipid and glucose metabolism in several tissues and cell types including liver, heart, kidneys, adipose tissue and macrophages. PPARα-activation suppresses chylomicron and increases HDL production by enterocytes [75,76]. In addition, its agonists promote secretion of macrophage cholesterol via stimulating expression of ABCA1 and LXR to increase reverse cholesterol transport [77].

Part Two: The Process of Cholesterol Esterified by LCAT

LCAT, a lipoprotein-associated enzyme, is a key player in the RCT pathway. LCAT has two different catalytic activities that account for its ability to esterify cholesterol. One is phospholipase A2 activity, and the other is its transesterification activity. It requires apoA-I and, to a lesser degree, other apolipoproteins, which most likely activate LCAT by modifying the presentation of its substrates, namely, phospholipids and cholesterol, on the surface of lipoproteins [78]. After FC efflux to pre-β HDL (the nascent, discoidal-shaped HDL), cholesterol in HDLs may be esterified by the enzymatic activity of LCAT. The LCAT reaction occurs in two steps. After binding to a lipoprotein, LCAT cleaves the fatty acid in the sn-2 position of phosphatidylcholine and transfers it onto a serine residue. Next, the fatty acid is transesterified to the 3-β-hydroxyl group on the A-ring of cholesterol to form cholesterol ester. Cholesteryl esters formed by LCAT partition, which are more hydrophobic than free cholesterol, are transferred from the surface of lipoproteins to the hydrophobic core. This process converts pre-β HDL to HDL$_2$ and HDL$_3$ particles, which are the major HDL species found in plasma and which represent larger, spherical-shaped α-migrating forms of HDL. LCAT is important in the process of RCT by generating a gradient of free cholesterol from cells to HDL [79]. This effect of LCAT prevents the back exchange of cholesterol by passive diffusion from HDL to peripheral cells and thus is believed to promote net removal of cholesterol from peripheral cells to HDL. Without ongoing esterification of cholesterol, the capacity of HDL to remove and bind additional cholesterol would eventually be diminished. Two lipases, endothelial lipase (EL) and HL, are the complete opposite of LCAT in HDL metabolism. HL and EL are members of the triglyceride lipase family, which also includes LPL [80,81]. EL has high phospholipase A1 activity and remodels HDL into small particles, whereas HL is more effective in hydrolyzing triglycerides [82]. Although HL causes a remodeling of HDL into smaller particles, it also promotes the release of lipid-poor apoA-I [83]. The combined functions of HL and EL have a significant effect on plasma HDL-C levels [84–87]. HL- and EL-deficient mice were studied to demonstrate that the magnitude of macrophage-derived [$^3$H] cholesterol in feces was increased, which promoted macrophage-to-feces RCT, and its ability to protect against LDL oxidation was enhanced [88].

Part Three: The Exchange of CE (Cholesteryl Esters) Mediated by CETP

CETP is a hydrophobic glycoprotein that is synthesized in several tissues but mainly in the liver. It facilitates the exchange of cholesteryl esters and triglycerides between HDL and apoB-containing particles (LDL, IDL, VLDL) and represents a major branching point for RCT [7,8]. This results in the recycling of cholesterol and an increasing attenuation in blood circulation, with the potential to go back into the artery wall. Two major observations were made using many novel methods, such as innovative X-ray crystallographic, electron microscopic (EM), and bioinformatics observations: (i) CETP connects with or forms bridges between two lipoproteins, e.g., HDL and LDL, with resultant neutral lipid transfer; and (ii) CETP appears to contain a hydrophobic tunnel along its entire long axis capable of neutral lipid transfer. These observations may clearly explain the relationships between CETP interactions with lipoproteins and the lipid transfer processes [89]. Most CEs derived from LCAT do not return to the liver via the HDL SR-B1 pathway but, rather, through more atherogenic pathways. CETP mediates the transfer of most CE from HDL to VLDL or to other more atherogenic intermediate-density lipoproteins and remnants, and the transfer of triglycerides from VLDL-1 to HDL results in larger, relatively triglyceride-enriched LDL species [90]. Transfer of CE from HDL directly to LDL by CETP could also be antiatherogenic if the LDL is cleared by the liver LDL receptor. Another transfer protein in this part of RCT, PLTP, transfers phospholipids between VLDL and HDL [91]. PLTP is one of the main modulators of plasma HDL size, composition and function [92] and one of the major modulators of HDL metabolism in plasma [93]. The level of HDL or HDL production is dramatically decreased in PLTP$^{-/-}$ mice [94,95], and PLTP deficiency attenuates plaque accumulation in different atherosclerotic models. This indicates that PLTP plays an important role in atherogenesis, and its function goes well beyond that of transferring phospholipids between lipoprotein particles [96,97].

Part Four: Catabolism of HDL Cholesterol in Biliary Pathway

After efflux, cholesterol in HDLs may be esterified by the enzymatic activity of LCAT whereupon HDLs can deliver the excess cholesterol from peripheral cells back to the liver in in distinct ways: HDL cholesteryl esters, but not the protein component of HDLs, are selectively taken up into the liver via SR-BI. Ultimately, cholesterol is excreted from the liver into the bile, either directly as free cholesterol or after conversion into bile acids, and eliminated from the body via the feces. In humans, HDL-C can be metabolized by the liver via another pathway: CETP exchanges of HDL CE for triglycerides in apoB-containing lipoproteins, followed by hepatic uptake mediated by LDL-r. An LDL-r deficiency in mice substantially decreases selective HDL CE uptake by liver and adrenals. Thus, LDL-r expression has a substantial impact on HDL metabolism in mice [98].

Via the classic RCT pathway, excessive cholesterol collected from peripheral tissues, which is delivered back to the liver, is followed by biliary secretion and elimination via the feces. In addition to the traditional RCT-mediated biliary pathway, in the last few years, direct trans-intestinal excretion of plasma-derived cholesterol (TICE) was shown to contribute substantially to fecal neutral sterol (FNS) excretion in mice, describing the transport of cholesterol from blood to the intestinal lumen directly via enterocytes. The TICE pathway was called a nonhepatobiliary-related route, which has been shown to have a high degree of correlation with the main contributors Niemann-Pick disease, type C1/2 (NPC1/2), ABCG5/G8, LDL-r, and LXR [99–101]. The application of PPAR δ agonist and LXR agonists, have been shown to stimulate the process of TICE [102]. In the RCT pathway, HDL plays an important role. In contrast, there is evidence from animal experiments that HDL plays an essential role in TICE [103].

### 2.2.2. Antioxidant Properties of High-Density Lipoprotein

LDL is one of the main causes of AS. Oxidation of LDL yields a more pro-atherogenic particle, and numerous studies have found that HDLs are capable of impeding oxidative changes in LDL. HDL exhibits potent antioxidant activity, which may arise from synergy in the inactivation of oxidized LDL lipids by enzymatic and nonenzymatic mechanisms, in part reflecting distinct intrinsic physicochemical properties [104]. The anti-oxidative properties of HDL critically involve HDL-associated enzymes, such as paraoxonase 1 (PON1), lipoprotein-associated phospholipase A2 (Lp-PLA2), and LCAT, which have been reported to hydrolyze oxidized phospholipids into lyso-phosphatidylcholine [105–109]. In addition, HDL carries glutathione selenoperoxidase, which can reduce lipid peroxide (LOOH) to the corresponding hydroxides and thereby detoxify them [110]. ApoA-I can remove oxidized lipids from LDL, suggesting that HDL can function as an acceptor of oxidized lipids. In cell culture experiments, apoA-I removes lipids from LDL and thereby makes LDL resistant to vascular cell-mediated oxidation and prevents oxidized LDL-induced monocyte adherence and chemotaxis [108]. Other HDL apolipoproteins, such as apoA-II, apoA-IV, apoE, and apoJ, also function as antioxidants *in vitro*.

### 2.2.3. HDL in the Endothelium

Traditionally, the endothelium has been considered to be an inert component of the vessel wall. Injury to the endothelium results in deleterious alterations of endothelial physiology, also referred to as endothelial dysfunction, which represents a key early step in the development of an atherosclerotic lesion and is implicated in the malignant development that follows. It has become evident that HDL from healthy subjects can exert direct potential atheroprotective effects on endothelial cells and can positively affect several endothelial functions in the regulation of vascular tone, inflammation [111] and endothelial oxidant stress [112], which is associated with activation of NO synthesis by HDL. NO is an endothelium-derived signaling molecule that activates guanylate cyclase in vascular smooth muscle cells (SMCs) to induce relaxation, which is generated by a constitutive eNOS. Notably, HDLs also beneficially affect the vasculature by promoting endothelial cell survival. In endothelial cells, HDLs still maintain anti-thrombotic functions, mainly inhibiting thrombin-induced tissue factor, mediating the extrinsic coagulation pathway, and stimulating the activation of the anticoagulant proteins C and S.

### 2.2.4. Anti-inflammatory Properties of HDL

HDL and apoA I are believed to protect against the development of AS, in part, which is closely related to their anti-inflammatory function. The anti-inflammatory properties of HDL encompass suppression of macrophage inflammatory cytokine production and inhibition of the expression of endothelial cell adhesion molecules that promote the entry of monocytes and neutrophils into arteries. HDLs potently block murine experimental endotoxinemia, indicating that HDLs bind lipopolysaccharides (LPS), which leads to lower systemic proinflammatory cytokine levels and improved survival rates [113]. Up-regulation of adhesion molecules (E-selectin, ICAM-1, and VCAM-1) and secretion of pro-inflammatory mediators (monocyte chemoattractant 1, MCP-1) lead to the activation of the vascular endothelium and induce AS.

## 3. Novel Therapies with HDL as the Key Components

### 3.1. Novel Pharmacotherapeutic Strategies in Increasing HDL

### 3.1.1. Infusions of Special HDL

One approach to increase the serum levels of HDL is to infuse reconstituted HDL (rHDL) or recombinant HDL particles into the circulation rather than by increasing HDL indirectly by modulating HDL metabolism. CSL-111 is a reconstituted HDL-particle comprising both human apoA-1 and soybean phosphatidylcholine. CSL-112 is being evaluated in phase II trials to improve the weakness of CSL-111. There is a benefit of CSL-111/CSL-112 infusions to increase HDL-C levels and up-regulation of cellular cholesterol efflux in patients with acute coronary syndrome (ACS). Some trials have demonstrated that infusions of recombinant HDL-containing apoA-I Milano and CER-001 significantly reduce or stabilize mean plaque atheroma volume and still exert greater anti-inflammatory effects in LDL-r$^{-/-}$ mice [114–116]. However, most clinical studies have demonstrated that rapid clearance of CER-001 leads to a requirement for repeated administration due to the inability to achieve effective plasma concentrations [117]. Pegylation of apoA-I in rHDL markedly increases its plasma half-life and enhances its antiatherogenic properties *in vivo* [118].

### 3.1.2. Autologous Delipidated HDL

Another novel approach to HDL therapeutics is to raise the levels of HDL particles by intravenous infusion with the use of autologous delipidated HDL. Preclinical evaluation of selective delipidated HDL in a nonhuman primate model of dyslipidemia achieved a significant 6.9% reduction in aortic atheroma volume, as assessed by intravenous ultrasound (IVUS) [119]. In a small clinical study in which 28 patients with ACS received five weekly infusions of delipidated HDL, the decreased total atheroma volume by 5.2% from baseline may be associated with selectively increased preβ-HDL [120].

### *3.2. Novel Pharmacotherapeutic Strategies in Four Steps of RCT*

### 3.2.1. First Part of RCT

#### ApoA-I Mimetic Peptides

ApoA-I mimetics, which are short synthetic peptides, mimic the amphipathic a-helix of apoA-I. Because the first apoA-I mimetic consisted of 18 amino acids, additional improved peptides were generated by increasing the number of phenylalanine residues on the hydrophobic face (referred to as 2F, 3F, 4F, 5F, 6F, and 7F) of the polypeptide. Two different types of apoA-I mimetic peptide 4F were used in clinical trials, including L-4F (the 4F peptide synthesized from L-amino acids) and D-4F (the 4F peptide synthesized from all D-amino acids). Compared with high plasma L-4F levels not improving HDL anti-inflammatory function, a single dose of D-4F was found to have a significant effect on the inflammatory index of HDL with modest oral bioavailability [10]. ABCA1-independent lipid efflux played a major role in this aspect of RCT. Recently, 5A, an asymmetric bihelical peptide based on 2F, had increased ABCA1-dependent cholesterol efflux and decreased hemolysis compared with its parent compound [121]. Furthermore, ATI-5261 synthetic peptide, like apoA-I mimetic peptides, successfully enhances cholesterol efflux from macrophages and reduces aortic AS, which exerts its effects similar to that of the role of HDL in reverse cholesterol transport in mice. Reservelogix-208 (RVX-208), an apoA-I upregulator that increases endogenous synthesis of apoA-I, has a unique mechanism of action related to epigenetically influencing the accessibility of apoA-I gene by the transcription machinery. Encouragingly, there is a progressive discovery that nanolipid particles containing multivalent peptides (HDL-like nanoparticles) promoted efficient cellular cholesterol efflux and were functionally superior to those derived from monomeric apoA-I mimetic peptides [122].

#### Regulators of ABCA1 and ABCG1

LXR and miR-33 (similars: miR-758, miR-26 and miR-106b) have been demonstrated to regulate macrophages ABCA1 and ABCG1. The activation of the former can upregulate the expression of two genes, which promote macrophage cholesterol efflux and augment intestinal HDL generation, while the opposite is true of the latter. In the animal experiments, LXR agonists exhibit a significant decrease of atherosclerotic plaque formation and exert anti-inflammatory characteristics [123]. The negative effects caused by hepatic LXR agonists, such as a fatty liver, will be overcome by more intestine-specific LXRα agonists that are still under preclinical investigation [124]. As an alternative to LXR agonism, gene- silencing approaches involving microRNA 33 are being explored to control gene expression at the post-transcriptional level [125]. An antisense oligonucleotide has been developed that effectively silences miR-33 (miRNA-33a/b) and increases ABCA1 and ABCG1 expression in macrophages. Studies in nonhuman primates showed that antisense oligonucleotides to miRNA-33a/b that induce ABCA1 expression as well as the expression of other proteins involved in fatty acid metabolism result in a 50% increase in HDL-C and a decrease in VLDL [126]. Recent studies in PPARγ-activated macrophages demonstrated that treatment with miR-613 leads to inhibition of cholesterol efflux by downregulating LXRα and ABCA1 [127]. This study sheds new insights into the possibility that the miR-613 inhibitor may serve as a novel therapy for the treatment of cholesterol metabolism diseases. Similarly, PPAR agonists also increase ABCA1 gene expression and target the LXR gene. PPARα (K-877), PPARγ (INT131) and PPARα/δ (GFT505) modulators play pivotal roles in ABCA1 gene expression and apoA-I secretion, and increase HDL-C in patients with hyperlipidemia [126,127]. Growing evidence suggests that MBX-8025, a promising PPARδ agonist targeting the ABCA1 gene and LXRα gene, with a greater binding potency, is under development. Additional dual PPARα/γ and PPARβ/δ agonists that are under development may play an important role in RCT pathway [78].

### 3.2.2. Second Part of RCT

LCAT Agonists

LCAT plays a major role in this part of RCT, and mediates the esterification of free cholesterol located at the surface of lipoprotein particles. Several drug development approaches have recently been initiated for modulating LCAT activity. Recombinant human LCAT (rhLCAT) efficiently promotes the process of RCT, which induces a marked increase in HDL-C levels and the maturation of small preb-HDL into α-migrating particles in LCAT-deficient plasma [128]. Currently, recombinant LCAT (ACP-501) in Phase 1 clinical trials was shown to rapidly and substantially elevate the formation of cholesteryl esters, representing the promotion of RCT. It is reported that Apo AI-derived peptides present the ability to modulate LCAT activity, with a significant increase in 3H-cholesteryl ester production in a higher proportion at 0.3 mg/mL [129]. An early apoA-I mimetic peptide (ETC-642) promotes LCAT activation and has entered clinical development [130].

### 3.2.3. Third Part of RCT

CETP Inhibitor

**Anacetrapib.** The Determining Efficacy and Tolerability of Anacetrapib has shown a 138% increase in HDL-C, a 40% reduction in LDL-C, and a 36% decrease in Lp(a) of patients with CAD. This results from the ability of anacetrapib to inhibit both heterotypic and homotypic CE transfer. A large phase III study, Randomized EValuationof the Effects of Anacetrapib Through Lipid-modification (REVEAL), will be completed in January 2017 [131].

**Evacetrapib.** Evacetrapib has a similar structure and mechanism of action as torcetrapib but, based on $IC_{50}$ values, appears to be a more potent CETP inhibitor than torcetrapib or anacetrapib [132]. It is the most recently developed CETP-inhibitor currently undergoing a phase III CVD outcome trial. In a phase II study, high doses of evacetrapib do not elevate blood pressure in rats and do not induce aldosterone or cortisol biosynthesis in a human adrenal cortical carcinoma cell line [133]. The Assessment of Clinical Effects of Cholesteryl Ester Transfer Protein Inhibition With Evacetrapib in Patients at a High-Risk for Vascular Outcomes (ACCELERATE) phase III CVD outcome trial is recruiting.

**BAY 60-5521/TA-8995.** BAY 60-5521 and TA-8995 are two potential inhibitors of CETP. The clinical results suggest that these two agents are safe and well tolerated, with effective inhibition of CETP activity and increased high HDL-C [134,135].

**CETP ASO.** Antisense oligonucleotide inhibitor of CETP (CETP ASO) is associated with reductions in CETP mRNA and also shows an enhanced effect on macrophage RCT. It is reported that the CETP ASO does not act through binding and inactivating CETP associated with the HDL particle. Instead, it specifically targets and degrades CETP mRNA. In hyperlipidemic, CETP transgenic, LDL-r$^{-/-}$ mice, CETP ASO provided comparable reductions in HDL cholesterol, decreases in CETP activity with an enhanced effect on macrophage RCT, and results in less accumulation of aortic cholesterol [136]. CETP ASO could produce a unique therapeutic profile, distinct from the current CETP drugs being evaluated in late-stage clinical trials.

### 3.2.4. Fourth Part of RCT

SR-BI Activators

SR-BI, the HDL receptor, is expressed on hepatocytes and facilitates selective absorption of cholesterol ester from HDL. The discovery and development of traditional drugs that can modulate SR-BI expression and/or activity are being investigated through high-throughput screening of chemical compound libraries. Endogenous or exogenous agents (such as natural products) and the detailed high-resolution molecular structure of SR-BI protein, the regulators of SR-BI activity, should provide new insights for the progress of non-genetic therapies to modulate HDL metabolism *in vivo*.

As an alternative to traditional drugs, gene therapeutics are being extensively explored. In general, miRNA-125a and miRNA-455 can bind to specific sites in the 3′ UTR of SR-BI mRNA and act as endogenous attenuators of SR-BI protein expression [137]. It is clear that activation of miRNA-125a and miRNA-455 shows a negative relationship with SR BI-mediated selective HDL uptake and SR-BI-supported steroid hormone synthesis. Therefore, inhibition of miRNA-125a and miRNA-455 may enhance SR-BI expression to alter the course of the atherogenic sequence by increasing plasma HDL-C flux. Furthermore, in a recent report, the level of SR-BI expression was repressed by miR-185, miR-96, and miR-223, individually, in HepG2 cells [138]. Inhibitions of these miRNAs were associated with increased SR-BI expression and reduced AS by facilitating RCT and cholesterol removal from the body.

### 3.3. Natural Drugs Associated with HDL at the Future Stage

Nature is an irrefutable source of inspiration for the modern man in many aspects. The observation and understanding of nature have allowed the development of new materials, new sources of energies, new drugs, *etc.* Specifically, natural products provide a great contribution to the development of new agents for the treatment of hyperlipidemic diseases. Compared with chemical drugs or other synthetically created products, natural products could more closely mimic physiological processes and may be more effective in treatment of cardiovascular diseases. There are many natural drugs for the treatment of AS associated with HDL, such as polysaccharide, sesamin, anthocyanins, 24(S)-Saringosterol and many others. Chitosan and its related products, which are polysaccharides, are used to enhance natural products and may represent promising strategies in the near future.

### 3.3.1. Chitosan

Chitosan (CTS), the only positive alkaline polysaccharide extracted from shrimp or crab shells and exhibiting biocompatibility and a nontoxic nature, has the potential for hypolipidemic and weight loss activities. In 1980, the first article demonstrating that proper supplementation of chitosan to the diet seemed to be effective in lowering plasma cholesterol in rats was published [139]. And the meta-analysis by Baker *et al.*, based on six randomized placebo-controlled clinical trials in hypercholesterolemic patients has showed that chitosan only induced a decrease in total cholesterol level but without any significant changes in LDL-cholesterol, HDL-cholesterol or triglyceride concentrations [140]. Since then, the research of CTS in treatment of hyperlipemia has been extensively undertaken. It is reported that because CTS is a biodegradable carbohydrate polymer, it has been prepared in a variety of dosages for use in delivery systems in previous studies [141–143], and these new preparations were found to have a more effective capability of improving hyperlipidemia in rats [144,145].

### 3.3.2. Chitosan Oligosaccharides

Chitosan oligosaccharides (COS) are depolymerized products of chitosan by either chemical or enzymatic hydrolysis. COS is a polymer of glucosamine with a number of bioactive properties. Recent studies have suggested that COS inhibits adipogenesis and promotes RCT through altered expression of a number of key regulators of lipid metabolism, including leptin [146], SR-BI and CYP7A1. It is reported that COS plays a positive role in the RCT pathway *in vivo* and stimulates hepatic LDL-r expression, inducing the LDL cholesterol lowering. Furthermore, chitosan oligosaccharides have anti-inflammatory functions similar to that of HDL, which downregulates the expression of E-selectin, ICAM-1 and IL-8 induced by LPS in endothelial cells through blockade of p38 MAPK and PI3K/Akt signaling pathways [147,148].

### 4. Discussion

HDL particles have various effects *in vitro* and *in vivo* that may protect arteries from chemical or biological hazards or facilitate repair of injuries. Nevertheless, HDL has not yet been successfully exploited for therapy. One possible reason may be the complexity of HDL particles, resulting in the physiological heterogeneity that contributes to the antiatherogenic functions of HDL. Moreover, for more than 50 years, HDL-C has been known as an independent clinical biomarker of cardiovascular risk. However, the failures of drugs to increase HDL-C drive scientists to search for new biomarkers to replace it, such as HDL functions, HDL size, and HDL-P. Among these, HDL functions will be a reliable diagnostic biomarker for the identification, personalized treatment stratification, and monitoring of patients at increased cardiovascular risk. The functionality of HDLs has two important roles. First, it promotes reverse cholesterol transport; second, it modulates inflammation and oxidation. HDL-based interventions in promoting the RCT process may be more promising. Meanwhile, some natural products with functions similar to that of HDL also play unique roles in the RCT process and will enter a new era in treatment of CVD in the near future (Table 2).

**Table 2.** Overview of Classes of HDL-based therapies. HDL-based therapies are an innovative approach against atherosclerosis. In this review, they can be summarized as the following three categories: strategies increasing HDL, strategies of RCT in four steps and natural drugs. ●: primary; ♦: secondary.

| Strategies Increasing HDL | | Strategies of RCT in Four Steps | | | Natural Drugs | |
|---|---|---|---|---|---|---|
| ●Infusions of special HDL | ♦ rHDL: CSL-111 CER-001 CSL-112 | ●First part of RCT | | ●Second part of RCT | ●Polysaccharide: | Chitosan; Chitosan oligosaccharides |
| | | ApoA-I mimetic peptides; ApoA-I upregulator | | LCAT Agonists; ♦ rLCAT: ACP-501 | | |
| | | LXR agonists; Mir-33; PPAR modulators | | | ●Anthocyanins | |
| | | ●Third part of RCT | | ●Fourth part of RCT | | |
| ●Autologous delipidated HDL | ♦ CETP inhibitor: | Anacetrapib; Evacetrapib; BAY 60-5521; TA-8995; CETP ASO | ♦ SR-BI activators: | Traditional drugs; Gene therapeutics | ●Sesamin | |
| | | | | | ●24(S)-Saringosterol | |
| | | | | | ●Others | |

**Acknowledgments:** This project was financially supported by the National Science Foundation of China (no. 81173107), the Science and Technology Planning Project of Guangdong, China (2013B021100018, 2013B090600050, 2015A010101318).

**Conflicts of Interest:** The authors declare no conflict of interests.

## References

1. Roger, V.L.; Go, A.S.; Lloyd-Jones, D.M.; Benjamin, E.J.; Berry, J.D.; Borden, W.B.; Bravata, D.M.; Dai, S.; Ford, E.S.; Fox, C.S.; *et al.* Executive summary: Heart disease and stroke statistics-2012 update: A report from the American heart association. *Circulation* **2012**, *125*, 188–197. [PubMed]

2. Gordon, T.; Castelli, W.P.; Hjortland, M.C.; Kannel, W.B.; Dawber, T.R. High density lipoprotein as a protective factor against coronary heart disease. The framingham study. *Am. J. Med.* **1977**, *62*, 707–714. [CrossRef]

3. Baigent, C.; Blackwell, L.; Emberson, J.; Holland, L.E.; Reith, C.; Bhala, N.; Peto, R.; Barnes, E.H.; Keech, A.; Simes, J.; *et al.* Efficacy and safety of more intensive lowering of LDL cholesterol: A meta-analysis of data from 170,000 participants in 26 randomised trials. *Lancet* **2010**, *376*, 1670–1681. [PubMed]

4. Barter, P.; Gotto, A.M.; LaRosa, J.C.; Maroni, J.; Szarek, M.; Grundy, S.M.; Kastelein, J.J.; Bittner, V.; Fruchart, J.C. HDL cholesterol, very low levels of LDL cholesterol, and cardiovascular events. *N. Engl. J. Med.* **2007**, *357*, 1301–1310. [CrossRef] [PubMed]

5. Dumitrescu, L.; Goodloe, R.; Bradford, Y.; Farber-Eger, E.; Boston, J.; Crawford, D.C. The effects of electronic medical record phenotyping details on genetic association studies: HDL-C as a case study. *Biodata Min.* **2015**, *8*, 15. [CrossRef] [PubMed]

6. Remaley, A.T.; Norata, G.D.; Catapano, A.L. Novel concepts in HDL pharmacology. *Cardiovasc. Res.* **2014**, *103*, 423–428. [CrossRef] [PubMed]

7. Schwartz, G.G.; Olsson, A.G.; Abt, M.; Ballantyne, C.M.; Barter, P.J.; Brumm, J.; Chaitman, B.R.; Holme, I.M.; Kallend, D.; Leiter, L.A.; *et al.* Effects of dalcetrapib in patients with a recent acute coronary syndrome. *N. Engl. J. Med.* **2012**, *367*, 2089–2099. [CrossRef] [PubMed]

8. Barter, P.J.; Caulfield, M.; Eriksson, M.; Grundy, S.M.; Kastelein, J.J.; Komajda, M.; Lopez-Sendon, J.; Mosca, L.; Tardif, J.C.; Waters, D.D.; *et al.* Effects of torcetrapib in patients at high risk for coronary events. *N. Engl. J. Med.* **2007**, *357*, 2109–2122. [CrossRef] [PubMed]

9. Marsche, G.; Saemann, M.D.; Heinemann, A.; Holzer, M. Inflammation alters HDL composition and function: Implications for HDL-raising therapies. *Pharmacol. Ther.* **2013**, *137*, 341–351. [CrossRef] [PubMed]

10. Shah, P.K. Atherosclerosis: Targeting endogenous apo AI—A new approach for raising HDL. *Nat. Rev. Cardiol.* **2011**, *8*, 187–188. [CrossRef] [PubMed]

11. Holleboom, A.G.; Jakulj, L.; Franssen, R.; Decaris, J.; Vergeer, M.; Koetsveld, J.; Luchoomun, J.; Glass, A.; Hellerstein, M.K.; Kastelein, J.J.; *et al. In vivo* tissue cholesterol efflux is reduced in carriers of a mutation in APOA1. *J. Lipid Res.* **2013**, *54*, 1964–1971. [CrossRef] [PubMed]

12. Umemoto, T.; Han, C.Y.; Mitra, P.; Averill, M.M.; Tang, C.; Goodspeed, L.; Omer, M.; Subramanian, S.; Wang, S.; den Hartigh, L.J.; *et al.* Apolipoprotein AI and high-density lipoprotein have anti-inflammatory effects on adipocytes via cholesterol transporters: ATP-binding cassette A-1, ATP-binding cassette G-1, and scavenger receptor B-1. *Circ. Res.* **2013**, *112*, 1345–1354. [CrossRef] [PubMed]

13. Navab, M.; Reddy, S.T.; van Lenten, B.J.; Fogelman, A.M. HDL and cardiovascular disease: Atherogenic and atheroprotective mechanisms. *Nat. Rev. Cardiol.* **2011**, *8*, 222–232. [CrossRef] [PubMed]

14. Hewing, B.; Parathath, S.; Barrett, T.; Chung, W.K.; Astudillo, Y.M.; Hamada, T.; Ramkhelawon, B.; Tallant, T.C.; Yusufishaq, M.S.; Didonato, J.A.; *et al.* Effects of native and myeloperoxidase-modified apolipoprotein A-I on reverse cholesterol transport and atherosclerosis in mice. *Arterioscler. Thromb. Vasc. Biol.* **2014**, *34*, 779–789. [CrossRef] [PubMed]

15. Rader, D.J.; Alexander, E.T.; Weibel, G.L.; Billheimer, J.; Rothblat, G.H. The role of reverse cholesterol transport in animals and humans and relationship to atherosclerosis. *J. Lipid Res.* **2009**, *50*, S189–S194. [CrossRef] [PubMed]

16. Hartman, J.; Frishman, W.H. Inflammation and atherosclerosis: A review of the role of interleukin-6 in the development of atherosclerosis and the potential for targeted drug therapy. *Cardiol. Rev.* **2014**, *22*, 147–151. [CrossRef] [PubMed]

17. Glass, C.K.; Witztum, J.L. Atherosclerosis. The road ahead. *Cell* **2001**, *104*, 503–516. [CrossRef]

18. Libby, P. Inflammation in atherosclerosis. *Nature* **2002**, *420*, 868–874. [CrossRef] [PubMed]

19. Prasad, V.; Bonow, R.O. The cardiovascular biomarker conundrum: Challenges and solutions. *JAMA* **2011**, *306*, 2151–2152. [CrossRef] [PubMed]

20. The AIM-HIGH Investigators. The role of niacin in raising high-density lipoprotein cholesterol to reduce cardiovascular events in patients with atherosclerotic cardiovascular disease and optimally treated low-density lipoprotein cholesterol: baseline characteristics of study participants. The Atherothrombosis Intervention in Metabolic syndrome with low HDL/high triglycerides: Impact on Global Health outcomes (AIM-HIGH) trial. *Am. Heart J.* **2011**, *161*, 538–543.

21. Boden, W.E.; Probstfield, J.L.; Anderson, T.; Chaitman, B.R.; Desvignes-Nickens, P.; Koprowicz, K.; McBride, R.; Teo, K.; Weintraub, W. Niacin in patients with low HDL cholesterol levels receiving intensive statin therapy. *N. Engl. J. Med.* **2011**, *365*, 2255–2267. [PubMed]

22. Li, C.; Zhang, W.; Zhou, F.; Chen, C.; Zhou, L.; Li, Y.; Liu, L.; Pei, F.; Luo, H.; Hu, Z.; *et al.* Cholesteryl ester transfer protein inhibitors in the treatment of dyslipidemia: A systematic review and meta-analysis. *PLoS ONE* **2013**, *8*, e77049. [CrossRef] [PubMed]

23. Boekholdt, S.M.; Arsenault, B.J.; Hovingh, G.K.; Mora, S.; Pedersen, T.R.; Larosa, J.C.; Welch, K.M.; Amarenco, P.; Demicco, D.A.; Tonkin, A.M.; *et al.* Levels and changes of HDL cholesterol and apolipoprotein A-I in relation to risk of cardiovascular events among statin-treated patients: A meta-analysis. *Circulation* **2013**, *128*, 1504–1512. [CrossRef] [PubMed]

24. Keene, D.; Price, C.; Shun-Shin, M.J.; Francis, D.P. Effect on cardiovascular risk of high density lipoprotein targeted drug treatments niacin, fibrates, and CETP inhibitors: Meta-analysis of randomised controlled trials including 117,411 patients. *BMJ* **2014**, *349*, g4379. [CrossRef] [PubMed]

25. Voight, B.F.; Peloso, G.M.; Orho-Melander, M.; Frikke-Schmidt, R.; Barbalic, M.; Jensen, M.K.; Hindy, G.; Holm, H.; Ding, E.L.; Johnson, T.; *et al.* Plasma HDL cholesterol and risk of myocardial infarction: a mendelian randomisation study. *Lancet* **2012**, *380*, 572–580. [CrossRef]

26. Saely, C.H.; Vonbank, A.; Drexel, H. HDL cholesterol and residual risk of first cardiovascular events. *Lancet* **2010**, *376*, 1738–1739. [CrossRef]

27. Mora, S.; Glynn, R.J.; Boekholdt, S.M.; Nordestgaard, B.G.; Kastelein, J.J.; Ridker, P.M. On-treatment non-high-density lipoprotein cholesterol, apolipoprotein B, triglycerides, and lipid ratios in relation to residual vascular risk after treatment with potent statin therapy: JUPITER (justification for the use of statins in prevention: An intervention trial evaluating rosuvastatin). *J. Am. Coll. Cardiol.* **2012**, *59*, 1521–1528. [PubMed]

28. Rye, K.A.; Bursill, C.A.; Lambert, G.; Tabet, F.; Barter, P.J. The metabolism and anti-atherogenic properties of HDL. *J. Lipid Res.* **2009**, *50*, S195–S200. [CrossRef] [PubMed]

29. Barter, P.; Kastelein, J.; Nunn, A.; Hobbs, R. High density lipoproteins (HDLs) and atherosclerosis: The unanswered questions. *Atherosclerosis* **2003**, *168*, 195–211. [CrossRef]

30. Arsenault, B.J.; Lemieux, I.; Despres, J.P.; Gagnon, P.; Wareham, N.J.; Stroes, E.S.; Kastelein, J.J.; Khaw, K.T.; Boekholdt, S.M. HDL particle size and the risk of coronary heart disease in apparently healthy men and women: the EPIC-Norfolk prospective population study. *Atherosclerosis* **2009**, *206*, 276–281. [CrossRef] [PubMed]

31. Azevedo, C.H.; Wajngarten, M.; Prete, A.C.; Diament, J.; Maranhao, R.C. Simultaneous transfer of cholesterol, triglycerides, and phospholipids to high-density lipoprotein in aging subjects with or without coronary artery disease. *Clinics* **2011**, *66*, 1543–1548. [PubMed]

32. Pascot, A.; Lemieux, I.; Prud'Homme, D.; Tremblay, A.; Nadeau, A.; Couillard, C.; Bergeron, J.; Lamarche, B.; Despres, J.P. Reduced HDL particle size as an additional feature of the atherogenic dyslipidemia of abdominal obesity. *J. Lipid Res.* **2001**, *42*, 2007–2014. [PubMed]

33. Du, X.M.; Kim, M.J.; Hou, L.; le Goff, W.; Chapman, M.J.; van Eck, M.; Curtiss, L.K.; Burnett, J.R.; Cartland, S.P.; Quinn, C.M.; *et al.* HDL particle size is a critical determinant of ABCA1-mediated macrophage cellular cholesterol export. *Circ. Res.* **2015**, *116*, 1133–1142. [PubMed]

34. Kim, D.S.; Burt, A.A.; Rosenthal, E.A.; Ranchalis, J.E.; Eintracht, J.F.; Hatsukami, T.S.; Furlong, C.E.; Marcovina, S.; Albers, J.J.; Jarvik, G.P. HDL-3 is a superior predictor of carotid artery disease in a case-control cohort of 1725 participants. *J. Am. Heart Assoc.* **2014**, *3*, e902. [CrossRef] [PubMed]

35. Ridker, P.M.; Genest, J.; Boekholdt, S.M.; Libby, P.; Gotto, A.M.; Nordestgaard, B.G.; Mora, S.; MacFadyen, J.G.; Glynn, R.J.; Kastelein, J.J. HDL cholesterol and residual risk of first cardiovascular events after treatment with potent statin therapy: An analysis from the JUPITER trial. *Lancet* **2010**, *376*, 333–339. [CrossRef]

36. Mora, S.; Glynn, R.J.; Ridker, P.M. High-density lipoprotein cholesterol, size, particle number, and residual vascular risk after potent statin therapy. *Circulation* **2013**, *128*, 1189–1197. [CrossRef] [PubMed]
37. El, H.K.; Arsenault, B.J.; Franssen, R.; Despres, J.P.; Hovingh, G.K.; Stroes, E.S.; Otvos, J.D.; Wareham, N.J.; Kastelein, J.J.; Khaw, K.T.; *et al.* High-density lipoprotein particle size and concentration and coronary risk. *Ann. Intern. Med.* **2009**, *150*, 84–93.
38. Mora, S.; Otvos, J.D.; Rifai, N.; Rosenson, R.S.; Buring, J.E.; Ridker, P.M. Lipoprotein particle profiles by nuclear magnetic resonance compared with standard lipids and apolipoproteins in predicting incident cardiovascular disease in women. *Circulation* **2009**, *119*, 931–939. [CrossRef] [PubMed]
39. Ala-Korpela, M.; Soininen, P.; Savolainen, M.J. Letter by Ala-Korpela et al regarding article, "Lipoprotein particle profiles by nuclear magnetic resonance compared with standard lipids and apolipoproteins in predicting incident cardiovascular disease in women". *Circulation* **2009**, *120*, e149–e150. [CrossRef] [PubMed]
40. Hutchins, P.M.; Ronsein, G.E.; Monette, J.S.; Pamir, N.; Wimberger, J.; He, Y.; Anantharamaiah, G.M.; Kim, D.S.; Ranchalis, J.E.; Jarvik, G.P.; *et al.* Quantification of HDL particle concentration by calibrated ion mobility analysis. *Clin. Chem.* **2014**, *60*, 1393–1401. [CrossRef] [PubMed]
41. Dahlback, B.; Nielsen, L.B. Apolipoprotein M—A novel player in high-density lipoprotein metabolism and atherosclerosis. *Curr. Opin. Lipidol.* **2006**, *17*, 291–295. [CrossRef] [PubMed]
42. Elsoe, S.; Christoffersen, C.; Luchoomun, J.; Turner, S.; Nielsen, L.B. Apolipoprotein M promotes mobilization of cellular cholesterol *in vivo. Biochim. Biophys. Acta* **2013**, *1831*, 1287–1292. [CrossRef] [PubMed]
43. Wolfrum, C.; Poy, M.N.; Stoffel, M. Apolipoprotein M is required for prebeta-HDL formation and cholesterol efflux to HDL and protects against atherosclerosis. *Nat. Med.* **2005**, *11*, 418–422. [CrossRef] [PubMed]
44. Su, W.; Jiao, G.; Yang, C.; Ye, Y. Evaluation of apolipoprotein M as a biomarker of coronary artery disease. *Clin. Biochem.* **2009**, *42*, 365–370. [CrossRef] [PubMed]
45. Borup, A.; Christensen, P.M.; Nielsen, L.B.; Christoffersen, C. Apolipoprotein M in lipid metabolism and cardiometabolic diseases. *Curr. Opin. Lipidol.* **2015**, *26*, 48–55. [CrossRef] [PubMed]
46. Brea, D.; Sobrino, T.; Blanco, M.; Fraga, M.; Agulla, J.; Rodriguez-Yanez, M.; Rodriguez-Gonzalez, R.; Perez, D.L.O.N.; Leira, R.; Forteza, J.; *et al.* Usefulness of haptoglobin and serum amyloid A proteins as biomarkers for atherothrombotic ischemic stroke diagnosis confirmation. *Atherosclerosis* **2009**, *205*, 561–567. [CrossRef] [PubMed]
47. King, V.L.; Thompson, J.; Tannock, L.R. Serum amyloid A in atherosclerosis. *Curr. Opin. Lipidol.* **2011**, *22*, 302–307. [CrossRef] [PubMed]
48. Liuzzo, G.; Biasucci, L.M.; Gallimore, J.R.; Grillo, R.L.; Rebuzzi, A.G.; Pepys, M.B.; Maseri, A. The prognostic value of C-reactive protein and serum amyloid a protein in severe unstable angina. *N. Engl. J. Med.* **1994**, *331*, 417–424. [CrossRef] [PubMed]
49. Delanghe, J.R.; Langlois, M.R.; de Bacquer, D.; Mak, R.; Capel, P.; van Renterghem, L.; de Backer, G. Discriminative value of serum amyloid A and other acute-phase proteins for coronary heart disease. *Atherosclerosis* **2002**, *160*, 471–476. [CrossRef]
50. Lepedda, A.J.; Nieddu, G.; Zinellu, E.; de Muro, P.; Piredda, F.; Guarino, A.; Spirito, R.; Carta, F.; Turrini, F.; Formato, M. Proteomic analysis of plasma-purified VLDL, LDL, and HDL fractions from atherosclerotic patients undergoing carotid endarterectomy: Identification of serum amyloid A as a potential marker. *Oxid. Med. Cell. Longev.* **2013**, *2013*, 385214. [CrossRef] [PubMed]
51. McMahon, M.; Grossman, J.; FitzGerald, J.; Dahlin-Lee, E.; Wallace, D.J.; Thong, B.Y.; Badsha, H.; Kalunian, K.; Charles, C.; Navab, M.; *et al.* Proinflammatory high-density lipoprotein as a biomarker for atherosclerosis in patients with systemic lupus erythematosus and rheumatoid arthritis. *Arthritis Rheum.* **2006**, *54*, 2541–2549. [CrossRef] [PubMed]
52. Fung, E.T.; Thulasiraman, V.; Weinberger, S.R.; Dalmasso, E.A. Protein biochips for differential profiling. *Curr. Opin. Biotechnol.* **2001**, *12*, 65–69. [CrossRef]
53. Issaq, H.J.; Veenstra, T.D.; Conrads, T.P.; Felschow, D. The SELDI-TOF MS approach to proteomics: Protein profiling and biomarker identification. *Biochem. Biophys. Res. Commun.* **2002**, *292*, 587–592. [CrossRef] [PubMed]
54. Watanabe, J.; Chou, K.J.; Liao, J.C.; Miao, Y.; Meng, H.H.; Ge, H.; Grijalva, V.; Hama, S.; Kozak, K.; Buga, G.; *et al.* Differential association of hemoglobin with proinflammatory high density lipoproteins in atherogenic/hyperlipidemic mice. A novel biomarker of atherosclerosis. *J. Biol. Chem.* **2007**, *282*, 23698–23707. [CrossRef] [PubMed]

55. Frohlich, J.; Al-Sarraf, A. Cholesterol efflux capacity and atherosclerosis. *N. Engl. J. Med.* **2011**, *364*, 1474–1475. [PubMed]

56. Uto-Kondo, H.; Ayaori, M.; Ogura, M.; Nakaya, K.; Ito, M.; Suzuki, A.; Takiguchi, S.; Yakushiji, E.; Terao, Y.; Ozasa, H.; *et al.* Coffee consumption enhances high-density lipoprotein-mediated cholesterol efflux in macrophages. *Circ. Res.* **2010**, *106*, 779–787. [CrossRef] [PubMed]

57. Oram, J.F.; Lawn, R.M.; Garvin, M.R.; Wade, D.P. ABCA1 is the cAMP-inducible apolipoprotein receptor that mediates cholesterol secretion from macrophages. *J. Biol. Chem.* **2000**, *275*, 34508–34511. [CrossRef] [PubMed]

58. Santamarina-Fojo, S.; Peterson, K.; Knapper, C.; Qiu, Y.; Freeman, L.; Cheng, J.F.; Osorio, J.; Remaley, A.; Yang, X.P.; Haudenschild, C.; *et al.* Complete genomic sequence of the human ABCA1 gene: analysis of the human and mouse ATP-binding cassette A promoter. *Proc. Natl. Acad. Sci. USA* **2000**, *97*, 7987–7992. [CrossRef] [PubMed]

59. Wang, N.; Lan, D.; Chen, W.; Matsuura, F.; Tall, A.R. ATP-binding cassette transporters G1 and G4 mediate cellular cholesterol efflux to high-density lipoproteins. *Proc. Natl. Acad. Sci. USA* **2004**, *101*, 9774–9779. [CrossRef] [PubMed]

60. Nakamura, K.; Kennedy, M.A.; Baldan, A.; Bojanic, D.D.; Lyons, K.; Edwards, P.A. Expression and regulation of multiple murine ATP-binding cassette transporter G1 mRNAs/isoforms that stimulate cellular cholesterol efflux to high density lipoprotein. *J. Biol. Chem.* **2004**, *279*, 45980–45989. [CrossRef] [PubMed]

61. Yancey, P.G.; Bortnick, A.E.; Kellner-Weibel, G.; de la Llera-Moya, M.; Phillips, M.C.; Rothblat, G.H. Importance of different pathways of cellular cholesterol efflux. *Arterioscler. Thromb. Vasc. Biol.* **2003**, *23*, 712–719. [CrossRef] [PubMed]

62. Dikkers, A.; Freak, D.B.J.; Annema, W.; Groen, A.K.; Tietge, U.J. Scavenger receptor BI and ABCG5/G8 differentially impact biliary sterol secretion and reverse cholesterol transport in mice. *Hepatology* **2013**, *58*, 293–303. [CrossRef] [PubMed]

63. Von Eckardstein, A.; Nofer, J.R.; Assmann, G. High density lipoproteins and arteriosclerosis. Role of cholesterol efflux and reverse cholesterol transport. *Arterioscler. Thromb. Vasc. Biol.* **2001**, *21*, 13–27. [CrossRef] [PubMed]

64. Rosenson, R.S.; Brewer, H.J.; Davidson, W.S.; Fayad, Z.A.; Fuster, V.; Goldstein, J.; Hellerstein, M.; Jiang, X.C.; Phillips, M.C.; Rader, D.J.; *et al.* Cholesterol efflux and atheroprotection: Advancing the concept of reverse cholesterol transport. *Circulation* **2012**, *125*, 1905–1919. [CrossRef] [PubMed]

65. Joy, T.; Hegele, R.A. The end of the road for CETP inhibitors after torcetrapib? *Curr. Opin. Cardiol.* **2009**, *24*, 364–371. [CrossRef] [PubMed]

66. Wang, N.; Tall, A.R. Regulation and mechanisms of ATP-binding cassette transporter A1-mediated cellular cholesterol efflux. *Arterioscler. Thromb. Vasc. Biol.* **2003**, *23*, 1178–1184. [CrossRef] [PubMed]

67. Curtiss, L.K.; Valenta, D.T.; Hime, N.J.; Rye, K.A. What is so special about apolipoprotein AI in reverse cholesterol transport? *Arterioscler. Thromb. Vasc. Biol.* **2006**, *26*, 12–19. [CrossRef] [PubMed]

68. Kontush, A.; Chapman, M.J. Functionally defective high-density lipoprotein: A new therapeutic target at the crossroads of dyslipidemia, inflammation, and atherosclerosis. *Pharmacol. Rev.* **2006**, *58*, 342–374. [CrossRef] [PubMed]

69. Freeman, S.R.; Jin, X.; Anzinger, J.J.; Xu, Q.; Purushothaman, S.; Fessler, M.B.; Addadi, L.; Kruth, H.S. ABCG1-mediated generation of extracellular cholesterol microdomains. *J. Lipid Res.* **2014**, *55*, 115–127. [CrossRef] [PubMed]

70. Song, G.J.; Kim, S.M.; Park, K.H.; Kim, J.; Choi, I.; Cho, K.H. SR-BI mediates high density lipoprotein (HDL)-induced anti-inflammatory effect in macrophages. *Biochem. Biophys. Res. Commun.* **2015**, *457*, 112–118. [CrossRef] [PubMed]

71. Wang, X.; Collins, H.L.; Ranalletta, M.; Fuki, I.V.; Billheimer, J.T.; Rothblat, G.H.; Tall, A.R.; Rader, D.J. Macrophage ABCA1 and ABCG1, but not SR-BI, promote macrophage reverse cholesterol transport *in vivo*. *J. Clin. Investig.* **2007**, *117*, 2216–2224. [CrossRef] [PubMed]

72. Daniil, G.; Zannis, V.I.; Chroni, A. Effect of apoA-I Mutations in the capacity of reconstituted HDL to promote ABCG1-mediated cholesterol efflux. *PLoS ONE* **2013**, *8*, e67993. [CrossRef] [PubMed]

73. Westerterp, M.; Murphy, A.J.; Wang, M.; Pagler, T.A.; Vengrenyuk, Y.; Kappus, M.S.; Gorman, D.J.; Nagareddy, P.R.; Zhu, X.; Abramowicz, S.; *et al.* Deficiency of ATP-binding cassette transporters A1 and G1 in macrophages increases inflammation and accelerates atherosclerosis in mice. *Circ. Res.* **2013**, *112*, 1456–1465. [CrossRef] [PubMed]

74. Bultel, S.; Helin, L.; Clavey, V.; Chinetti-Gbaguidi, G.; Rigamonti, E.; Colin, M.; Fruchart, J.C.; Staels, B.; Lestavel, S. Liver X receptor activation induces the uptake of cholesteryl esters from high density lipoproteins in primary human macrophages. *Arterioscler. Thromb. Vasc. Biol.* **2008**, *28*, 2288–2295. [CrossRef] [PubMed]

75. Hanf, R.; Millatt, L.J.; Cariou, B.; Noel, B.; Rigou, G.; Delataille, P.; Daix, V.; Hum, D.W.; Staels, B. The dual peroxisome proliferator-activated receptor α/δ agonist GFT505 exerts anti-diabetic effects in db/db mice without peroxisome proliferator-activated receptor gamma-associated adverse cardiac effects. *Diabetes Vasc. Dis. Res.* **2014**, *11*, 440–447. [CrossRef] [PubMed]

76. Colin, S.; Briand, O.; Touche, V.; Wouters, K.; Baron, M.; Pattou, F.; Hanf, R.; Tailleux, A.; Chinetti, G.; Staels, B.; *et al.* Activation of intestinal peroxisome proliferator-activated receptor-α increases high-density lipoprotein production. *Eur. Heart J.* **2013**, *34*, 2566–2574. [CrossRef] [PubMed]

77. Sahebkar, A.; Chew, G.T.; Watts, G.F. New peroxisome proliferator-activated receptor agonists: Potential treatments for atherogenic dyslipidemia and non-alcoholic fatty liver disease. *Expert Opin. Pharmacother.* **2014**, *15*, 493–503. [CrossRef] [PubMed]

78. Rousset, X.; Shamburek, R.; Vaisman, B.; Amar, M.; Remaley, A.T. Lecithin cholesterol acyltransferase: An anti- or pro-atherogenic factor? *Curr. Atheroscler. Rep.* **2011**, *13*, 249–256. [CrossRef] [PubMed]

79. Soran, H.; Hama, S.; Yadav, R.; Durrington, P.N. HDL functionality. *Curr. Opin. Lipidol.* **2012**, *23*, 353–366. [CrossRef] [PubMed]

80. Olivecrona, G.; Olivecrona, T. Triglyceride lipases and atherosclerosis. *Curr. Opin. Lipidol.* **2010**, *21*, 409–415. [CrossRef] [PubMed]

81. Chatterjee, C.; Sparks, D.L. Hepatic lipase, high density lipoproteins, and hypertriglyceridemia. *Am. J. Pathol.* **2011**, *178*, 1429–1433. [CrossRef] [PubMed]

82. Yasuda, T.; Ishida, T.; Rader, D.J. Update on the role of endothelial lipase in high-density lipoprotein metabolism, reverse cholesterol transport, and atherosclerosis. *Circ. J.* **2010**, *74*, 2263–2270. [CrossRef] [PubMed]

83. Annema, W.; Tietge, U.J. Role of hepatic lipase and endothelial lipase in high-density lipoprotein-mediated reverse cholesterol transport. *Curr. Atheroscler. Rep.* **2011**, *13*, 257–265. [CrossRef] [PubMed]

84. Ishida, T.; Choi, S.; Kundu, R.K.; Hirata, K.; Rubin, E.M.; Cooper, A.D.; Quertermous, T. Endothelial lipase is a major determinant of HDL level. *J. Clin. Investig.* **2003**, *111*, 347–355. [CrossRef] [PubMed]

85. Ruel, I.L.; Couture, P.; Cohn, J.S.; Bensadoun, A.; Marcil, M.; Lamarche, B. Evidence that hepatic lipase deficiency in humans is not associated with proatherogenic changes in HDL composition and metabolism. *J. Lipid Res.* **2004**, *45*, 1528–1537. [CrossRef] [PubMed]

86. Lambert, G.; Amar, M.J.; Martin, P.; Fruchart-Najib, J.; Foger, B.; Shamburek, R.D.; Brewer, H.J.; Santamarina-Fojo, S. Hepatic lipase deficiency decreases the selective uptake of HDL-cholesteryl esters *in vivo*. *J. Lipid Res.* **2000**, *41*, 667–672. [PubMed]

87. Jaye, M.; Lynch, K.J.; Krawiec, J.; Marchadier, D.; Maugeais, C.; Doan, K.; South, V.; Amin, D.; Perrone, M.; Rader, D.J. A novel endothelial-derived lipase that modulates HDL metabolism. *Nat. Genet.* **1999**, *21*, 424–428. [PubMed]

88. Escola-Gil, J.C.; Chen, X.; Julve, J.; Quesada, H.; Santos, D.; Metso, J.; Tous, M.; Jauhiainen, M.; Blanco-Vaca, F. Hepatic lipase- and endothelial lipase-deficiency in mice promotes macrophage-to-feces RCT and HDL antioxidant properties. *Biochim. Biophys. Acta* **2013**, *1831*, 691–697. [CrossRef] [PubMed]

89. Zhang, L.; Yan, F.; Zhang, S.; Lei, D.; Charles, M.A.; Cavigiolio, G.; Oda, M.; Krauss, R.M.; Weisgraber, K.H.; Rye, K.A.; *et al.* Structural basis of transfer between lipoproteins by cholesteryl ester transfer protein. *Nat. Chem. Biol.* **2012**, *8*, 342–349. [CrossRef] [PubMed]

90. Chapman, M.J.; le Goff, W.; Guerin, M.; Kontush, A. Cholesteryl ester transfer protein: At the heart of the action of lipid-modulating therapy with statins, fibrates, niacin, and cholesteryl ester transfer protein inhibitors. *Eur. Heart J.* **2010**, *31*, 149–164. [CrossRef] [PubMed]

91. Rao, R.; Albers, J.J.; Wolfbauer, G.; Pownall, H.J. Molecular and macromolecular specificity of human plasma phospholipid transfer protein. *Biochemistry* **1997**, *36*, 3645–3653. [CrossRef] [PubMed]

92. Yu, Y.; Guo, S.; Feng, Y.; Feng, L.; Cui, Y.; Song, G.; Luo, T.; Zhang, K.; Wang, Y.; Jiang, X.C.; *et al.* Phospholipid transfer protein deficiency decreases the content of S1P in HDL via the loss of its transfer capability. *Lipids* **2014**, *49*, 183–190. [CrossRef] [PubMed]

93. Albers, J.J.; Vuletic, S.; Cheung, M.C. Role of plasma phospholipid transfer protein in lipid and lipoprotein metabolism. *Biochim. Biophys. Acta* **2012**, *1821*, 345–357. [CrossRef] [PubMed]

94. Jiang, X.C.; Bruce, C.; Mar, J.; Lin, M.; Ji, Y.; Francone, O.L.; Tall, A.R. Targeted mutation of plasma phospholipid transfer protein gene markedly reduces high-density lipoprotein levels. *J. Clin. Investig.* **1999**, *103*, 907–914. [CrossRef] [PubMed]

95. Yazdanyar, A.; Quan, W.; Jin, W.; Jiang, X.C. Liver-specific phospholipid transfer protein deficiency reduces high-density lipoprotein and non-high-density lipoprotein production in mice. *Arterioscler. Thromb. Vasc. Biol.* **2013**, *33*, 2058–2064. [CrossRef] [PubMed]

96. Jiang, X.C.; Qin, S.; Qiao, C.; Kawano, K.; Lin, M.; Skold, A.; Xiao, X.; Tall, A.R. Apolipoprotein B secretion and atherosclerosis are decreased in mice with phospholipid-transfer protein deficiency. *Nat. Med.* **2001**, *7*, 847–852. [CrossRef] [PubMed]

97. Luo, Y.; Shelly, L.; Sand, T.; Reidich, B.; Chang, G.; Macdougall, M.; Peakman, M.C.; Jiang, X.C. Pharmacologic inhibition of phospholipid transfer protein activity reduces apolipoprotein-B secretion from hepatocytes. *J. Pharmacol. Exp. Ther.* **2010**, *332*, 1100–1106. [CrossRef] [PubMed]

98. Rinninger, F.; Heine, M.; Singaraja, R.; Hayden, M.; Brundert, M.; Ramakrishnan, R.; Heeren, J. High density lipoprotein metabolism in low density lipoprotein receptor-deficient mice. *J. Lipid Res.* **2014**, *55*, 1914–1924. [CrossRef] [PubMed]

99. Van der Velde, A.E.; Vrins, C.L.; van den Oever, K.; Kunne, C.; Oude, E.R.; Kuipers, F.; Groen, A.K. Direct intestinal cholesterol secretion contributes significantly to total fecal neutral sterol excretion in mice. *Gastroenterology* **2007**, *133*, 967–975. [CrossRef] [PubMed]

100. Van der Velde, A.E.; Brufau, G.; Groen, A.K. Transintestinal cholesterol efflux. *Curr. Opin. Lipidol.* **2010**, *21*, 167–171. [CrossRef] [PubMed]

101. Blanchard, C.; Moreau, F.; Cariou, B.; Le May, C. Trans-intestinal cholesterol excretion (TICE): A new route for cholesterol excretion. *Med. Sci.* **2014**, *30*, 896–901.

102. Vrins, C.L.; van der Velde, A.E.; van den Oever, K.; Levels, J.H.; Huet, S.; Oude, E.R.; Kuipers, F.; Groen, A.K. Peroxisome proliferator-activated receptor delta activation leads to increased transintestinal cholesterol efflux. *J. Lipid Res.* **2009**, *50*, 2046–2054. [CrossRef] [PubMed]

103. Vrins, C.L.; Ottenhoff, R.; van den Oever, K.; de Waart, D.R.; Kruyt, J.K.; Zhao, Y.; van Berkel, T.J.; Havekes, L.M.; Aerts, J.M.; van Eck, M.; *et al.* Trans-intestinal cholesterol efflux is not mediated through high density lipoprotein. *J. Lipid Res.* **2012**, *53*, 2017–2023. [CrossRef] [PubMed]

104. Kontush, A.; Chantepie, S.; Chapman, M.J. Small, dense HDL particles exert potent protection of atherogenic LDL against oxidative stress. *Arterioscler. Thromb. Vasc. Biol.* **2003**, *23*, 1881–1888. [CrossRef] [PubMed]

105. Kontush, A.; Chapman, M.J. Antiatherogenic function of HDL particle subpopulations: Focus on antioxidative activities. *Curr. Opin. Lipidol.* **2010**, *21*, 312–318. [CrossRef] [PubMed]

106. Mackness, B.; Mackness, M. The antioxidant properties of high-density lipoproteins in atherosclerosis. *Panminerva Med.* **2012**, *54*, 83–90. [PubMed]

107. Vohl, M.C.; Neville, T.A.; Kumarathasan, R.; Braschi, S.; Sparks, D.L. A novel lecithin-cholesterol acyltransferase antioxidant activity prevents the formation of oxidized lipids during lipoprotein oxidation. *Biochemistry* **1999**, *38*, 5976–5981. [CrossRef] [PubMed]

108. Navab, M.; Hama, S.Y.; Anantharamaiah, G.M.; Hassan, K.; Hough, G.P.; Watson, A.D.; Reddy, S.T.; Sevanian, A.; Fonarow, G.C.; Fogelman, A.M. Normal high density lipoprotein inhibits three steps in the formation of mildly oxidized low density lipoprotein: Steps 2 and 3. *J. Lipid Res.* **2000**, *41*, 1495–1508. [PubMed]

109. Turunen, P.; Jalkanen, J.; Heikura, T.; Puhakka, H.; Karppi, J.; Nyyssonen, K.; Yla-Herttuala, S. Adenovirus-mediated gene transfer of Lp-PLA2 reduces LDL degradation and foam cell formation *in vitro*. *J. Lipid Res.* **2004**, *45*, 1633–1639. [CrossRef] [PubMed]

110. Chen, N.; Liu, Y.; Greiner, C.D.; Holtzman, J.L. Physiologic concentrations of homocysteine inhibit the human plasma GSH peroxidase that reduces organic hydroperoxides. *J. Lab. Clin. Med.* **2000**, *136*, 58–65. [CrossRef] [PubMed]

111. Besler, C.; Heinrich, K.; Rohrer, L.; Doerries, C.; Riwanto, M.; Shih, D.M.; Chroni, A.; Yonekawa, K.; Stein, S.; Schaefer, N.; *et al.* Mechanisms underlying adverse effects of HDL on eNOS-activating pathways in patients with coronary artery disease. *J. Clin. Investig.* **2011**, *121*, 2693–2708. [CrossRef] [PubMed]

112. Sorrentino, S.A.; Besler, C.; Rohrer, L.; Meyer, M.; Heinrich, K.; Bahlmann, F.H.; Mueller, M.; Horvath, T.; Doerries, C.; Heinemann, M.; *et al.* Endothelial-vasoprotective effects of high-density lipoprotein are impaired in patients with type 2 diabetes mellitus but are improved after extended-release niacin therapy. *Circulation* **2010**, *121*, 110–122. [CrossRef] [PubMed]

113. Levine, D.M.; Parker, T.S.; Donnelly, T.M.; Walsh, A.; Rubin, A.L. *In vivo* protection against endotoxin by plasma high density lipoprotein. *Proc. Natl. Acad. Sci. USA* **1993**, *90*, 12040–12044. [CrossRef] [PubMed]

114. Nissen, S.E.; Tsunoda, T.; Tuzcu, E.M.; Schoenhagen, P.; Cooper, C.J.; Yasin, M.; Eaton, G.M.; Lauer, M.A.; Sheldon, W.S.; Grines, C.L.; *et al.* Effect of recombinant ApoA-I Milano on coronary atherosclerosis in patients with acute coronary syndromes: A randomized controlled trial. *JAMA* **2003**, *290*, 2292–2300. [CrossRef] [PubMed]

115. Ibanez, B.; Giannarelli, C.; Cimmino, G.; Santos-Gallego, C.G.; Alique, M.; Pinero, A.; Vilahur, G.; Fuster, V.; Badimon, L.; Badimon, J.J. Recombinant HDL(Milano) exerts greater anti-inflammatory and plaque stabilizing properties than HDL(wild-type). *Atherosclerosis* **2012**, *220*, 72–77. [CrossRef] [PubMed]

116. Tardy, C.; Goffinet, M.; Boubekeur, N.; Ackermann, R.; Sy, G.; Bluteau, A.; Cholez, G.; Keyserling, C.; Lalwani, N.; Paolini, J.F.; *et al.* CER-001, a HDL-mimetic, stimulates the reverse lipid transport and atherosclerosis regression in high cholesterol diet-fed LDL-receptor deficient mice. *Atherosclerosis* **2014**, *232*, 110–118. [CrossRef] [PubMed]

117. Tardif, J.C.; Ballantyne, C.M.; Barter, P.; Dasseux, J.L.; Fayad, Z.A.; Guertin, M.C.; Kastelein, J.J.; Keyserling, C.; Klepp, H.; Koenig, W.; *et al.* Effects of the high-density lipoprotein mimetic agent CER-001 on coronary atherosclerosis in patients with acute coronary syndromes: A randomized trial. *Eur. Heart J.* **2014**, *35*, 3277–3286. [CrossRef] [PubMed]

118. Murphy, A.J.; Funt, S.; Gorman, D.; Tall, A.R.; Wang, N. Pegylation of high-density lipoprotein decreases plasma clearance and enhances antiatherogenic activity. *Circ. Res.* **2013**, *113*, e1–e9. [CrossRef] [PubMed]

119. Waksman, R.; Torguson, R.; Kent, K.M.; Pichard, A.D.; Suddath, W.O.; Satler, L.F.; Martin, B.D.; Perlman, T.J.; Maltais, J.A.; Weissman, N.J.; *et al.* A first-in-man, randomized, placebo-controlled study to evaluate the safety and feasibility of autologous delipidated high-density lipoprotein plasma infusions in patients with acute coronary syndrome. *J. Am. Coll. Cardiol.* **2010**, *55*, 2727–2735. [CrossRef] [PubMed]

120. Sacks, F.M.; Rudel, L.L.; Conner, A.; Akeefe, H.; Kostner, G.; Baki, T.; Rothblat, G.; de la Llera-Moya, M.; Asztalos, B.; Perlman, T.; *et al.* Selective delipidation of plasma HDL enhances reverse cholesterol transport *in vivo*. *J. Lipid Res.* **2009**, *50*, 894–907. [CrossRef] [PubMed]

121. Van Capelleveen, J.C.; Brewer, H.B.; Kastelein, J.J.; Hovingh, G.K. Novel therapies focused on the high-density lipoprotein particle. *Circ. Res.* **2014**, *114*, 193–204. [CrossRef] [PubMed]

122. Zhao, Y.; Imura, T.; Leman, L.J.; Curtiss, L.K.; Maryanoff, B.E.; Ghadiri, M.R. Mimicry of high-density lipoprotein: Functional peptide-lipid nanoparticles based on multivalent peptide constructs. *J. Am. Chem. Soc.* **2013**, *135*, 13414–13424. [CrossRef] [PubMed]

123. Joseph, S.B.; Castrillo, A.; Laffitte, B.A.; Mangelsdorf, D.J.; Tontonoz, P. Reciprocal regulation of inflammation and lipid metabolism by liver X receptors. *Nat. Med.* **2003**, *9*, 213–219. [CrossRef] [PubMed]

124. Lo, S.G.; Murzilli, S.; Salvatore, L.; D'Errico, I.; Petruzzelli, M.; Conca, P.; Jiang, Z.Y.; Calabresi, L.; Parini, P.; Moschetta, A. Intestinal specific LXR activation stimulates reverse cholesterol transport and protects from atherosclerosis. *Cell Metab.* **2010**, *12*, 187–193.

125. Bartel, D.P. MicroRNAs: Genomics, biogenesis, mechanism, and function. *Cell* **2004**, *116*, 281–297. [CrossRef]

126. Rayner, K.J.; Esau, C.C.; Hussain, F.N.; McDaniel, A.L.; Marshall, S.M.; van Gils, J.M.; Ray, T.D.; Sheedy, F.J.; Goedeke, L.; Liu, X.; *et al.* Inhibition of miR-33a/b in non-human primates raises plasma HDL and lowers VLDL triglycerides. *Nature* **2011**, *478*, 404–407. [CrossRef] [PubMed]

127. Zhao, R.; Feng, J.; He, G. miR-613 regulates cholesterol efflux by targeting LXRα and ABCA1 in PPARβ activated THP-1 macrophages. *Biochem. Biophys. Res. Commun.* **2014**, *448*, 329–334. [CrossRef] [PubMed]

128. Simonelli, S.; Tinti, C.; Salvini, L.; Tinti, L.; Ossoli, A.; Vitali, C.; Sousa, V.; Orsini, G.; Nolli, M.L.; Franceschini, G.; *et al.* Recombinant human LCAT normalizes plasma lipoprotein profile in LCAT deficiency. *Biologicals* **2013**, *41*, 446–449. [CrossRef] [PubMed]

129. Aguilar-Espinosa, S.L.; Mendoza-Espinosa, P.; Delgado-Coello, B.; Mas-Oliva, J. Lecithin cholesterol acyltransferase (LCAT) activity in the presence of Apo-AI-derived peptides exposed to disorder-order conformational transitions. *Biochem. Biophys. Res. Commun.* **2013**, *441*, 469–475. [CrossRef] [PubMed]

130. Barylski, M.; Toth, P.P.; Nikolic, D.; Banach, M.; Rizzo, M.; Montalto, G. Emerging therapies for raising high-density lipoprotein cholesterol (HDL-C) and augmenting HDL particle functionality. *Best Pract. Res. Clin. Endocrinol. Metab.* **2014**, *28*, 453–461. [CrossRef] [PubMed]

131. Barter, P.J.; Rye, K.A. Cholesteryl ester transfer protein inhibition as a strategy to reduce cardiovascular risk. *J. Lipid Res.* **2012**, *53*, 1755–1766. [CrossRef] [PubMed]

132. Mohammadpour, A.H.; Akhlaghi, F. Future of cholesteryl ester transfer protein (CETP) inhibitors: A pharmacological perspective. *Clin. Pharmacokinet.* **2013**, *52*, 615–626. [CrossRef] [PubMed]

133. Cao, G.; Beyer, T.P.; Zhang, Y.; Schmidt, R.J.; Chen, Y.Q.; Cockerham, S.L.; Zimmerman, K.M.; Karathanasis, S.K.; Cannady, E.A.; Fields, T.; *et al.* Evacetrapib is a novel, potent, and selective inhibitor of cholesteryl ester transfer protein that elevates HDL cholesterol without inducing aldosterone or increasing blood pressure. *J. Lipid Res.* **2011**, *52*, 2169–2176. [CrossRef] [PubMed]

134. Ford, J.; Lawson, M.; Fowler, D.; Maruyama, N.; Mito, S.; Tomiyasu, K.; Kinoshita, S.; Suzuki, C.; Kawaguchi, A.; Round, P.; *et al.* Tolerability, pharmacokinetics and pharmacodynamics of TA-8995, a selective cholesteryl ester transfer protein (CETP) inhibitor, in healthy subjects. *Br. J. Clin. Pharmacol.* **2014**, *78*, 498–508. [CrossRef] [PubMed]

135. Boettcher, M.F.; Heinig, R.; Schmeck, C.; Kohlsdorfer, C.; Ludwig, M.; Schaefer, A.; Gelfert-Peukert, S.; Wensing, G.; Weber, O. Single dose pharmacokinetics, pharmacodynamics, tolerability and safety of BAY 60–5521, a potent inhibitor of cholesteryl ester transfer protein. *Br. J. Clin. Pharmacol.* **2012**, *73*, 210–218. [CrossRef] [PubMed]

136. Bell, T.R.; Graham, M.J.; Lee, R.G.; Mullick, A.E.; Fu, W.; Norris, D.; Crooke, R.M. Antisense oligonucleotide inhibition of cholesteryl ester transfer protein enhances RCT in hyperlipidemic, CETP transgenic, LDLr$^{-/-}$ mice. *J. Lipid Res.* **2013**, *54*, 2647–2657. [CrossRef] [PubMed]

137. Hu, Z.; Shen, W.J.; Kraemer, F.B.; Azhar, S. MicroRNAs 125a and 455 repress lipoprotein-supported steroidogenesis by targeting scavenger receptor class B type I in steroidogenic cells. *Mol. Cell. Biol.* **2012**, *32*, 5035–5045. [CrossRef] [PubMed]

138. Wang, L.; Jia, X.J.; Jiang, H.J.; Du, Y.; Yang, F.; Si, S.Y.; Hong, B. MicroRNAs 185, 96, and 223 repress selective high-density lipoprotein cholesterol uptake through posttranscriptional inhibition. *Mol. Cell. Biol.* **2013**, *33*, 1956–1964. [CrossRef] [PubMed]

139. Sugano, M.; Fujikawa, T.; Hiratsuji, Y.; Nakashima, K.; Fukuda, N.; Hasegawa, Y. A novel use of chitosan as a hypocholesterolemic agent in rats. *Am. J. Clin. Nutr.* **1980**, *33*, 787–793. [PubMed]

140. Baker, W.L.; Tercius, A.; Anglade, M.; White, C.M.; Coleman, C.I. A meta-analysis evaluating the impact of chitosan on serum lipids in hypercholesterolemic patients. *Ann. Nutr. Metab.* **2009**, *55*, 368–374. [CrossRef] [PubMed]

141. Su, Z.Q.; Wu, S.H.; Zhang, H.L.; Feng, Y.F. Development and validation of an improved Bradford method for determination of insulin from chitosan nanoparticulate systems. *Pharm. Biol.* **2010**, *48*, 966–973. [CrossRef] [PubMed]

142. Tan, S.; Gao, B.; Tao, Y.; Guo, J.; Su, Z.Q. Antiobese effects of capsaicin-chitosan microsphere (CCMS) in obese rats induced by high fat diet. *J. Agric. Food Chem.* **2014**, *62*, 1866–1874. [CrossRef] [PubMed]

143. Chen, J.; Huang, G.D.; Tan, S.R.; Guo, J.; Su, Z.Q. The preparation of capsaicin-chitosan microspheres (CCMS) enteric coated tablets. *Int. J. Mol. Sci.* **2013**, *14*, 24305–24319. [CrossRef] [PubMed]

144. Tao, Y.; Zhang, H.L.; Hu, Y.M.; Wan, S.; Su, Z.Q. Preparation of chitosan and water-soluble chitosan microspheres via spray-drying method to lower blood lipids in rats fed with high-fat diets. *Int. J. Mol. Sci.* **2013**, *14*, 4174–4184. [CrossRef] [PubMed]

145. Pan, H.; Guo, J.; Su, Z. Advances in understanding the interrelations between leptin resistance and obesity. *Physiol. Behav.* **2014**, *130*, 157–169. [CrossRef] [PubMed]

146. Zhang, H.L.; Tao, Y.; Guo, J.; Hu, Y.M.; Su, Z.Q. Hypolipidemic effects of chitosan nanoparticles in hyperlipidemia rats induced by high fat diet. *Int. Immunopharmacol.* **2011**, *11*, 457–461. [CrossRef] [PubMed]

*Int. J. Mol. Sci.* **2015**, *16*, 17245–17272

147. Li, Y.; Xu, Q.; Wei, P.; Cheng, L.; Peng, Q.; Li, S.; Yin, H.; Du, Y. Chitosan oligosaccharides downregulate the expression of E-selectin and ICAM-1 induced by LPS in endothelial cells by inhibiting MAP kinase signaling. *Int. J. Mol. Med.* **2014**, *33*, 392–400. [PubMed]

148. Liu, H.T.; Huang, P.; Ma, P.; Liu, Q.S.; Yu, C.; Du, Y.G. Chitosan oligosaccharides suppress LPS-induced IL-8 expression in human umbilical vein endothelial cells through blockade of p38 and Akt protein kinases. *Acta Pharmacol. Sin.* **2011**, *32*, 478–486. [CrossRef] [PubMed]

International Journal of
*Molecular Sciences*

MDPI

*Article*

# Atherosclerotic Calcification Detection: A Comparative Study of Carotid Ultrasound and Cone Beam CT

Fisnik Jashari [1], Pranvera Ibrahimi [1,*], Elias Johansson [1,2], Jan Ahlqvist [3], Conny Arnerlöv [4], Maria Garoff [3], Eva Levring Jäghagen [3], Per Wester [1] and Michael Y. Henein [1]

[1] Department of Public Health and Clinical Medicine, Umeå University, 90187 Umeå, Sweden; fisnik.jashari@medicin.umu.se (F.J.); elias.johansson@umu.se (E.J.); per.wester@medicin.umu.se (P.W.); michael.henein@medicin.umu.se (M.Y.H.)

[2] Department of Pharmacology and Clinical Neuroscience, Umeå University, 90187 Umeå, Sweden

[3] Department of Odontology, Umeå University, 90187 Umeå, Sweden; jan.ahlqvist@umu.se (J.A.); maria.garoff@umu.se (M.G.); eva.levring.jaghagen@umu.se (E.L.J.)

[4] Department of Surgical and Perioperative Sciences, Umeå University, 90187 Umeå, Sweden; conny.arnerlov@vll.se

* Correspondence: pranvera.ibrahimi@medicin.umu.se; Tel.: +46-90-785-1431; Fax: +46-90-137-633

Academic Editor: Kurt A. Jellinger
Received: 30 June 2015; Accepted: 13 August 2015; Published: 21 August 2015

**Abstract:** Background and Aim: Arterial calcification is often detected on ultrasound examination but its diagnostic accuracy is not well validated. The aim of this study was to determine the accuracy of carotid ultrasound B mode findings in detecting atherosclerotic calcification quantified by cone beam computed tomography (CBCT). Methods: We analyzed 94 carotid arteries, from 88 patients (mean age $70 \pm 7$ years, 33% females), who underwent pre-endarterectomy ultrasound examination. Plaques with high echogenic nodules and posterior shadowing were considered calcified. After surgery, the excised plaques were examined using CBCT, from which the calcification volume ($mm^3$) was calculated. In cases with multiple calcifications the largest calcification nodule volume was used to represent the plaque. Carotid artery calcification by the two imaging techniques was compared using conventional correlations. Results: Carotid ultrasound was highly accurate in detecting the presence of calcification; with a sensitivity of 88.2%. Based on the quartile ranges of calcification volumes measured by CBCT we have divided plaque calcification into four groups: <8; 8–35; 36–70 and >70 $mm^3$. Calcification volumes $\geq 8$ were accurately detectable by ultrasound with a sensitivity of 96%. Of the 21 plaques with <8 $mm^3$ calcification volume; only 13 were detected by ultrasound; resulting in a sensitivity of 62%. There was no difference in the volume of calcification between symptomatic and asymptomatic patients. Conclusion: Carotid ultrasound is highly accurate in detecting the presence of calcified atherosclerotic lesions of volume $\geq 8$ $mm^3$; but less accurate in detecting smaller volume calcified plaques. Further development of ultrasound techniques should allow better detection of early arterial calcification.

**Keywords:** carotid atherosclerosis; ultrasound; calcification

## 1. Introduction

Carotid atherosclerosis is an important cause of ischaemic stroke [1] and patients with severe stenosis benefit significantly from carotid endarterectomy (CEA) [2]. However, it is well recognised that plaque features add extra diagnostic accuracy for better risk stratification of such patients [3]. According to the conventional atherosclerotic cascade, calcium formation in the carotid wall could be an early pathology that occurs well before significant plaque stenosis and hence is described as subclinical atherosclerosis. Although carotid calcification is detected in 50%–60% of cases [4], often in significantly stenotic lesions (>50%) [5], its association with cerebrovascular events is uncertain. Indeed, controversies exist as to the effect of calcification on plaque nature, with studies suggesting that it both increases stability [6] and is a marker for vulnerability, irrespective of the degree of stenosis [7]. Recently, a systematic review suggested that symptomatic plaques have less calcification compared to asymptomatic plaques [8].

Historically, carotid calcification is detected by plain roentgen imaging. With the development of ultrasound examination of vascular pathology, carotid 2D grey scale has become a routine investigation in vascular laboratories. While ultrasound may detect the presence of calcification, it does not have the accurate means for quantifying its extent. Recently, computed tomography (CT) has become the modality of choice in assessing arterial wall calcification. Even calcification volumes of 1 mm$^3$ can be detected and quantified by cone beam CT (CBCT) [9]. However, any CT investigation carries the risk of significant radiation, as well as limited availability. The aim of this study was to determine whether ultrasound (US) B mode could detect carotid calcification quantified by CBCT.

## 2. Results

### 2.1. Clinical Characteristics

Among the 88 patients included in the study, the mean age was 70 ± 7 years and 33% were females. Baseline characteristics of the patients are presented in Table 1. Seventy-three patients had symptomatic carotid stenosis and the remaining 15 patients had asymptomatic carotid stenosis. Six of the symptomatic patients underwent bilateral carotid endarterectomy (CEA); a stenosis severity of 50%–69% was found in six, stenosis severity of 70%–99% in 85 and near-occlusion in three arteries.

**Table 1.** Baseline characteristics of the study population.

| Baseline | Study Population (*n* = 88) |
| --- | --- |
| Age (years), mean (SD) Females, *n* (%) | 70 (7), 29 (33) |
| Systolic blood pressure (mmHg), mean (SD) | 147 (22.6) |
| Diastolic blood pressure (mmHg), mean (SD) | 78 (12) |
| Total cholesterol (mmol/L), mean (SD) | 4.61 (1.03) |
| LDL (mmol/L), mean (SD) | 2.60 (0.92) |
| HDL (mmol/L), mean (SD) | 1.27 (0.48) |
| Creatinine (µmol/L), mean (SD) | 84 (25) |
| HBA1c (mmol/mol), mean (SD) | 52.5 (12.7) |
| Symptomatic carotid stenosis, *n* (%) | 73 (83) |
| Prior myocardial infarction, *n* (%) | 14 (16) |
| Current angina pectoris, *n* (%) | 7 (8) |
| Heart failure, *n* (%) | 1 (1.1) |
| Previous stroke (>6 months to the present evaluation), *n* (%) | 14 (16) |
| Claudication (lower extremity artery disease), *n* (%) | 10 (11.4) |
| Any previous revascularization for ischemia, *n* (%) | 24 (27.3) |
| Current smoker, *n* (%) | 9 (10.2) |
| Diabetes, *n* (%) | 29 (33) |
| Lipid lowering medicine, *n* (%) | 82 (93.2) |
| Platelet inhibiting or anticoagulation medicine, *n* (%) | 88 (100) |
| Blood pressure reducing medicine, *n* (%) | 84 (95.5) |

### 2.2. Carotid Calcification: Ultrasound vs. Cone Beam CT (CBCT)

The pre-operative ultrasound examination detected calcification in 82 (87.2%) of the 94 carotid arteries. Calcification volume was quantified with CBCT in 94 CEA specimens and confirmed the presence of calcium in 93 (98.9%). There was no statistically significant difference between the 50%–69% and the 70%–99% degree of stenosis groups (Table 2). Furthermore, there was no difference in carotid calcification between different subgroups according to age, gender or other risk factors (Table 2).

Carotid ultrasound was highly accurate in detecting the presence of calcification, with a sensitivity of 88.2%. Using quartiles of calcification volumes measured by CBCT, plaque calcification extent was divided into four groups: <8, 8–35, 36–70 and >70 mm$^3$ (Figure 1). Calcification volumes $\geq$8 mm$^3$ were accurately detectable by ultrasound with a sensitivity of 96%. Of the 21 plaques with <8 mm$^3$ calcification volume, only 13 were detected by ultrasound, resulting in a sensitivity of 62% (Table 3). There was only one case that was negative for calcification on CBCT but this was also negative on ultrasound (Table 3).

**Table 2.** Subgroup analyses.

| Subgroups | Arteries $n$ | <8 mm$^3$ Calcification Volumes on CBCT, $n$ (%) | $p$ |
|---|---|---|---|
| Age | | | |
| <75 | 68 | 18 (26.5) | |
| $\geq$75 | 26 | 4 (15.3) | 0.21 |
| Gender | | | |
| Female | 30 | 7 (23.3) | |
| Male | 64 | 15 (23.4) | 0.99 |
| Symptomatic | 79 | 20 (25.3) | |
| Asymptomatic | 15 | 2 (13.3) | 0.31 |
| Current smoker | | | |
| Yes | 10 | 4 (40.0) | |
| No | 84 | 18 (21.4) | 0.19 |
| Diabetes | | | |
| Yes | 31 | 8 (25.8) | |
| No | 63 | 14 (22.2) | 0.60 |
| Previous stroke | | | |
| Yes | 17 | 2 (11.8) | |
| No | 77 | 20 (26.0) | 0.21 |
| Previous MI | | | |
| Yes | 15 | 3 (20.0) | |
| No | 79 | 19 (24.0) | 0.73 |
| Statin therapy | | | |
| Yes | 87 | 21 (24.1) | |
| No | 7 | 1 (14.3) | 0.55 |
| Degree of stenosis | | | |
| 50%–69% | 6 | 0 (0) | |
| 70%–99% | 85 | 20 (23.5) | |
| Near-occlusion | 3 | 2 (66.7) | 0.08 |

**Table 3.** Accuracy of Doppler ultrasound in detecting carotid calcification.

| CBCT\Ultrasound | Total Arteries, $n$ | Calcification on CBCT | Calcification on US | US Sensitivity | US Specificity |
|---|---|---|---|---|---|
| Calcification | 94 | 93 | 82 | 88.20% | 100% |
| Calcification $\geq$8 mm$^3$ | 72 | 72 | 69 | 96% | N/A |
| Calcification <8 mm$^3$ | 22 | 21 | 13 | 62% | 100% |

N/A: Not applicable.

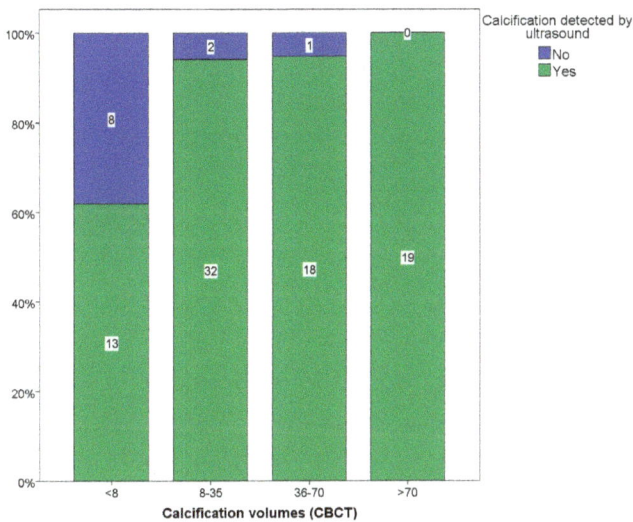

**Figure 1.** Quartiles of calcification volumes detected and not detected by ultrasound.

### 2.3. Carotid Calcification vs. Symptoms

There was no statistical difference in calcification, evaluated by US, between symptomatic and asymptomatic patients. Carotid arteries of symptomatic patients were calcified in 89% of cases compared to 93% in asymptomatic individuals ($p = 0.51$). We then explored whether patients with dispersed calcification within the plaque were more often associated with symptoms compared to those with all the calcification gathered in a solitary nodule, but there was no difference between the groups ($p = 0.32$). There also no difference in calcification volume quartiles (<8, 8–35, 36–70 and >70 mm$^3$) between symptomatic and asymptomatic patients ($p = 0.35$). In addition, we compared calcification volumes <8 mm$^3$ on CBCT between different risk factor subgroups but there was no significant difference in volume between those with and without the risk factor (Table 2).

### 2.4. Reproducibility Analyses

The inter-observer agreement for the presence of calcification in the carotid artery stenotic lesions evaluated by US was good ($K = 0.905$, $p < 0.001$).

### 3. Discussion

#### 3.1. Findings

Our results show that carotid US is accurate in detecting the presence of calcification in a group of patients with significant stenosis recruited for carotid endarterectomy (CEA). US sensitivity was 96% in identifying calcification volume $\geq 8$ mm$^3$ by CBCT. However, calcification volumes <8 mm$^3$ were inconsistently detected, with a sensitivity of only 62%. In addition, there was no difference for calcification between symptomatic and asymptomatic patients, neither between other risk factors and degree of stenosis.

#### 3.2. Data Interpretation

Calcium formation can be found in any arterial bed. It may be present in the lumen as a calcified plaque cap or may also invade the intima or media [10]. It has been shown that the amount of calcification quantified by carotid CT correlated with respective histology sections [11,12]. Carotid US is an easy and economically favorable method to study different plaque features and their potential association with symptoms [13]. Although it was previously reported that carotid plaque echogenicity, evaluated by US, may not adequately reflect the degree of carotid plaque calcification [14], our study is the first to demonstrate the accuracy of carotid US in identifying calcification deposits and showing that non-detectable calcification is of lower volume and is dispersed throughout the plaque. It was reported that calcification of the carotid plaque is associated with adverse outcomes, such as ipsilateral ischemic stroke, 30 days after carotid artery stenting [15]. Thus, CEA could be feasible in such cases. As many centers now rely mainly on the carotid US findings before intervention, identification of calcification may guide towards better patient risk stratification.

Our findings show that US was accurate in identifying carotid plaques, which after excision proved to have a calcification volume of at least 8 mm$^3$. This accuracy was seen in both the two technologies used to assess carotid calcification, CBCT, which directly quantified the calcification volume of extirpated carotid plaques, and US, which qualitatively evaluated plaque echogenicity and posterior shadowing as indirect evidence for plaque calcification. The calcium volume detectable by US ($\geq 8$ mm$^3$) is relatively significant when compared with normal intima-media thickness, which is usually less than 1 mm, highlighting the accuracy of the US examination in detecting sizable carotid masses. In contrast, the same method failed to detect carotid calcification below a volume of 8 mm$^3$ with reliable accuracy. In addition, our findings did not show any statistical difference in the presence of calcification between symptomatic and asymptomatic patients, in keeping with our previous findings [9].

The exact mechanism behind carotid calcification as a potential risk for cerebrovascular events and in predicting strokes still remains controversial [5,6,16]. A systematic review suggested that clinically symptomatic plaques have a lower degree of calcification than asymptomatic plaques [7]. This review had some potential limitations due to the wide range of methods used, highlighting the need for a well-validated, accurate and reproducible technique for calcium quantification. The relationship between arterial calcification and vascular events is more thoroughly studied in the setting of coronary artery disease because of the widely available CT scans performed on patients in cardiology units. Many studies have already shown that the extent of coronary calcification correlates with future development of cardiac events and disease burden [17]. This of course does not refute the presence of a subgroup of patients with calcific coronary artery disease in whom extensive calcification, based on Agatston score classification, is associated with recurrent angina but stable plaques [10,18]. Finally, it has recently been shown that statin therapy increases plaque echogenicity [19] and coronary artery calcification [20]. In our cohort 87.3% of patients were using statin therapy, limiting us in carrying out a reliable comparison between groups.

### 3.3. Clinical Implications

Our findings support the use of US optimum grey scale echogenicity and plaque shadowing as an accurate manifestation of sizable calcification of a minimum volume of 8 mm$^3$. US does not appear well suited for the detection of smaller calcium deposits, although they might play a role in plaque vulnerability. Although this method of calcification detection by US is subjective in nature, it establishes a potential foundation for the future development of quantitative models, which could guide towards improved identification of plaque characteristics as a step towards achieving plaque stabilization through optimum treatment. This approach, however, is not sufficiently accurate to detect early calcification pathology based on this biased patient selection. Further development of US plaque studies using methods which quantify echogenicity (grey scale median) could enable detection of earlier stages of calcification and reduce the need for advanced CT examinations and the resultant radiation for patients.

*3.4. Study Limitations*

Because almost all (98.9%) carotid plaques were calcified, as assessed by CBCT, evaluation of US positive and negative predictive value was not possible. We cannot be sure if all calcified material was excised during the CEA procedure but we are working on the assumption that the largest plaque with the largest calcium volume was removed and assessed by CBCT. Furthermore, we cannot be sure that the calcification quantified by CBCT was the same hyperechogenic calcium nodule seen by US in cases with multiple calcifications but we made a further assumption that the nodule with the largest volume was the same nodule as caused the posterior shadowing. In addition, the number of patients with asymptomatic carotid stenosis was small compared with the symptomatic patients, therefore weakening our conclusion in this subgroup. The exact role of stenosis severity in determining accuracy of US in detecting calcification cannot be established since most of our patients had significant (>50%) stenosis.

## 4. Experimental Section

*4.1. Subjects*

This is a secondary analysis of the Doppler US examinations of the Panorama arm of the SPACE study [5]. In the main study, we included 100 patients with significant carotid stenosis ($\geq$50%) who were all eligible for CEA, and who received a pre-operative carotid US examination using the conventional protocol. Patients' age, gender, recent symptoms and pre-existing co-morbidities, as well as severity of carotid stenosis based on Doppler velocities, were documented. All 100 patients underwent CEA: 94 unilaterally and six bilaterally. Except for five plaques from unilateral CEA, the remainder of the excised plaques was collected, making a total of 101 plaques from 95 patients. This study was approved by the Regional Ethical Board in Umeå (07-004M) and is in compliance with the Helsinki Declaration.

*4.2. Ultrasound Examination*

All patients were examined by experienced ultrasound operators ($n$ = 9) and findings were confirmed by a second opinion before reporting. A conventional US system (Acuson Sequoia; Siemens Medical Solutions, Mountain View, CA, USA) equipped with an 8L5 (frequency bandwidth 5–8 MHz) linear-array transducer was used for all carotid examinations. A visible plaque on B-mode with increased Doppler flow velocity was graded according to validated criteria [21], aimed to reproduce North American Symptomatic Carotid Endarterectomy Trial (NASCET) type stenosis for angiography.

*4.3. Carotid Ultrasound Analysis*

Two observers blinded to the CBCT results of the excised plaques reviewed all stored US images. A plaque was defined as calcified if there was a hyperechogenic spot with posterior shadowing (Figure 2). Where there was disagreement between the two observers ($n$ = 2), the examination was re-evaluated by both observers and consensus was reached.

Seven patients were excluded, five because of suboptimal image quality and two because there was no calcification in the internal carotid artery (ICA) plaque but a distinct calcification in the external carotid artery (ECA). Since it was unclear whether the ECA would be included in the extirpated plaque, it was impossible to make a reasonable assessment without breaking the blinding. The remaining 88 patients constituted the study cohort.

*4.4. Carotid Endarterectomy (CEA)*

All operations were performed under general anesthesia. The common carotid artery and the internal and external carotid arteries were carefully exposed. A longitudinal incision was made in the common carotid artery and this extended into the internal carotid artery and to the healthy artery distal to the plaque. The intimal thickening of the common carotid artery was divided and the plaque was gently dissected from the artery, leaving the adventitial layer intact and taking care to minimize trauma to the plaque. The plaque was further removed from the external carotid artery, along with dissection of the plaque and its calcification until its end in the internal carotid artery. After removal, the plaque was placed in a plastic tube and immediately stored in −20 °C and transferred over to a −80 °C freezer until CBCT analysis.

## 4.5. CBCT

Excised plaques were examined by CBCT (Cone Beam Computed Tomography, 3D Accuitomo 170, J Morita MFG Corporation, Kyoto, Japan; 60 kV, 1 mA, 360°). Due to differences in plaque size, volumes of 4 × 4 or 6 × 6 cm were used and they had a resolution of 0.08–0.125 mm voxel size. The reconstructions were made with 0.5 mm slice thickness and 0.5 mm increment. To counteract partial volume, effect thresholds for maximum window-level were halved in all reconstructions. Volume measurements were performed with a software program (General Electric Company, Barrington, IL, USA, advantage workstation 4.3, Volume Viewer 2) (Figure 1). All calcifications were measured in $mm^3$, the resolution of the volume reconstructions were 1 $mm^3$. In plaques with multiple calcifications, the largest calcification nodule volume was used to represent the plaque (Figure 2).

## 4.6. Statistics

All statistical analyses were performed using IBM SPSS Statistics 22. Quartile analyses were used for categorizing calcification volume into four groups. Kappa values were calculated to determine the accuracy of US in detecting calcification. Chi-square test was used to compare categorical and ordinal variables with a pre-selected significance level of $p < 0.05$.

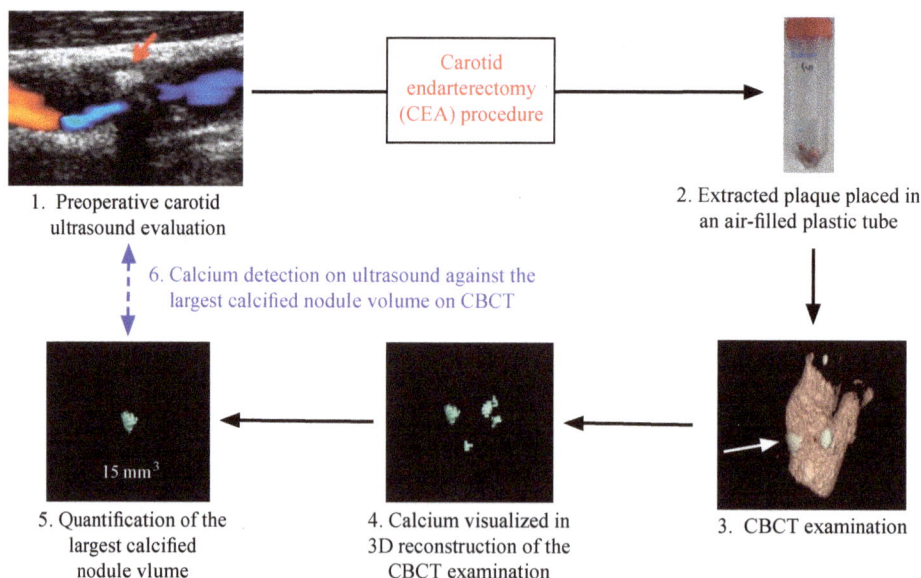

**Figure 2.** Flowchart of the work process of the study, starting with preoperative ultrasound examination, Cone Beam Computed Tomography (CBCT) evaluation of the carotid specimens removed after carotid endarterectomy (CEA) to images reconstruction and data analysis. Red arrow: hyperechoic spot with posterior shadowing (calcification) on ultrasound. White arrow: calcification on CBCT image.

## 5. Conclusions

Carotid US is highly accurate in detecting the presence of calcified atherosclerotic lesions of CBCT volume of more than 8 mm$^3$. However, it was less accurate in detecting smaller volume calcified plaques, with relatively high false negativity. Calcification was very common in both symptomatic and asymptomatic patients. Further development of US techniques should allow improved detection of early arterial calcification. Since calcification is a dynamic process, finding a method that avoids radiation could be of importance for longitudinal assessment of calcification progression and its effects on plaque stability.

**Author Contributions:** Fisnik Jashari and Elias Johansson conceived and designed the study; Fisnik Jashari and Pranvera Ibrahimi performed ultrasound data analysis, Jan Ahlqvist, Eva Levring Jäghagen and Maria Garoff performed the CBCT 3D image reconstruction and calcification volumes measurement; Fisnik Jashari and Pranvera Ibrahimi statically analyzed the data; Michael Y. Henein, Per Wester, Conny Arnerlöv critically revised the manuscript and helped with results interpretation. Fisnik Jashari wrote the paper.

**Conflicts of Interest:** The authors declare no conflict of interest.

## References

1. Grau, A.J.; Weimar, C.; Buggle, F.; Heinrich, A.; Goertler, M.; Neumaier, S.; Glahn, J.; Brandt, T.; Hacke, W.; Diener, H.-C.; *et al.* Risk factors, outcome, and treatment in subtypes of ischemic stroke: The German Stroke Data Bank. *Stroke J. Cereb. Circ.* **2001**, *32*, 2559–2566. [CrossRef]

2. Chaturvedi, S.; Bruno, A.; Feasby, T.; Holloway, R.; Benavente, O.; Cohen, S.N.; Cote, R.; Hess, D.; Saver, J.; Spence, J.D.; *et al.* Carotid endarterectomy—An evidence-based review: Report of the Therapeutics and Technology Assessment Subcommittee of the American Academy of Neurology. *Neurology* **2005**, *65*, 794–801. [CrossRef] [PubMed]

3. Hashimoto, H.; Tagaya, M.; Niki, H.; Etani, H. Computer-assisted analysis of heterogeneity on B-mode imaging predicts instability of asymptomatic carotid plaque. *Cerebrovasc. Dis.* **2009**, *28*, 357–364. [CrossRef] [PubMed]

4. Slevin, M.; Wang, Q.; Font, M.A.; Luque, A.; Juan-Babot, O.; Gaffney, J.; Kumar, P.; Kumar, S.; Badimon, L.; Krupinski, J. Atherothrombosis and plaque heterology: Different location or a unique disease? *Pathobiol. J. Immunopathol. Mol. Cell. Boil.* **2008**, *75*, 209–225. [CrossRef] [PubMed]

5. Garoff, M.; Johansson, E.; Ahlqvist, J.; Jaghagen, E.L.; Arnerlöv, C.; Wester, P. Detection of calcifications in panoramic radiographs in patients with carotid stenoses ≥50%. *Oral Surg. Oral Med. Oral Pathol. Oral Radiol.* **2014**, *117*, 385–391. [CrossRef] [PubMed]

6. Huang, H.; Virmani, R.; Younis, H.; Burke, A.P.; Kamm, R.D.; Lee, R.T. The impact of calcification on the biomechanical stability of atherosclerotic plaques. *Circulation* **2001**, *103*, 1051–1056. [CrossRef] [PubMed]

7. Nandalur, K.R.; Baskurt, E.; Hagspiel, K.D.; Finch, M.; Phillips, C.D.; Bollampally, S.R.; Kramer, C.M. Carotid artery calcification on CT may independently predict stroke risk. *AJR Am. J. Roentgenol.* **2006**, *186*, 547–552. [CrossRef] [PubMed]

8. Kwee, R.M. Systematic review on the association between calcification in carotid plaques and clinical ischemic symptoms. *J. Vasc. Surg.* **2010**, *51*, 1015–1025. [CrossRef] [PubMed]

9. Garoff, M.; Johansson, E.; Ahlqvist, J.; Arnerlöv, C.; Levring Jäghagen, E.; Wester, P. Calcium quantity in carotid plaques: Detection in panoramic radiographs and association with degree of stenosis. *Oral Surg. Oral Med. Oral Pathol. Oral Radiol.* **2015**, *120*, 269–274. [CrossRef] [PubMed]

10. Ibrahimi, P.; Jashari, F.; Nicoll, R.; Bajraktari, G.; Wester, P.; Henein, M.Y. Coronary and carotid atherosclerosis: How useful is the imaging? *Atherosclerosis* **2013**, *231*, 323–333. [CrossRef] [PubMed]

11. Ababneh, B.; Rejjal, L.; Pokharel, Y.; Nambi, V.; Wang, X.; Tung, C.H.; Han, R.I.; Talyor, A.A.; Kougias, P.; Lumsden, A.B.; *et al.* Distribution of calcification in carotid endarterectomy tissues: Comparison of micro-computed tomography imaging with histology. *Vasc. Med.* **2014**, *19*, 343–350. [CrossRef] [PubMed]

12. Miralles, M.; Merino, J.; Busto, M.; Perich, X.; Barranco, C.; Vidal-Barraquer, F. Quantification and characterization of carotid calcium with multi-detector CT-angiography. *Eur. J. Vasc. Endovasc. Surg.* **2006**, *32*, 561–567. [CrossRef] [PubMed]

13. Lal, B.K.; Hobson, R.W., 2nd.; Pappas, P.J.; Kubicka, R.; Hameed, M.; Chakhtoura, E.Y.; Jamil, Z.; Padber, F.T., Jr.; Haser, P.B.; Durán, W.N. Pixel distribution analysis of B-mode ultrasound scan images predicts histologic features of atherosclerotic carotid plaques. *J. Vasc. Surg.* **2002**, *35*, 1210–1217. [CrossRef] [PubMed]

14. Denzel, C.; Lell, M.; Maak, M.; Höckl, M.; Balzer, K.; Müller, K.M.; Fellner, C.; Fellner, F.A.; Lang, W. Carotid artery calcium: Accuracy of a calcium score by computed tomography—An *in vitro* study with comparison to sonography and histology. *Eur. J. Vasc. Endovasc. Surg.* **2004**, *28*, 214–220. [CrossRef] [PubMed]

15. Setacci, C.; Chisci, E.; Setacci, F.; Iacoponi, F.; de Donato, G.; Rossi, A. Siena carotid artery stenting score: A risk modelling study for individual patients. *Stroke* **2010**, *41*, 1259–1265. [CrossRef] [PubMed]

16. Elias-Smale, S.E.; Odink, A.E.; Wieberdink, R.G.; Hofman, A.; Hunink, M.G.; Krestin, G.P.; Koudstaal, P.J.; Breteler, M.B.; van der Lugt, A.; Witteman, J.C.M. Carotid, aortic arch and coronary calcification are related to history of stroke: The Rotterdam study. *Atherosclerosis* **2010**, *212*, 656–660. [CrossRef] [PubMed]

17. Bajraktari, G.; Nicoll, R.; Ibrahimi, P.; Jashari, F.; Schmermund, A.; Henein, M.Y. Coronary calcium score correlates with estimate of total plaque burden. *Int. J. Cardiol.* **2013**, *167*, 1050–1052. [CrossRef] [PubMed]

18. Nicoll, R.; Henein, M.Y. Arterial calcification: Friend or foe? *Int. J. Cardiol.* **2013**, *167*, 322–327. [CrossRef] [PubMed]

19. Ibrahimi, P.; Jashari, F.; Bajraktari, G.; Wester, P.; Henein, M.Y. Ultrasound assessment of carotid plaque echogenicity response to statin therapy: A systematic review and meta-analysis. *Int. J. Mol. Sci.* **2015**, *16*, 10734–10747. [CrossRef] [PubMed]

20. Henein, M.; Granåsen, G.; Wiklund, U.; Schmermund, A.; Guerci, A.; Erbel, R.; Raggi, P. High dose and long-term statin therapy accelerate coronary artery calcification. *Int. J. Cardiol.* **2015**, *184*, 581–586. [CrossRef] [PubMed]

21. Hansen, F.; Bergqvist, D.; Lindblad, B.; Lindh, M.; Matzsch, T.; Lanne, T. Accuracy of duplex sonography before carotid endarterectomy—A comparison with angiography. *Eur. J. Vasc. Endovasc. Surg.* **1996**, *12*, 331–336. [CrossRef]

MDPI AG

St. Alban-Anlage 66

4052 Basel, Switzerland

Tel. +41 61 683 77 34

Fax +41 61 302 89 18

http://www.mdpi.com

*IJMS* Editorial Office

E-mail: ijms@mdpi.com

http://www.mdpi.com/journal/ijms